Pattern Recognition
and Machine Vision

RIVER PUBLISHERS SERIES IN INFORMATION SCIENCE AND TECHNOLOGY

Volume 6

Consulting Series Editor

KWANG-CHENG CHEN
National Taiwan University
Taiwan

Information science and technology enables 21st century into an Internet and multimedia era. Multimedia means the theory and application of filtering, coding, estimating, analyzing, detecting and recognizing, synthesizing, classifying, recording, and reproducing signals by digital and/or analog devices or techniques, while the scope of "signal" includes audio, video, speech, image, musical, multimedia, data/content, geophysical, sonar/radar, bio/medical, sensation, etc. Networking suggests transportation of such multimedia contents among nodes in communication and/or computer networks, to facilitate the ultimate Internet. Theory, technologies, protocols and standards, applications/ services, practice and implementation of wired/wireless networking are all within the scope of this series. We further extend the scope for 21st century life through the knowledge in robotics, machine learning, cognitive science, pattern recognition, quantum/biological/molecular computation and information processing, and applications to health and society advance.

- Communication/Computer Networking Technologies and Applications
- Queuing Theory, Optimization, Operation Research, Statistical Theory and Applications
- Multimedia/Speech/Video Processing, Theory and Applications of Signal Processing
- Computation and Information Processing, Machine Intelligence, Cognitive Science, and Decision

For a list of other books in this series, see final page.

Pattern Recognition and Machine Vision

In Honor and Memory of
Professor King-Sun Fu 傅京孫教授

Editor

Patrick Shen-Pei Wang

Northeastern University, USA and
East China Normal University (China)

LONDON AND NEW YORK

Published 2010 by River Publishers
River Publishers
Alsbjergvej 10, 9260 Gistrup, Denmark
www.riverpublishers.com

Distributed exclusively by Routledge
4 Park Square, Milton Park, Abingdon, Oxon OX14 4RN
605 Third Avenue, New York, NY 10158

First published in paperback 2024

Pattern Recognition and Machine Vision / by Patrick Shen-Pei Wang.

Routledge is an imprint of the Taylor & Francis Group, an informa business

Publisher's Note
The publisher has gone to great lengths to ensure the quality of this reprint but points out that some imperfections in the original copies may be apparent.

While every effort is made to provide dependable information, the publisher, authors, and editors cannot be held responsible for any errors or omissions.

ISBN: 978-87-92329-36-3 (hbk)
ISBN: 978-87-7004-551-3 (pbk)

King-Sun Fu

2 October 1930 (Nanking, China) – 29 April 1985 (DC, USA)

Table of Contents

vii

Preface

In recent years, there has been a growing interest in the field of Pattern Recognition and Machine Vision in academia and industry. Novel theories have been developed, as well as new designs of technology and systems, both in hardware and software. They are widely applied to our daily life to solve realistic problems in diverse areas including science, engineering, agriculture, e-commerce, education, robotics, government, hospital, games and animation, medical imaging analysis and diagnosis, military, and homeland security. The foundation of all these can be traced back to the late Professor King-Sun Fu (2 October 1930–29 April 1985), one of the fathers of Pattern Recognition, who was a visionary and founded the International Association for Pattern Recognition (IAPR) in 1978. Ever since then, even after 30 years, the world has witnessed the rapid growth and development of this field, and most people are affected by its applications in our daily lives.

Today, on the eve of his 80th birthday and 25th anniversary of the unfortunate and untimely passing of Professor Fu, we are proud to produce this volume of collected works by world renowned professionals and experts in Pattern Recognition and Machine Vision, in memory of the late Professor King-Sun Fu. We hope this book will help promote further its course of not only fundamental principles, systems and technologies, but also its vast applications to help solving our daily life problems.

There has been much praise about Professor Fu's extraordinary achievements on PRMV, including hundreds of paper and book publications, numerous prestigious awards, the founding of IEEE-PAMI and IAPR and an NSF Research center and so on. Professor Fu is known as one of the greatest computer scientists and engineers. But actually more than that, he was also a great artist and music lover. He believed in a liberal arts education, and played volleyball well. In his Lab at Purdue University, most, if not all his team researchers and students' popular entertainment and exercise was to play volleyball. He also enjoyed reading novels and poems. I remember when he came to Boston to give a seminar we had some discussions on his most recent work then, on "towards a unification theory of statistical and syntactic pattern recognition." He noticed on my desk a copy of my article "On Puccini's opera 'Turandot' and Chinese folk song 'Jasmine flower'." It immediately attracted his attention and interest to request a copy. Everywhere we met or communicated, be it in the USA, Canada, Europe or Asia, he always showed strong concerns not only about the progress of pattern recognition, but also about world affairs, human affairs and the environment. Although, regrettably, he did not finish his ambitious plan to establish a theory towards unification of statistical, syntactical and structural pattern recognition, pretty much like Einstein did not have time to finish his effort towards unification of forces, yet, Professor King-Sun Fu indeed was not only a distinguished professor, but also a great teacher from whom we indeed have learned much. No wonder, as Professor Thomas Huang put it, in his King-Sun Fu Prize laureate speech at 2002 ICPR, Quebec City, Canada, that "There is no so called last student of ProfessorFu. We all are his students." Yes, indeed, Professor Fu, you live in our hearts forever.

This book on *Pattern Recognition and Machine Vision* in memory of Professor King-Sun Fu is divided into five parts.

Part I: Fundamental Principles and Methods in Pattern Recognition begins with the paper by Joydeep Ghosh, Meghana Deodhar, and Gunjan Gupta, presenting several approaches to co-clustering, namely the simultaneous clustering of features and objects. Robust Overlapping Co-Clustering is highlighted by addressing the mining of dense, arbitrarily positioned, possibly overlapping co-clusters from large, noisy datasets. The paper by Minhua Huang and Robert M. Haralick presents a novel stable and reliable probabilistic graphical model for recognizing patterns in texts where time and memory complexities for predicating new texts are significantly less than those of existing models. Limin Cui, Yuan Yan Tang, Patrick S.P. Wang, and Fucheng Liao's chapter, Wavelet-Galerkin algorithm for solving the first kind of weak singular integral equations with a logarithmic kernel is presented, which uses the Tikhonov regularization method to solve the system of stiff equation. Alan Wee-Chung Liew and Hong Yan discuss how gene expression study is performed and how to analyze these data based on pattern recognition methods, which include microarray image segmentation, missing value imputation, gene expression data clustering and biclustering and spectral analysis of time series data. The paper by Marco Reisert, Nikos Canterakis and Hans Burkhardt presents foundations and recent advances in the field of invariance and covariance against group transformations with a focus on the Euclidean Group transformation. Different methodologies on how to obtain invariant features and covariant image transformations are discussed. Conrad Sanderson and Brian C. Lovell presents a fast, scalable identity inference algorithm providing recognition of faces that are partially aligned and exhibit variations in pose, expression, illumination, resolution and scale. The proposed algorithm is inspired by methods used in text processing as well as 2D Hidden Markov Models.

Part II: Fundamental Principles and Methods in Machine Vision begins the paper by Feng Chen and Yi-Ping Phoebe Chen which presents a review of pattern recognition methods in genomics including the investigation of DNA, RNA, protein, and relationships between them. Xiaoqing Ding and Junxia Gu present two exemplar based classifiers for action recognition based on 3D body joints. Multiple features are fused to improve performance. Josh Harguess and J.K. Aggarwal exploit the inherent symmetry of the human face by averaging the right and left half-face to produce the average-half-face. Results using the average-half-face on two- and three-dimensional face databases by utilizing several popular face recognition algorithms show an increased accuracy over using the original full face for face recognition. Anja Attig and Petra Perner propose a framework for model building in image processing by meta-learning, which is seen as a classification process during which the signal characteristics are mapped to the right image-processing parameters to ensure the best image-processing output. The mapping function is realized by case-based reasoning. Ajay Misher and Yiannis Aloimonos define segmentation as the task of segmenting regions containing given fixation points. Only one image region is segmented per object, containing the fixation point. Prior knowledge of the objects is utilized to build object models to facilitate segmentation of known objects from input images with clutter. Guan-Yu Chen and Yung-Cheng Chen show a new idea of creating a template to be used as an atlas, where the atlas has average anatomic shapes and their positions, instead of probability maps only.

Part III: Machine Learning and Image Recognition begins with the paper by Giovanni Maria Farinella and Sebastiano Battiato, which reviews the approaches to scene classification detailing the description and discussion of relevant peculiarities. Yun Fu and Thomas S. Huang review state-of-the-art manifold and subspace learning methods in pattern recognition applications including four subspace learning algorithms derived from manifold criteria, namely Locally Embedded Analysis, Discriminant Simplex Analysis, Correlation Embedding Analysis, and Correlation Tensor Analysis. The contribution by Xiaoyi Jiang and Horst Bunke presents the generalized median of a set of input objects based on an optimization framework followed by a variety of problems dealing with objects like vectors, contours, strings, graphs, clustering, and image segmentations. H. Allende, C. Moraga, R. Nanculef and R. Salas discuss an ensemble learning algorithm to build a predictive model by combining a set of learning hypotheses instead of laboriously designing the complete map between inputs and responses in a single step. D.W. Repperger, C.A. Phillips and J.M. Flach pursue the Cybernetics viewpoint for modeling and analyzing a human-machine system within the context of control and information theory.

Part IV: Systems and Basic Technologies begins with Jung-Ming Wang, Wan-Ya Liao, Sei-Wang Chen, and Chiou-Shann Fuh introducing a multi-view representation of a pedestrian characterized by a unit sphere called the viewsphere. Pedestrian views are clustered on the viewsphere into aspects, having the same characteristics as a silhouette. The paper by Xiaoxiao Niu, Ching Y. Suen, and Tien D. Bui presents a check processing system including the extraction of target regions, isolation of digits for amount and the legal amount sentence, date fields, and the recognition of various elements. Bin Fang, Jing Wen, Y.Y. Tang, and Patrick S.P. Wang review developments in automatic off-line signature verification frameworks including data acquisition, pre-processing, feature extraction, and pattern matching together with techniques and algorithms used at different stages. The paper by Chin-Hung Teng, Ho-Ling Hsu, and Wen-Hsing Hsu reviews biometric authentication, including issues in designing a large scale biometric system, a number of standards, the architecture and design criteria and a potential mechanism for web-based biometric authentication. Yuping Wang introduces experimental design methods as a branch of statistics that extracts the most information possible from the fewest samples and considers the use of experimental design methods and quantization to design efficient evolutionary operators.

Part V: Machine Learning and Pattern Recognition Applications begins with Yue Lu, Xiao Tu, Shujing Lu, and Patrick S.P. Wang presenting the use of pattern recognition technology for postal automation in China including image acquisition, postcode and address segmentation, and recognition in automatic letter sorting machines. Che-Wei Lee and Wen-Hsiang Tsai present a lossless data hiding method based on histogram shifting which employs a scheme of adaptive division of cover images into blocks to yield large data hiding capacities as well as high stego-image qualities. The paper by Sen-Shyong Fann, Nae-Jye Hwang, Sheng-Tai Liaw and Huey-Liang Hwang presents an innovative handwriting input device embedded in the thin film transistor (TFT) array of a display panel and recognition of a two-dimensional pattern taken from this input device is also addressed including how to determine the location of a light spot or shadow of a finger. The paper by Chi-Yueh Lin and Hsiao-Chuan Wang presents a speech recognizer based on a cascaded structure consisting of an attribute detection and conditional random field approach to make use of phonetic knowledge within the phoneme decoding process. Shaoyi Du and Nanning Zheng present a thorough discussion of rigid and non-rigid point set registration. Rigid registration is formulated and the typical iterative closest point algorithm is presented. The history of non-rigid point set registration is introduced followed by modeling and algorithms of non-rigid registration of point sets. Finally, last but not least, the chapter by Ping Guo and Xinyu Chen presents a comprehensive review currently employed approaches in image feature representation, especially the Bag of Visual-words method and probability latent models. Some learning methods to map image representation with its corresponding semantic contents are also discussed.

A short description of each chapter evidently is not adequate to present the excellent contributions by all authors. The high quality of River's publications, however, provides ample opportunity to let each chapter to speak for itself, and therefore readers are strongly encouraged to peruse individual chapters in detail. This monumental, milestone book is indeed very rich and stimulating, full of vibrant activities in theory, applications, and system technologies in Pattern Recognition and Machine Vision.

Producing this book could not have been done by a single individual alone. It is the result of dedicated team work. I like to take this opportunity to show my deepest appreciations to all contributors, without whom this book could never be published. I also want to thank all foreword writers (in alphabetic order): Professors Jake Aggarwal, Herbert Freeman, Brian Lovell, and Chingyi Suen, who are either current or past Presidents, Fellows, or King-Sun Fu Prize Laureates of IAPR, the largest and most prestigious organization of its kind. Thanks also go to River Publisher, for its strong support of this book project. Finally, I like to thank my wife Sue and sons George Da-Yuan and David Da-Wen, for their understanding, encouragement and prayer. Above all, I thank God for granting me life and soul, without which I can never have today.

Patrick Shen-Pei Wang
Northeastern University (USA) and
East China Normal University (China)

Foreword

It is a pleasure and a distinct honor to introduce the book edited by Professor Patrick Wang in memory of the late Professor King-Sun Fu. Professor Fu was a friend and a mentor to several of the authors of the various chapters including myself. This collection of papers is the newest book devoted to presenting the state of the research, developments and applications in pattern recognition, image processing, computer vision, and machine vision.

I begin this foreword with a historical note and a tribute to my mentor and friend Professor Fu. The book entitled *Handbook of Pattern Recognition and Image Processing* (Academic Press, 1986) was edited by Professors Tzay Young and King-Sun Fu. Professor Fu passed away in 1985 and the book was published posthumously. Professor Fu was a teacher, a researcher, and a scholar. at the time of Professor Fu's untimely death, he was one of the prominent leaders of the field of Pattern Recognition. Among his many lasting accomplishments two stand out. He was the founding editor of the *IEEE Transactions on Pattern Analysis and Machine Intelligence*. Today this transaction has the highest regard of researchers and developers. It is among the crown jewels of the publications of the IEEE. He also helped start the organization known as the International Association for Pattern Recognition (IAPR) in the early 1970s. Today IAPR has membership from 42 countries all over the globe. The family and friends of Professor Fu created the K.S. Fu Prize in his memory under the auspices of the IAPR. IAPR awards this prize every two years at its International Conference on Pattern Recognition popularly known as ICPR. It is noteworthy that seven winners of the King Sun Fu Prize (out of a total of 11 awarded so far) are among the authors in the book edited by Professors Tzay and Fu. The book attests to the vision of Professor Fu; he was a true visionary. It is a fitting tribute to Professor Fu that Drs. Chen, Pau, and Wang started a series of handbooks in the vein of the book edited by Young and Fu.

The first in the series is the book edited by Drs. Chen, Pau, and Wang (World Scientific, 1993) entitled *Handbook of Pattern Recognition and Com-*

puter Vision. The book was dedicated to the memory of Professor Fu and it contained 32 papers. The change in the title from "Image Processing" to "Computer Vision" reflected the emphasis and the rise of computer vision as an important area of research and development. Professor Young in his preface explains image processing by saying that "the general goal of image processing is to analyze images of a given scene and to recognize the content of the scene." This is not too different from computer vision. The second edition of this book (World Scientific, 1999) contained 34 papers and many papers from the earlier book were revised and expanded with new material. These two handbooks together constituted an important landmark on the Pattern Recognition and Computer Vision scene in the 1990s. Obviously, they did not cover everything but they did cover a lot of ground. The second edition contained a thought provoking foreword by the late Professor Rosenfeld. He alludes to the fact that most real world vision problems are mathematically ill-defined and that real world domains do not satisfy simple mathematical or probabilistic models. I agree with the observation at least in part. However, mathematics provides the foundation and structure for building better models leading to the design and implementation of reliable and real world systems. In fact computer/machine vision has been successful in creating working systems by paying attention to details and through the adequate use of computer memory and power. Today the successes of computer/machine vision are spread over a large spectrum of applications, ranging from documents, objects, people, activities and events.

The third edition (World Scientific, 2005) edited by Chen and Wang presented an entirely new collection of papers. It contained 33 papers and reflected the exciting areas of pattern recognition and computer vision. It presented basic methods in the areas of pattern recognition and computer vision. In addition, it emphasized application areas like document processing and biometrics. It included papers on some of the newer areas like human motion recognition.

The present book is titled *Pattern Recognition and Machine Vision*. Again, the title of the book reflects the emphasis of the book. The papers in this volume are driven by application rather than by methodology. It consists of 28 chapters that have been divided into five fairly homogenous parts. This collection of papers has several review papers, original contributions, and papers outlining the state of art in several application areas of pattern recognition and machine vision. It is a well balanced book with strong appeal for researchers and students alike.

I would like to end this foreword on a personal note. It has been a privilege to introduce this book as a tribute to Professor Fu. I am delighted to note that I am in the select group of three professors, Thomas S. Huang, Ching Y. Suen, and J.K. Aggarwal, who have papers in all of the five collections of papers mentioned in this foreword.

J.K. Aggarwal
Department of Electrical & Computer Engineering
The University of Texas at Austin

Foreword

I consider it a great honor to have been asked to write this foreword to a book issued in memory of King-Sun Fu who died 25 years ago. King-Sun Fu was the unquestioned leader of the field of pattern recognition and a dear friend to many of us who were his contemporaries. He was a key founder of the International Association for Pattern Recognition (IAPR), served as the first editor of the IAPR Newsletter, served as the first chairman of the IEEE Machine Pattern Analysis Technical Committee, and led the effort to have the IEEE Computer Society create the *Transactions on Pattern Analysis and Machine Intelligence*. He was extremely active in publishing books and technical papers in the area of pattern recognition and served as advisor to some 75 doctoral students at Purdue University. His untimely death in 1985 was a shock to the entire pattern recognition community. Shortly thereafter the International Association for Pattern Recognition set up a biennial prize to honor his memory. The prize is awarded only to those who have made an extraordinary contribution to the field of pattern recognition. To this day it is regarded as the "Nobel prize" of the field.

This book contains 28 articles written by persons who either knew King-Sun Fu personally or were drawn to pattern recognition by the challenges that it offers. The topics covered range over a broad area, spanning machine learning, image processing, and classical pattern recognition, and all can be said to have been influenced by King-Sun Fu's pioneering work. This book is a wonderful testimonial to this great person to whose legacy we owe so much.

Herbert Freeman
Professor Emeritus, Rutgers University

Foreword

It is widely acknowledged that no one deserves more credit for the founding of the International Association for Pattern Recognition than King Sun Fu. After attempting to incorporate the emerging field of pattern recognition into artificial intelligence at the 2nd International Joint Conference on Artificial Intelligence in London in 1971, King Sun Fu invited several persons to set up a committee for an international conference on pattern recognition.[1]

This initiative resulted in the First International Joint Conference on Pattern Recognition (IJCPR) which was held in Washington, D.C., from 30th October to 1st November, 1973. The success of this conference led to the second IJCPR held in Lyngby, near Copenhagen, Denmark, from the 13th to the 15th of August, 1974. Discussions at Lyngby suggested the formation of an entirely new international society which would be a kind of federation of national organizations in pattern recognition. The name of the organization would be "International Association for Pattern Recognition" as suggested by Herbert Freeman.

At the third IJCPR held at Hotel del Coronado, Coronado, California, the draft constitution of the IAPR was approved on the 8th of November, 1976 and the IAPR came into official existence in January, 1978. The association had the following inaugural executive officers:

President	K.S. Fu
Vice President	M. Aizerman
Secretary	C.J.D.M. Verhagen
Treasurer	H. Freeman

Today the IAPR is a vibrant organisation with 41 member organisations from all around the globe. It sponsors numerous pattern recognition conferences in many countries as well as organising the International Conference on Pattern Recognition (formerly IJCPR) conference series.

---

[1] Freeman, H., Detailed History of the IAPR, http://www.iapr.org/docs/IAPR-History.pdf (last visited: January 2010).

The IAPR owes a huge debt to King Sun Fu, so it was a tremendous blow when he died suddenly on 29 April 1985. Aged only 55, his early death was a loss to the whole community and was keenly felt around the world. He, more than anyone, created the organization, serving as its inaugural president, and was a leading figure in the scientific field. In 1988 the IAPR awarded the first K.S. Fu Prize to commemorate his achievements. The prize was to be given no more often than biennially to a living person "in recognition of a technical contribution of far-reaching significance and impact on the field of pattern recognition or its closely allied fields made at any time in the past." The K.S. Fu Prize would serve as the "Nobel Prize" for pattern recognition.

Now more than 25 years after his untimely death, it is appropriate to commemorate King Sun Fu with this edition of collected works representing aspects of the exciting field of pattern recognition which he did so much to establish. I sincerely regret that I did not have the honour of meeting King Sun Fu, but I hear much about him from friends and colleagues who knew him well. Some researchers of my generation tell me proudly that they took his classes at Purdue; older researchers knew him as a valued colleague and friend. Most will agree with Albert Einstein's sentiment, "If I have seen farther than others, it is because I was standing on the shoulders of giants." King Sun Fu was truly one such giant.

Brian C. Lovell
President of the IAPR, 2008–2010

Foreword

It is my great pleasure to write the preface of this book in honour and memory of the Late Professor King-Sun Fu (1930–1985).

I met Professor Fu in 1973 at the First International Joint Conference on Pattern Recognition organized by him in Washington D.C. In subsequent years, I met him on many occasions at different conferences, and at meetings of the international Chinese Language Computer Society of which he and I served as its President in different periods. Professor Fu was an exceptional individual with a warm heart and kindness. He worked very hard and had always been very helpful to his colleagues and students. During those years, I benefited a lot from his foresight and advice, e.g. recommending my book on *Computational Analysis of Mandarin* for publication by Birkhauser Verlag in Switzerland in 1979, serving as the External Examiner of my doctoral student in 1984, and discussing the future directions of Chinese computer input and output techniques. As can be seen from his extended biography, published works, and the list of Ph.D. dissertations supervised by him on pp. 291–294, 295–300, and 301–303 respectively in volume 8, May 1986 of the *IEEE Transactions on Pattern Analysis and Machine Intelligence*, the journal he founded in 1978, and the prestigious IAPR (International Association for Pattern Recognition, http://www.iapr.org) award in his name, there is no doubt that Professor Fu is the most respected scholar and pioneer in our field. In fact, his foundation of IAPR and ICPRs, his publications and students continue to exert a tremendous influence up to this date and for many more years to come.

I have known Professor Patrick S.P. Wang, editor of this book, for more than 20 years. He has co-founded the *International Journal of Artificial Intelligence and Pattern Recognition*, co-edited the *Handbook of Pattern Recognition and Image Processing*, and many books and papers on AI and pattern recognition. I am very pleased to see the completion of this new book dedicated to the memory of the late Professor K.S. Fu. I am especially happy to see that it contains the excellent contributions of many prominent researchers in this field, covering a wide spectrum of subjects ranging from theory to practice, which is indeed, a reminiscence of Professor Fu's effort in editing his handbook of Pattern Recognition and his keen interest on PR principles and applications.

I strongly recommend this comprehensive book to all academic researchers and practitioners in the fields of pattern recognition and computer vision.

Ching Y. Suen
Concordia Chair in AI & Pattern Recognition
Director – CENPARMI, Concordia University

PART I

FUNDAMENTAL PRINCIPLES
AND METHODS IN
PATTERN RECOGNITION

1

Detection of Dense Co-Clusters in Large, Noisy Datasets

Joydeep Ghosh, Meghana Deodhar and Gunjan Gupta

Department of Electrical and Computer Engineering, University of Texas at Austin, 1 University Station C0803, Austin, TX 78712, USA; E-mail: {ghosh, deodhar, ggupta}@ece.utexas.edu

Abstract

Clustering problems often involve datasets where only a part of the data is relevant to the problem. For example, in analysis of microarray data for detecting related genes, only a subset of the genes typically show cohesive expressions and that to only within a subset of the conditions/features. The existence of a large number of non-informative data points and features makes it challenging to hunt for coherent and meaningful clusters from such datasets. Additionally, clusters could exist in different subspaces of the feature space. For these reasons, a co-clustering algorithm that simultaneously clusters objects and features is often more suitable as compared to one that is restricted to traditional "one-sided" clustering. In this chapter, we discuss several approaches to finding useful co-clusters, using microarray datasets as the motivating application. In particular, we highlight Robust Overlapping Co-Clustering (ROCC) [8], a scalable and very versatile framework that addresses the problem of efficiently mining dense, arbitrarily positioned, possibly overlapping co-clusters from large, noisy datasets. ROCC is demonstrated to have several desirable properties that make it extremely well suited to a number of real life applications.

Patrick Shen-Pei Wang (Ed.), Pattern Recognition and Machine Vision – In Honor and Memory of Professor King-Sun Fu, 3–17.

Keywords: co-clustering, cluster mining, density based clustering, microarray data analysis.

1.1 Introduction

Though clustering is a classic procedure with a long history and extensive literature in the pattern recognition community [16], several complex datasets that have emerged from recent, real-life problems impose additional requirements or constraints that limit the applicability of traditional clustering techniques. These constraints may arise from a variety of data properties such as the size or complexity of the data or its streaming nature. They may also arise from application requirements, e.g., a desire for getting clusters of comparable size for actionability [13].

In this chapter we focus on one such constrained clustering scenario, namely, situations for which only a part of the data forms cohesive clusters. This setting is relevant for analysis of microarray data, which can be represented as a matrix, where each row corresponds to a gene or some other DNA sequence such as cDNA clones, and each column represents an experimental condition or sample (e.g. cancerous vs. normal tissue). The matrix entries are the normalized "expression" level relative to a reference level for a gene (row) for the corresponding condition (col). Additional details on microarray data and relevant analysis techniques can be found in the chapter by Liew and Yan. A certain biological process may be active in a subset of the experiments, and genes that have similar trends in expression levels for this subset of conditions are likely to be functionally related [17]. If we define a "co-cluster" as a set of rows (objects) which have relatively coherent behavior over a set of columns (features) [2], then potentially funtionally related genes can be identified by clustering in the correct "expression" subspace, i.e., identifying an appropriate co-cluster in the microarray data matrix. Note that typically only a small subset of the genes cluster well and the rest can be considered non-informative [14]. Moreover, genes can participate in multiple biological processes and hence fully belong to multiple clusters. Similarly, in a given sample or condition, multiple processes can be active. This means that the set of co-clusters correponding to all the active processes can be arbitrarily positioned and may overlap in rows and/or columns.

Problems addressed by eCommerce businesses, such as market basket analysis and fraud detection also involve huge, noisy datasets with coherent patterns occurring only in small pockets of the data. Moreover, for such data, it is also often the case that coherent clusters reside within different and possibly overlapping subsets of features. Additionally, it is possible that some features may not be relevant to any cluster. Thus one needs to identify the data points that cluster well while disregarding the others, and also, for each cluster, determine the subspace where the clustering is most evident.

Traditional clustering algorithms like k-means or approaches such as feature clustering [9] do not allow clusters existing in different subsets of the feature space to be detected easily. This observation has led to a variety of subspace clustering methods. For example, *co-clustering* simultaneously clusters the data along multiple axes. In the case of microarray data analysis, where this is also called biclustering, the procedure simultaneously clusters the genes as well as the experiments [6] and can hence detect clusters existing in different subspaces of the feature space. This particular application is additionally complicated since the relevant co-clusters may be arbitrarily positioned in the data matrix, could be overlapping (indicating genes that are fully members of multiple biological processes) and are obfuscated by the presence of a large number of irrelevant points (irrelevant genes or conditions). Thus the most effective clustering method for such settings should be able to discover dense, arbitrarily positioned and overlapping co-clusters in the data, while simultaneously pruning away non-informative objects and features.

1.2 Approaches to Detecting Dense (Co-)Clusters

Perhaps the most natural approach to cluster only part of the data is to define clusters around local density peaks, and discard data that is not close to these dense regions. The classic method based on this philosophy is Wishart's modal analysis [27]. More recent density based clustering algorithms such as DBSCAN [11], OPTICS [1], Bregman Bubble Clustering [14] and Hierarchical Density Shaving [15] also use the notion of local density to cluster only a relevant subset of the data into multiple dense clusters. However, all of these approaches are developed for one-sided clustering only, where the data points are clustered based on their similarity across the entire set of features. In contrast, both co-clustering (also known as biclustering) and subspace clustering approaches that are considered next, locate clusters in subspaces of the feature space. The literature in both areas is recent but explosive, so we refer to the surveys and comparative studies in [20, 22, 23] as good starting points.

Co-clustering was first applied to gene expression data by Cheng and Church [6], who used a greedy search heuristic to generate arbitrarily positioned, overlapping co-clusters, based on a homogeneity constraint. This was a pioneering work that spawned tremendous interest in biclustering approaches for bioinformatics. However, their iterative insertion and deletion based algorithm is expensive, since it identifies individual co-clusters sequentially rather than all at once. The algorithm also causes random perturbations to the data while masking discovered biclusters, which reduces the clustering quality. The plaid model approach [18] improves upon this by directly modeling overlapping clusters, but still cannot identify multiple co-clusters simultaneously. These algorithms are not very general as they assume additive Gaussian noise models. Neither can they effectively handle missing data. The xMotif algorithm [21] is an iterative search method that identifies submatrices with genes that have near constant expression levels across a subset of experiments. Since functionally related genes may have quite different (though correlated) expression levels, most generated biclusters are not very interesting for microarray data analysis.

In addition to the greedy, iterative algorithms discussed above, notable deterministic algorithms such as BiMax [23] and OPSM [4] have also been proposed. The BiMax approach is based on a binary data model, which results in a number of co-clusters that is exponential in the number of genes and experiments. Its attraction is the simplicity of the binary model, but the exponential growth makes it impractical in case of large datasets. The order preserving sub matrix algorithm (OPSM) looks for submatrices in which the expression levels of all the genes induce the same linear ordering of the experiments. This algorithm although very accurate, is designed to identify only a single co-cluster. A recent extension to OPSM [29] finds multiple, overlapping co-clusters in noisy datasets, but is very expensive in the number of features.

Bregman Co-Clustering (BCC), proposed by Banerjee et al. [2], is a highly efficient, generalized framework for partitional co-clustering [20] that works with any distance measure that is a Bregman divergence, or equivalently any noise distribution from the regular exponential family. The BCC framework is however restricted to grid-based, partitional co-clustering and assigns every point in the data matrix to exactly one co-cluster, i.e., the co-clustering is exhaustive and exclusive.

Parsons et al. [22] present a survey of subspace clustering algorithms, which includes bottom-up grid based methods like CLIQUE and iterative top-down algorithms like PROCLUS. However, most of them are computationally intensive, need extensive tuning to get meaningful results and identify uniform clusters with very similar values rather than clusters with coherent trends or patterns. The pCluster model [25] and the more recent reg-cluster model [28] generalize subspace clustering and aim to identify arbitrary scaling and shifting co-regulations patterns. Subspace clustering approaches have the additional attraction that they are robust to irrelevant features, and each cluster if often more interpretable as it is defined by only a subset of the features. However, the current approaches are typically pattern-based and heuristic, without a principled cost function. Moreover they typically do not scale well due to high complexity in the number of features.

1.3 Robust Overlapping Co-clustering

Many of the limitations of the approaches described in the previous section are avoided by the recently proposed discuss Robust Overlapping Co-clustering (ROCC) method for discovering dense, arbitrarily positioned co-clusters in large, possibly high-dimensional datasets [8]. ROCC is robust in the presence of noisy and irrelevant objects as well as features, which the algorithm automatically detects and prunes during the clustering process. It is based on a systematically developed objective function, which is minimized by an iterative procedure that provably converges to a locally optimal solution. ROCC is also robust to the noise model of the data and can be tailored to use the most suitable distance measure for the data, selected from a large class of distance measures known as Bregman divergences. Therefore, we concentrate on this approach for the rest of the chapter.

The overall objective of ROCC is achieved in two steps. In the first step, the Bregman co-clustering algorithm is adapted to automatically prune away non-informative data points and perform feature selection by eliminating non-discriminative features and hence cluster only the relevant part of the dataset. This step finds co-clusters arranged in a grid structure, but only a predetermined number of rows and columns are assigned to the co-clusters. Note however that this result cannot be achieved by simply removing some rows/columns from the BCC result. An agglomeration step then appropriately merges similar co-clusters to discover dense, arbitrarily positioned, overlapping co-clusters. Figure 1.1 contrasts the nature of the co-clusters identified by ROCC with those found by BCC and illustrates the way in which they are conceptually derived from the partitional model of BCC.

The ROCC framework has the following key features that distinguish it from previously proposed co-clustering algorithms:

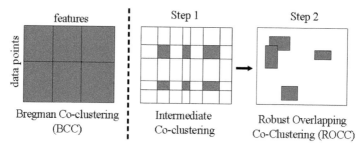

Figure 1.1 Nature of clusters identified by BCC and ROCC. Shaded areas represent clustered elements, rearranged according to cluster labels, while non-shaded areas denote discarded values

1. The ability to mine the most coherent co-clusters from large and noisy datasets.
2. Detection of arbitrarily positioned and possibly overlapping co-clusters in a principled manner by iteratively minimizing a suitable cost function.
3. Generalization to all Bregman divergences, including squared Euclidean distance, commonly used for clustering microarray data and I-divergence, commonly used for text data clustering [10].
4. The ability to naturally deal with missing data values, without introducing random perturbations or bias in the data.
5. Efficient detection of all co-clusters simultaneously rather than sequentially, enabling scalability to large and high-dimensional datasets.

None of the co-clustering algorithms [2, 6, 28, 29] discussed in Section 1.2 have all of the above properties. The contribution of ROCC is significant, since as described in Section 1.1 there exist several applications where all these properties are necessary for discovering meaningful patterns.

1.3.1 Problem Definition

We begin with the formulation of the first step of the ROCC algorithm. Let m be the total number of rows (data points) and n the total number of columns (features). The data can be represented as an $m \times n$ matrix Z of data points and features. Let s_r and s_c be the specified number of rows and columns, respectively, to be retained after pruning. If the exact values are not known, it is sufficient to set s_r and s_c conservatively to large values since the algorithm (Section 1.3.2) does a second round of pruning as needed. Our aim is to simultaneously cluster s_r rows and s_c columns of Z into a grid of k row clusters and l column clusters. The co-clusters will hence be comprised of $s_r \times s_c$

entries selected from the $m \times n$ entries of Z. Let $\mathcal{K}$ and $\mathcal{L}$ denote the sets consisting of the s_r clustered rows and the s_c clustered columns respectively. Let ρ be a mapping from the s_r rows $\in \mathcal{K}$ to the k row clusters and γ be a mapping from the s_c columns $\in \mathcal{L}$ to the l column clusters. Let squared Euclidean distance be the selected distance measure.[1] We want to find a co-clustering defined by (ρ, γ) and sets $\mathcal{K}$ and $\mathcal{L}$ for the specified s_r and s_c that minimize the following objective function

$$\sum_{g=1}^{k}\sum_{h=1}^{l} \sum_{u\in\mathcal{K}:\rho(u)=g} \sum_{v\in\mathcal{L}:\gamma(v)=h} w_{uv}(z_{uv} - \hat{z}_{uv})^2, \tag{1.1}$$

where z_{uv} is the original value in row u, column v of the matrix, assigned to row cluster g and column cluster h and $\hat{z}_{uv}$ is the value approximated within co-cluster g-h. w_{uv} is the non-negative weight associated with matrix entry z_{uv}, which allows the algorithm to deal with missing values and data uncertainties. For example, the weights for known values can be set to 1 and missing values can be effectively ignored by setting their weights to 0. The objective function is hence the element-wise squared error between the original and the approximated value, summed only over the clustered elements $(s_r \times s_c)$ of the matrix Z. The value $\hat{z}_{uv}$ can be approximated in several ways, depending on the type of summary statistics that each co-cluster preserves. Banerjee et al. [2] identify six possible sets of summary statistics, of increasing complexity, that one might be interested in preserving in the reconstructed matrix $\hat{Z}$, which lead to six different co-clustering schemes. Two of these approximation schemes for $\hat{z}_{uv}$ are described in Section 1.3.2.

In the next step of ROCC, the goal is to agglomerate similar co-clusters to recover the arbitrarily positioned co-clusters. In order to agglomerate co-clusters, we first define a distance measure between two candidate co-clusters ($cc1$ and $cc2$) as follows. Let cc denote the co-cluster formed by the union of the rows and columns in $cc1$ and $cc2$. The matrix entries $\hat{z}_{uv}$ in cc are approximated using the selected approximation scheme. The average element-wise error e for cc is computed as $e = \frac{1}{N}\sum_{z_{uv}\in cc}(z_{uv} - \hat{z}_{uv})^2$, where N is the number of elements in cc. The error e is defined to be the distance between $cc1$ and $cc2$.

---

[1] A more general description, which allows any Bregman divergence as the loss function, is given in Section 1.3.2.

1.3.2 ROCC Algorithm

Solving Step 1 of the ROCC problem. A co-clustering (ρ, γ), that minimizes the objective function (1.1), can be obtained by an iterative algorithm. The objective function can be expressed as a sum of row or column errors, computed over the s_r rows and s_c columns assigned to co-clusters. If row u is assigned to row cluster g, the row error is the error summed over the appropriate s_c elements in the row, i.e., if $\rho(u) = g$, then $E_u(g) = \sum_{h=1}^{l} \sum_{v \in \mathcal{L}:\gamma(v)=h} w_{uv}(z_{uv} - \hat{z}_{uv}(g))^2$. For a fixed γ, the best choice of the row cluster assignment for row u is the g that minimizes this error, i.e., $\rho^{\text{new}}(u) = \arg{}_g \min E_u(g)$. After computing the best row cluster assignment for all the m rows, the top s_r rows with minimum error are selected to participate in the current row clusters. A similar approach is used to assign columns to column clusters. Note that the rows/columns that are not included in the current s_r/s_c rows/columns assigned to co-clusters are still retained since they could be included in the co-clusters in future iterations.

Given the current row and column cluster assignments (ρ, γ), the values $\hat{z}_{uv}$ within each co-cluster have to be updated by recomputing the required co-cluster statistics based on the approximation scheme. This problem is identical to the Minimum Bregman Information (MBI) problem presented in [2] for updating the matrix reconstruction $\hat{Z}$. Solving the MBI problem for this update is guaranteed to decrease the objective function.

This iterative procedure is described in Figure 1.2. Step 1(i) decreases the objective function due to the property of the MBI solution, while Steps 1(ii) and 1(iii) directly decrease the objective function. The objective function hence decreases at every iteration. Since this function is bounded from below by zero, the algorithm is guaranteed to converge to a locally optimal solution.

Solving Step 2 of the pOCC Problem. We now provide a heuristic to hierarchically agglomerate similar co-clusters. The detailed steps are:

(i) *Pruning co-clusters.* Since the desired number of co-clusters is expected to be significantly smaller than the number of co-clusters at this stage of the algorithm, co-clusters with the largest error values can be filtered out in this step. Filtering also reduces the computation effort required by the following merging step. If one has no idea of the final number of co-clusters, a simple and efficient filtering heuristic is to select the error cut-off value as the one at which the sorted co-cluster errors show the largest increase between consecutive values. The co-clusters with errors greater than the cut-off are filtered out. Alternatively, if the final number of co-clusters to be found is pre-specified, it can be used to prune away an appropriate number of co-clusters with the largest errors.

(ii) *Merging similar co-clusters.* This step involves hierarchical, pairwise agglomeration of the co-clusters left at the end of the pruning step (Step 2(i)) to recover the true co-clusters. Each agglomeration identifies the "closest" pair of co-clusters that can be well represented by a single co-cluster model and are thus probably part of the same original co-cluster, and merges them to form a new co-cluster.[2] "Closest" here is in terms of the smallest value of distance as defined in Section 1.3.1. The rows and columns of the new co-cluster consist of the union of the rows and columns of the two merged co-clusters. Merging co-clusters in this manner allows co-clusters to share rows and columns and hence allows partial overlap between co-clusters. If the number of co-clusters to be identified is pre-specified, one can stop merging when this number is reached. If not, merging is continued all the way until only a single co-cluster (or a reasonably small number of co-clusters) is left. The increase in the distance between successively merged co-clusters is then computed and the set of co-clusters just before the largest increase is selected as the final solution.

Overall ROCC meta-algorithm. Here we put together the procedures described in Sections 1.3.2 and 1.3.2 and present the complete ROCC algorithm. The key idea is to over-partition the data into small co-clusters arranged in a grid structure and then agglomerate similar, partitioned co-clusters to recover the desired co-clusters. The iterative procedure (Section 1.3.2) is run with large enough values for the number of row and column clusters (k and l). Similarly, the s_r and s_c input parameters are set to sufficiently large values. Since the pruning step (Step 2(i) in Section 1.3.2) takes care of discarding less coherent co-clusters, setting $s_r \geq s_r^{\text{true}}$ and $s_c \geq s_c^{\text{true}}$ is sufficient. The resulting $k \times l$ clusters are then merged as in hierarchical agglomerative clustering until a suitable stopping criterion is reached. The pseudo-code for the complete algorithm is illustrated in Figure 1.2.

Approximation schemes. The ROCC algorithm can use each of the six schemes (co-clustering bases) listed by Banerjee et al. [2] for approximating the matrix entries $\hat{z}_{uv}$. For concreteness, we illustrate

---

[2] A variant of this algorithm can be derived by adopting Ward's method [26] to agglomerate co-clusters. Empirically we found little difference between the two approaches.

two specific approximation schemes with squared Euclidean distance, which give rise to block co-clusters and pattern-based co-clusters respectively.[3] The meta-algorithm in Figure 1.2 uses C to refer to the selected co-clustering basis.

Block co-clusters. Let the co-cluster row and column indices be denoted by sets U and V respectively. In this case, a matrix entry is approximated as $\hat{z}_{uv} = z_{UV}$, where

$$z_{UV} = \frac{1}{|U||V|} \sum_{u \in U, v \in V} z_{uv}$$

is the mean of all the entries in the co-cluster.

Pattern-based co-clusters. z_{uv} is approximated as $\hat{z}_{uv} = z_{uV} + z_{Uv} - z_{UV}$, where

$$z_{uV} = \frac{1}{|V|} \sum_{v \in V} z_{uv}$$

is the mean of the entries in row u whose column indices are in V and

$$z_{Uv} = \frac{1}{|U|} \sum_{u \in U} z_{uv}$$

is the mean of the entries in column v whose row indices are in U. This approximation can identify co-clusters that show a coherent trend or pattern in the data values, making it suitable for clustering gene expression data [7].

Distance Measures. In Section 1.3.1 we developed the objective function (1.1) assuming squared Euclidean distance as the distance measure. The objective function and the iterative procedure to minimize it can be generalized to all Bregman divergences [2]. The selected Bregman divergence is denoted by d_ϕ in Figure 1.2.

ROCC with pressurization The iterative minimization procedure in Step 1 of the ROCC algorithm begins with random initialization for ρ and γ, which could lead to poor local minima. A better local minimum can be achieved by applying an extension of the pressurization technique used by BBC [14]. The adopted strategy is to begin by clustering all the data and iteratively shaving

---

[3] These co-cluster definitions correspond to basis 2 and basis 6 defined by the BCC framework [2] respectively.

Algorithm: ROCC
Input: $Z_{m \times n}, s_r, s_c, k, l$, basis C, d_ϕ
Output: Set of co-clusters

Step 1
Begin with a random co-clustering (ρ, γ)
Repeat
Step (i): Update co-cluster models, $\forall [g]_1^k, [h]_1^l,$
Update statistics for co-cluster (g, h) based on
basis C to compute new $\hat{z}$ values

Step (ii): Update ρ
(iia). $\forall [u]_1^m, \rho(u) = \arg{}_g \min \sum_{h=1}^l \sum_{v \in \mathscr{L}: \gamma(v)=h} w_{uv} d_\phi(z_{uv}, \hat{z}_{uv}(g))$
(iib). $\mathscr{K} = s_r$ rows with least error from among the m rows

Step (iii): Update γ
(iiia). $\forall [v]_1^n, \gamma(v) = \arg{}_h \min \sum_{g=1}^k \sum_{u \in \mathscr{K}: \rho(u)=g} w_{uv} d_\phi(z_{uv}, \hat{z}_{uv}(h))$
(iiib). $\mathscr{L} = s_c$ columns with least error from among the n columns
until convergence

Step 2: Post-process (see text for details)
(i) Prune co-clusters with large errors.
(ii) Merge similar co-clusters until stopping criterion is reached.
return identified co-clusters.

Figure 1.2 Pseudo-code for ROCC meta-algorithm

off data points and features till s_r rows and s_c columns are left. Let $s_r^{\text{press}}(j)$ and $s_c^{\text{press}}(j)$ denote the number of data points and features to be clustered using the Step 1 procedure (Figure 1.2) in the jth iteration of pressurization. $s_r^{\text{press}}(1)$ and $s_c^{\text{press}}(1)$ are initialized to m and n respectively, after which these parameters are decayed exponentially till $s_r^{\text{press}}(j) = s_r$ and $s_c^{\text{press}}(j) = s_c$. The rate of decay is controlled by parameters β_{row} and β_{col}, which lie between 0 and 1. At iteration j, $s_r^{\text{press}}(j) = s_r + \lfloor (m - s_r) * \beta_{\text{row}}^{j-1} \rfloor$ and $s_c^{\text{press}}(j) = s_c + \lfloor (m - s_c) * \beta_{\text{col}}^{j-1} \rfloor$. The intuition is that by beginning with all the data being clustered and then slowly reducing the fraction of data clustered, co-clusters can move around considerably from their initial positions to enable the discovery of small, coherent patterns.

1.4 Application: Finding Dense Co-clusters in Microarray Data

This section presents an evaluation of the performance of ROCC on two yeast microarray datasets, the Lee dataset [19] and the Gasch dataset [12]. The Lee dataset consists of gene expression values of 5612 yeast genes across 591 experiments and can be obtained from the Stanford Microarray Database (`http://genome-www5.stanford.edu/`). The Gasch dataset consists of the expression values of 6151 yeast genes under 173 environmental stress conditions and is available at `http://genome-www.stanford.edu/yeast_stress/`. Since the ground truth for both datasets is available only in the form of pairwise linkages between the genes that are known to be functionally related, we compare the quality of the co-clusters identified by different co-clustering algorithms by computing the overlap lift [14] for the genes in each co-cluster. Overlap lift measures how many times more correct links are predicted as compared to random chance and is related to a normalized version of the proportion of disconnected genes measure used in [23]. On these datasets, the aim is to find the most coherent and biologically useful 150 to 200 co-clusters.

Figure 1.3 compares the performance of ROCC with prominent co-clustering algorithms, i.e., Cheng and Church's Biclustering algorithm, the OPSM algorithm [4], the BiMax algorithm [23], and the BCC algorithm on the Lee and Gasch microarray datasets. Through extensive experimentation, Prelic et al. [23] show that the OPSM and the BiMax algorithms outperform other well known co-clustering algorithms like Samba [24], ISA [5] and xMotif [21] on real microarray data. The BiMax and OPSM results were generated using the BicAT software(`http://www.tik.ee.ethz.ch/sop/bicat/`) [3]. Since BCC clusters all the data, pruning is carried out by a post-processing step. Both ROCC and BCC use squared Euclidean distance and find pattern-based co-clusters. The ROCC, BCC and Biclustering results are averaged over 10 trials, while OPSM and BiMax are deterministic.

Figure 1.3 shows that on both datasets, ROCC does much better than the other co-clustering approaches in terms of the overlap lift of the gene clusters. The figure also displays above each bar, the percentage of the data matrix entries clustered by the corresponding algorithm. On the Lee dataset, it is interesting that although ROCC clusters a much larger fraction of the data matrix entries than Biclustering, OPSM and BiMax, the co-clusters are of superior quality. A more detailed discussion on the experimental evaluation of ROCC is presented in [8].

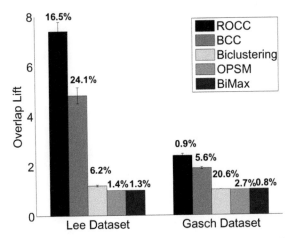

Figure 1.3 Comparison of ROCC with other co-clustering algorithms on the Lee and Gasch datasets

1.5 Concluding Remarks

In this chapter, we discussed the problem of identifying multiple dense co-clusters in noisy data and motivated this setting with the requirements of microarray data analysis. Several approaches to this problem were discussed, and a promising framework, Robust Overlapping Co-clustering was examined in detail. ROCC is robust to the presence of irrelevant data points and features, and discovers coherent co-clusters very accurately as illustrated in Section 1.4. Moreover, though ROCC requires several input parameters to be supplied, i.e., s_r, s_c, k and l, it is relatively very robust to the choice of these parameters because of the post-processing steps as detailed in Section 1.3.2. It will be worthile to further investigate the applicability of suitable instances of the ROCC framework to clustering problems in different domains like text mining and market basket analysis.

Acknowledgments

This research was supported by NSF grant IIS 0713142. We are grateful to Hyuk Cho and Inderjit Dhillon from the CS department at the University of Texas at Austin for insightful discussions and help with empirical studies.

References

[1] M. Ankerst, M.M. Breunig, H.-P. Kriegel, and J. Sander. OPTICS: Ordering points to identify the clustering structure. In *Proc. SIGMOD '99*, pages 49–60, 1999.

[2] A. Banerjee, I. Dhillon, J. Ghosh, S. Merugu, and D. Modha. A generalized maximum entropy approach to Bregman co-clustering and matrix approximation. *Journal of Machine Learning Research*, 8:1919–1986, 2007.

[3] S. Barkow, S. Bleuler, A. Prelic, P. Zimmermann, and E. Zitzler. Bicat: A biclustering analysis toolbox. *Bioinformatics*, 22(10):1282–1283, 2006.

[4] A. Ben-Dor, B. Chor, R. Karp, and Z. Yakhini. Discovering local structure in gene expression data: the order-preserving submatrix problem. In *Proc. Research in Comp. Mol. Bio. '02*, pages 49–57, 2002.

[5] S. Bergmann, J. Ihmels, and N. Barkai. Iterative signature algorithm for the analysis of large-scale gene expression data. *Phys. Rev. E. Stat. Nonlin. Soft Matter Phys.*, 67, 2003.

[6] Y. Cheng and G.M. Church. Biclustering of expression data. In *Proc. Intell. Syst. Mol. Bio. '00*, pages 93–103, 2000.

[7] H. Cho and I. Dhillon. Co-clustering of human cancer microarrays using minimum sum-squared residue co-clustering. *IEEE/ACM Trans. on Comp. Bio. and Bioinfo.*, 5:385–400, 2008.

[8] M. Deodhar, H. Cho, G. Gupta, J. Ghosh, and I. Dhillon. Robust overlapping co-clustering. Department of ECE, University of Texas at Austin, IDEAL-TR09, page Downloadable from http://www.lans.ece.utexas.edu/papers/techreports/deodhar08ROCC.pdf, 2009.

[9] I. Dhillon, S. Mallela, and R. Kumar. A divisive information-theoretic feature clustering algorithm for text classification. *Journal of Machine Learning Research*, 3:1265–1287, 2003.

[10] I. Dhillon, S. Mallela, and D. Modha. Information-theoretic co-clustering. In *Proc. KDD '03*, pages 89–98, 2003.

[11] M. Ester, H. Kriegel, J. Sander, and X. Xu. A density-based algorithm for discovering clusters in large spatial databases with noise. In *Proc. KDD '96*, 1996.

[12] A.P. Gasch, P.T. Spellman, C.M. Kao, O. Carmel-Harel, M.B. Eisen, G. Stotz, D. Botstein, and P.O. Brown. Genomic expression program in the response of yeast cells to environmental changes. *Molecular Cell Biology*, 11:4241–4257, 2000.

[13] J. Ghosh. Scalable clustering. In Nong Ye (Ed.), *The Handbook of Data Mining*, pages 247–277. Lawrence Erlbaum Assoc., 2003.

[14] G. Gupta and J. Ghosh. Bregman bubble clustering: A robust, scalable framework for locating multiple, dense regions in data. In *Proc. ICDM'06*, pages 232–243, 2006.

[15] G. Gupta, A. Liu, and J. Ghosh. Automated hierarchical density shaving: A robust, automated clustering and visualization framework for large biological datasets. *IEEE/ACM Trans. on Comp. Bio. and Bioinfo.*, 2008.

[16] A.K. Jain and R.C. Dubes. *Algorithms for Clustering Data*. Prentice Hall, New Jersey, 1988.

[17] D. Jiang, C. Tang, and A. Zhang. Cluster analysis for gene expression data: A survey. *IEEE Trans. Knowl. Data Eng.*, pages 1370–1386, 2004.

[18] L. Lazzeroni and A. B. Owen. Plaid models for gene expression data. *Statistica Sinica*, 12(1):61–86, 2002.

[19] I. Lee, S. Date, A. Adai, and E. Marcotte. A probabilistic functional network of yeast genes. *Science*, 306:1555–1558, 2004.

[20] S.C. Madeira and A.L. Oliveira. Biclustering algorithms for biological data analysis: A survey. *IEEE/ACM TCBB*, 1(1):24–45, 2004.

[21] T. Murali and S. Kasif. Extracting conserved gene expression motifs from gene expression data. *Pacific Symposium on Biocomputing*, 8:77–88, 2003.

[22] L. Parsons, E. Haque, and H. Liu. Subspace clustering for high dimensional data: a review. *SIGKDD Explor. Newsl.*, 6(1):90–105, 2004.

[23] A. Prelic, S. Bleuler, P. Zimmermann, A. Wille, et al. A systematic comparison and evaluation of biclustering methods for gene expression data. *Bioinformatics*, 22(9):1122–1129, 2006.

[24] A. Tanay, R. Sharan, and R. Shamir. Discovering statistically significant biclusters in gene expression data. *Bioinformatics*, 18:136–144, 2002.

[25] H. Wang, W. Wang, J. Yang, and P.S. Yu. Clustering by pattern similarity in large data sets. In *Proc. SIGMOD '02*, pages 394–405, 2002.

[26] J. Ward. Hierarchical grouping to optimize an objective function. *Journal of American Statistical Association*, 58(301):236–244, 1963.

[27] D. Wishart. Mode analysis: A generalization of nearest neighbour which reduces chaining effects. In *Proc. Colloquium in Numerical Taxonomy*, pages 282–308, 1968.

[28] X. Xu, Y. Lu, A. Tung, and W. Wang. Mining shifting-and-scaling co-regulation patterns on gene expression profiles. In *Proc. ICDE '06*, page 89, 2006.

[29] M. Zhang, W. Wang, and J. Liu. Mining approximate order preserving clusters in the presence of noise. In *Proc. ICDE '08*, pages 160–168, 2008.

2

Recognizing Patterns in Texts

Minhua Huang and Robert M. Haralick

Computer Science, Graduate Center, City University of New York,
New York, NY 10016, USA; E-mail: mhuang@gc.cuny.edu, haralick@aim.com

Abstract

We discuss a probabilistic graphical model for recognizing patterns in texts. It is derived from the probability function for a sequence of categories given a sequence of symbols under two reasonable conditional independence assumptions and represented by a product of combinations of conditional and marginal probability functions. The novelty of our model is that it has a mathematical representation which is completely different from existing graphical models such as CRFs, HMMs, and MEMMs. In our model, the global maximum probability is obtained by the local maximal probabilities. The time and memory complexities for predicating new texts are much less than these existing graphical models. It is also more stable and reliable. Up to now, we have used this model for recognizing NP chunks and senses of a polysemous word in sentences. This model has achieved very promising results on standard data sets. Moreover, we will use this model for extracting semantic roles of a verb in a sentence.

Keywords: probabilistic graphical model, clique, separator, NP chunking, word sense disambiguation, semantic role labelling.

2.1 Introduction

Knowing text patterns can help a machine to understand semantics of a sentence. These patterns can be viewed as syntactic patterns or semantic pat-

Patrick Shen-Pei Wang (Ed.), Pattern Recognition and Machine Vision – In Honor and Memory of Professor King-Sun Fu, 19–35.
© 2010 *River Publishers. All rights reserved.*

terns. For example, NP chunks (noun phrases) are syntactic patterns because they are defined by grammatical rules while senses of a polysemous word are semantic patterns because they can be identified by the contexts of the word. Here, we discuss a probabilistic graphical model for recognizing these patterns. The mathematical representation of our model is: $p(c_1, \ldots, c_N \mid s_1, \ldots, s_N) = \prod_{n=1}^{N} p(s_{n-1} \mid s_n, c_n) \, p(s_{n+1} \mid s_n, c_n) \, p(s_n \mid c_n) \, p(c_n)$. It is derived from the probability function of a sequence of categories $(c_1, \ldots, c_N)$ given a sequence of symbols $(s_1, \ldots, s_N)$ where the symbols carry the information in the lexicon and POS tags in a sentence. It has a different perspective compared with the existing graphical models such as CRFs [9], HMMs [16], MEMMs [13].

For identifying these patterns in texts, a sentence is partitioned into several phrases. In the case of the NP chunking or the semantic role labelling problems, these phrases can be NP chunks or semantic roles and OTHERs. In the case of the word sense disambiguation problem, the polysemous word is represented by its context. The context is a sequence of words made by the last words of each phrase or the particular phrase in which the polysemous word is embedded. By the model, we automatically assign the highest probability to each word of the sequence. We determine NP chunks by grouping consecutive words with the same particular category. We determine the category sense of the ambiguous word by selecting the most frequent category sense assigned to the context. We determine semantic roles by first, grouping consecutive words having the same assigned particular category, then categorizing these roles into different classes by designing a set of rules.

We test our model for identifying NP chunks with two data sets: the WSJ data set from the Penn Treebank and the CoNLL-2000 shared task data set. Our method achieves an average precision 97.7% and an average recall 98.7% on the first data set and an average precision 95.15% and an average recall 96.05% on the second data set. Moreover, we test our model for word sense disambiguation by the *line_serve_hard_interest* data sets. Our model achieves an average of precision 91.38% and an average of recall 91.08% for identifying the *project* sense of the word *line*; an average of precision 91.36% and an average of recall 90.07% for identifying the *supply_with_food* sense of the word *serve*; an average of precision 86.50% and an average of recall 91.43% for distinguishing the *difficult* sense of the word *hard*; an average of precision 89.50% and an average of recall 91.78% for identifying the *the_money_paid_for_the_use_of_money* sense of the word *interest*. We are going to test our model for semantic role labeling on the CoNLL-2005 shared task data set.

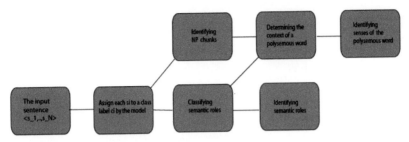

Figure 2.1 The relationship between NP chunking, semantic role labeling, and word sense disambiguation

2.2 The Method

2.2.1 An Example

Table 2.1 shows the input sentence *"He had deposited his paycheck to the local PNC bank last Saturday morning"* with its POS tags. In our method, in NP chunking and semantic role labelling, each word of the sentence is assigned to one of three different categories C_1, C_2, and C_3. While C_1 represents a word inside a block (a block can be a NP chunk or a semantic role), C_2 represents a word outside a block, C_3 represents a word starting a new block. NP chunks or semantic roles are formed by grouping successive words with the same category C_1 or starting with the category C_3 and followed by zero, one, or more consecutive C_1s. Moreover, different semantic roles are needed to be identified from the semantic roles obtained from this step by creating a method. In the word sense disambiguating, the context of the polysemous word *bank* is found by grouping the words corresponding to the last C_1 of consecutive C_1s or last C_2 of consecutive C_2s. Then, we assign C_1 and C_2 to each of the words in the context. While C_1 represents the fixed sense (financial sense) of the polysemous word 'bank', C_2 represents the others. We determine the sense of the polysemous word 'bank' by selecting the most frequent category which is C_1.

2.2.2 Describing the Task

Let L be a language, V be a vocabulary of L, and T be *POS* tags of V. Let S be a sequence of symbols associated with a sentence, $S = (s_1, \ldots, s_N)$, where $s_n = \langle w_n, t_n \rangle$, $w_n \in V$, $t_n \in T$. Let B be a block such as an NP chunk or a semantic role. Let C be a set of categories, $C = \{C_1, C_2, C_3\}$, where C_1 indicates the current symbol is in B, C_2 indicates the current symbol is not in

Table 2.1 An example of recognizing text patterns

Words in the sentence	POS tags	NP chunks	'bank' context	Semantic roles	NP chunks	The sense of 'bank'	Semantic roles
He	NNP	C_1	C_2	C_1	$\vert^1_1$		$\vert^{A_0}_{A_0}$
had	VBD	C_1		C_2			
deposited	VBN	C_2	C_1	C_2			Verb
his	PRP	C_1		C_2	$\vert^2$		$\vert^{A_1}$
paycheck	NNS	C_1	C_1	C_1	$\vert_2$		$\vert_{A_1}$, $\vert^{A_2}$
to	TO	C_2	C_1	C_3			
the	DT	C_1		C_1	$\vert^3$		$\vert_{A_2}$
local	JJ	C_1		C_1	$\vert_3$		$\vert_{A_2}$
PNC	NN	C_1		C_1	$\vert_3$		$\vert_{A_2}$
bank	NNS	C_1	C_1	C_1	$\vert_3$	Financial	$\vert_{A_2}$
last	JJ	C_3		C_3	$\vert^4$		$\vert^{A_3}$
Saturday	NNP	C_1		C_1	$\vert_4$		$\vert_{A_3}$
morning	NN	C_1	C_1	C_1	$\vert_4$		$\vert_{A_3}$

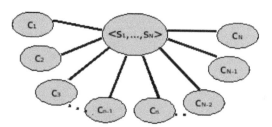

Figure 2.2 The graphical model of $p(c_1, \ldots, c_N \mid s_1, \ldots, s_N)$ under the assumption c_i is independent of $c_{j \neq i}$ given $(s_1, s_2, \ldots, s_N)$

B, and C_3 starts a new B. The tasks can be stated as given $S = (s_1, \ldots, s_N)$, we need to find

1. a sequence of categories, $(c_1, \ldots, c_N)$, $c_i \in C$, with the best description of S;
2. all the Bs based on $(c_1, \ldots, c_N)$, s.t. $B = \{B_1, \ldots, B_M\}$, s.t. $B_i \cap B_j = \phi$ and $B_i \subset S$ (see Section 2.2.6 for the definition of a block B).

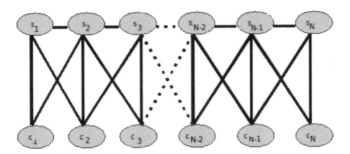

Figure 2.3 The probabilistic graphical model of $p(c_i \mid s_1, \ldots, s_N)$ under assumptions c_i is independent of $c_{j \neq i}$ given $(s_1, s_2, \ldots, s_N)$ and c_i is independent of $(s_1, \ldots, s_{i-2}, s_{i+2}, \ldots, s_N)$ given (s_{i-1}, s_i, s_{i+1})

2.2.3 Building Probabilistic Graphical Models

Given $S = (s_1, s_2, \ldots, s_N)$, $C = \{C_1, C_2, C_3\}$, for $s_i \in S$, we want to find $c_i \in C$, s.t.

$$(c_1, c_2, \ldots, c_N) = \underset{c_1, c_2, \ldots, c_N}{\text{argmax}} \; p(c_1, c_2, \ldots, c_N \mid s_1, s_2, \ldots, s_N). \quad (2.1)$$

According to the chain rule that follows from the definition of conditional probability

$$p(c_1, c_2, \ldots, c_N \mid s_1, s_2, \ldots, s_N)$$
$$= p(c_1 \mid c_2, \ldots, c_N, s_1, s_2, \ldots, s_N) p(c_2 \mid c_3, \ldots, c_N, s_1, s_2, \ldots, s_N) \cdots$$
$$p(c_N \mid s_1, s_2 \ldots, s_N).$$

Suppose c_i is independent of $c_{j \neq i}$ given $(s_1, s_2, \ldots, s_N)$. This means that the symbol sequence contains all the information with respect to the category chain associated with any word. From this point of view, $p(c_1, \ldots, c_N \mid s_1, \ldots, s_N)$ can be represented as shown in Figure 2.2 and Equation (2.2).

$$p(c_1, c_2, \ldots, c_N \mid s_1, s_2, \ldots, s_N) = \prod_{i=1}^{N} p(c_i \mid s_1, \ldots, s_N). \quad (2.2)$$

Assume c_i is independent of $(s_1, \ldots, s_{i-2}, s_{i+2}, \ldots, s_N)$ given (s_{i-1}, s_i, s_{i+1}). This means that all the information pertaining to the category class of the word i is in entities contained by the symbol associated with the word i, its predecessor word $i - 1$, and its successor word $i + 1$.

$$p(c_1, c_2, \ldots, c_N \mid s_1, s_2, \ldots, s_N) = \prod_{i=1}^{N} p(c_i \mid s_{i-1}, s_i, s_{i+1}). \quad (2.3)$$

The probability graphical model using these assumptions is shown in Figure 2.3. From this model, a set of $2N - 2$ cliques[1] is obtained:

$$\text{CIL} = \big\{ \{s_1, s_2, c_1\}, \{s_1, s_2, c_2\}, \{s_2, s_3, c_2\}, \ldots,$$

$$\{s_{N-1}, s_N, c_{N-1}\}, \{s_{N-1}, s_N, c_N\} \big\}.$$

Moreover, there is a corresponding set of $2N - 3$ separators:[2]

$$\text{SEP} = \big\{ \{s_1, s_2\}, \ldots, \{s_{N-1}, s_N\}, \{s_2, c_2\}, \ldots, \{s_{N-1}, c_{N-1}\} \big\}.$$

The junction tree[3] is formed as shown in Figure 2.4. The cliques are represented as nodes and separators are represented as edges. From this model, according to Bishop [2], $p(c_1, \ldots, c_N \mid s_1, \ldots, s_N)$ can be computed by the product of the probability of the cliques divided by the product of the probabilities of the separators. Hence:

$$p(c_1, \ldots c_N \mid s_1, \ldots, s_N) =$$

$$M_{s_1, \ldots, s_N} \prod_{n=1}^{N} p(s_{n-1} \mid s_n, c_n) p(s_{n+1} \mid s_n, c_n) p(s_n \mid c_n) p(c_n). \, (2.4)$$

$M_{s_1, \ldots, s_M}$ is a constant depending only on $s_1, \ldots, s_n$ and not depending on any c_n. It is defined by

$$M_{s_1, \ldots, s_N} = \frac{1}{p(s_1, s_2) p(s_2, s_3) \ldots p(s_{N-1}, s_N)}. \quad (2.5)$$

The derivation of Equation (2.4) can be found in the Appendix. Note that we define $p(s_0 \mid s_1, c_1) = p(s_{N+1} \mid s_N, c_N) = 1$.

---

[1] A clique is a maximal complete set of nodes.

[2] $\Gamma = \{\Gamma_1, \ldots, \Gamma_M\}$ is a set of separators, where $\Gamma_k = \Lambda_k \cap (\Lambda_1 \cup \ldots, \cup \Lambda_{k-1})$.

[3] A junction tree is a maximum spanning tree w.r.t separator size. A maximum spanning tree is a tree over V whose sum of edge weights has a maximum value. Here the edge weights are the sizes of the separators.

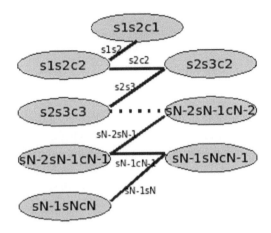

Figure 2.4 A junction tree for $p(c_1, \ldots c_N \mid s_1, \ldots, s_N)$

2.2.4 Making Decisions

Because each c_n is independent of each other in (2.4), its value can be determined individually. We can find c_n by $c_n = \text{argmax}_{c_n} p(s_{n-1} \mid s_n, c_n) p(s_{n+1} \mid s_n, c_n) \, p(s_n \mid c_n) p(c_n)$. Then, (2.1) can be rewritten as

$$(c_1, c_2, \ldots, c_N) = (\underset{c_1}{\text{argmax}}(p(s_2 \mid s_1, c_1) p(s_1 \mid c_1) p(c_1)),$$

$$\underset{c_2}{\text{argmax}}(p(s_1 \mid s_2, c_2) p(s_3 \mid s_2, c_2) p(s_2 \mid c_2) p(c_2)),$$

$$\ldots,$$

$$\underset{c_{N-1}}{\text{argmax}}(p(s_{N-2} \mid s_{N-1}, c_{N-1}) p(s_N \mid s_{N-1}, c_{N-1}) p(s_{N-1} \mid c_{N-1}) p(c_{N-1})),$$

$$\underset{c_N}{\text{argmax}}(p(s_{N-1} \mid s_N, c_N) p(s_N \mid c_N) p(c_N))). \tag{2.6}$$

2.2.5 Estimating Probabilities of the Model

We use a training set to estimate the probabilities for Equation (2.4). In our model, the probability of a current symbol being assigned to the class c in a sequence associated with a sentence is partially dependent on the probability of the previous symbol given the current symbol and the class c and the probability of the successive symbol given the current symbol and the class c. In this way, the adjacency between two neighboring symbols of an incoming sentence are preserved by the overlapping of our model.

Therefore, we can consider a group of k sentences for estimating the probabilities. We consider that the training set has K sentences. Each sentence k consists of N_k words, $\langle x_{k,1}, \ldots, x_{k,N_k} \rangle$, and the corresponding class labels $\langle y_{k,1}, \ldots, y_{k,N_k} \rangle$. Hence the training set is $\Psi = \{\psi_1, \psi_2, \ldots, \psi_K\}$, where $\psi_k = (\langle x_{k,1}, y_{k,1} \rangle, \ldots, \langle x_{k,N_k}, y_{k,N_k} \rangle)$.

Let t, w, z, c be random variables, where w designates a word, t designates a word before w, z designates a word after w, and c designates a class. Let $p_{w|c}(\alpha \mid \gamma)$ designate the conditional probability of a word being α given that its class is γ. Let $p_{t|w,c}(\alpha \mid \beta, \gamma)$ designate the conditional probability of the word previous to the current word being α given that the current word is β and its class is γ. Let $p_{z|w,c}(\alpha \mid \beta, \gamma)$ designate the conditional probability of the word after the current word being α given that the current word is β and its class is γ.

Let

$$I = \{(k, n) \mid k = 1, \ldots, K, n = 1, \ldots, N_k\}.$$

$p_{w|c}(\alpha \mid \gamma)$ can be estimated by

$$\hat{p}_{w|c}(\alpha \mid \gamma) = \frac{\#\{(k, n) \in I \mid \alpha = x_{k,n}, \gamma = y_{k,n}\}}{\#\{(k, n) \in I \mid \gamma = y_{k,n}\}}. \tag{2.7}$$

$p_{t|w,c}(\alpha \mid \beta, \gamma)$ can be estimated by

$$\hat{p}_{t|w,c}(\alpha \mid \beta, \gamma) = \frac{\#\{(k, n) \in I \mid \alpha = x_{k,n-1}, \beta = x_{k,n}, \gamma = y_{k,n}\}}{\#\{(k, n) \in I \mid \beta = x_{k,n}, \gamma = y_{k,n}\}}. \tag{2.8}$$

$p_{z|w,c}(\alpha \mid \beta, \gamma)$ can be estimated by

$$\hat{p}_{z|w,c}(\alpha \mid \beta, \gamma) = \frac{\#\{(k, n) \in I \mid \alpha = x_{k,n+1}, \beta = x_{k,n}, \gamma = y_{k,n}\}}{\#\{(k, n) \in I \mid \beta = x_{k,n}, \gamma = y_{k,n}\}}. \tag{2.9}$$

2.2.6 Determining an NP Chunk or a Semantic Role

We have determined $(c_1, \ldots, c_N)$ by (2.6). To form an NP chunk or a semantic role, we group together the consecutive words assigned with the same category. Let $S = (s_1, \ldots, s_N)$ be a sequence of symbols associating with a sentence. Let $(c_1, \ldots, c_N)$ be a sequence of category labels with the best description of S. Hence, from the method, we have $\mathcal{T} = (T_1, \ldots, T_N) = (\langle s_1, c_1 \rangle, \ldots, \langle s_N, c_N \rangle)$. Let $C = \{C_1, C_2, C_3\}$ be a set of class labels, where C_1 represents a symbol is in an NP chunk or semantic role, C_2 represents a symbol is not in an NP chunk or semantic role, C_3 represents a symbol starts in a new NP chunk or semantic role.

We define $\mathcal{B}$ to be a block if and only if:

1. for some $i < j$, $\mathcal{B} = (\langle s_i, c_i \rangle, \langle s_{i+1}, c_{i+1} \rangle, \ldots, \langle s_j, c_j \rangle)$
2. $c_i \in \{C_1, C_3\}$.
3. $c_n = C_1$, $n = i + 1, \ldots, j$.
4. B is maximal; $\mathcal{B}' \supseteq \mathcal{B}$ and $\mathcal{B}'$ satisfying (1), (2), and (3) implies $\mathcal{B}' = \mathcal{B}$.

Starting from the first label of the sequence, we try to find the succeeding label with the same class label. If it is the same, then we group these two together and try to find the next one. Otherwise, we break the chain and find the new chain.

2.2.7 Determining the Sense of a Polysemous Word

For this task, we consider that the polysemous word is represented by a symbol sequence. This sequence may be the original symbol sequence, or a sequence of head words with their POS tags extracted from NP chunking task, or a sequence of head words with their POS tags extracted from semantic role labeling task. An example of such sequence can be found in Table 2.1, where each head word is the last word of an NP chunk or a $\neg$ NP chunk.

From the model of Equation (2.4), for an input sequence $S = \langle s_1, \ldots, s_N \rangle$, we have obtained $(\langle s_1, c_1 \rangle, \ldots, \langle s_N, c_N \rangle)$, $c_i \in C = \{0, 1\}$. 1 represents a given fixed sense of the polysemous word while 0 represents others. We assign the class 1 for the polysemous word w_t if and only if:

$$\#\{c_n = 1 \mid n = 1, \ldots, N, s_n \in S\}$$
$$> \#\{c_m = 0 \mid m = 1, \ldots, N, s_m \in S\} \qquad (2.10)$$

2.3 Empirical Results

We define features based on Equation (2.4) as follows:

$$f(s_i, c_i) = p(c_i)q(s_i \mid c_i)r(s_{i-1} \mid s_i, c_i)t(s_{i+1} \mid s_i, c_i). \qquad (2.11)$$

We form the training set by including 90% instances and the testing set by including 10% instances of the whole data set. In this way, we do it iteratively 10 times to select the different training sets and testing sets. The evaluation

Table 2.2 The results on the CoNLL-2000 data

Measurement	Lexicon + POS tags	POS tags	Lexicon
	%	%	%
P_{re}	95.15	92.27	86.42
R_{ec}	96.05	93.76	93.35
F_{me}	95.59	92.76	89.75

metric we have used are precision p_{re}, recall R_{ec}, and f-measure

$$f_{me} = \frac{2 * p_{re} * R_{ec}}{P_{re} + R_{ec}}.$$

2.3.1 Identifying NP Chunks Using CoNLL-2000 Shared Task Data Set

We have conducted three different tests on the CoNLL-2000 shared task data set by choosing different values of features. From these selections, we examine the probabilities in order to find the one which contributes the best performance. In the first test, we include all lexicon and POS tags from the data set. We have tested our model according to descriptions in Section 2.3. We averaged the results that we received. The average precision is 95.15%, the average recall is 96.05%, and the average f-measure is 95.59%. In the second test, the lexicon is excluded. We only include the POS tags. In the third test, all POS tags are excluded and only the lexicon is included. The results are shown in Table 2.2. By comparing the three results on the CoNLL-2000 shared task data, we have noticed that if the model is built only on the lexical information, it has the lowest performance of f-measure 89.75%. The model's performance improved 3% in f-measure if it is constructed by POS tags. The model achieves the best performance of 95.59% in f-measure if we are considering both lexicons and POS tags.

2.3.2 Identifying NP Chunks Using WSJ Data Set from Penn Treebank

The second data set on which we have experimented is the WSJ data of Penn Treebank. The main reason for using this data set is that we want to see whether the performance of our model can be improved when it is built on more data. We build our model on a training set which is seven times larger than the CoNLL-2000 shared task training data set (Section 2.3.3). The performance of our method for the data is listed in Table 2.3. The average

Table 2.3 The test results on the WSJ data from the Penn Treebank

Training 800 files	Testing 100 files	P_{re}	R_{ec}	F_{me}
200–999	1100–1199	0.9806	0.9838	0.9822
	1200–1299	0.9759	0.9868	0.9814
	1300–1399	0.9794	0.9863	0.9828
	1400–1499	0.9771	0.9868	0.9817
	1500–1599	0.9768	0.9858	0.9814
	1600–1699	0.9782	0.9877	0.9829
	1700–1799	0.9770	0.9877	0.9824
	1800–1899	0.9771	0.9848	0.9809
	1900–1999	0.9774	0.9863	0.9819
	2000–2099	0.9735	0.9886	0.9806
μ		0.9773	0.9865	0.9818
σ		0.0019	0.0014	0.0008

precision is increased 2.7% from 95.15 to 97.73%. The average recall is increased 2.8% from 96.05 to 98.65%. The average f-measure is increased 2.7 from 95.59 to 98.2%.

2.3.3 Identifying Sense of Polysemous Words Using *line_interest_hard_serve* Data Sets

We test our model for WSD on the data sets *line*, *hard*, *serve*, and *interest*. The senses' descriptions and instances' distributions can be found in [11] and [3]. Because of the limitations of number of instances (a sentence having the polysemous word in it) for each sense in the corpora, we select the first three senses for each polysemous word in our test. Again, the values of features are made by lexicon + POS tags. We test our model based the descriptions of Section 2.3. The test results are shown in Table 2.4.

We have noticed that, with the same instances for a polysemous noun, adjective, or verb, our model achieves the best f-measure for polysemous nouns and the worst result for polysemous adjectives. For example, the average $f_{me} > 91\%$ if the number of instances > 1270 for the polysemous nouns *line* and *interest*. However, in order to keep the same f-measure value, the polysemous word *serve* needs to have more instance: 1841 instances. The polysemous adjective needs to have about 3350 instances to reach the average $f_{me} = 88.76\%$. Moreover, in any case, our model achieves an average $f_{ma} = 80\%$ if the number of instances is reduced to about 400. We conclude that the performance of our model on WSD is dependent on the number of instances in the training set: the larger the better.

Table 2.4 The results on *line, hard, serve, interest* data

Word	Sense Description	# of Instance	f_{me} %
line	*project*	2218	92.24
noun	*phone*	429	85.22
	text	404	81.95
hard	*difficult*	3345	88.76
adj	*not soft*	502	83.75
	physical not soft	376	80.05
serve	*supply with food*	1841	91.04
verb	*hold an office*	1272	87.48
	function as something	853	82.52
interest	*money paid for the use of money*	1272	91.45
	a share in		
noun	*a company*	500	88.85
	readiness to		
	give attention	361	79.95

2.4 Related Researches

Currently existing graphical models for NLP are HMMs [13] [16], MEMMs [13], and CRFs [9, 18]. These models are built under different conditional independence assumptions for obtaining the sequence $\langle c_1, \ldots, c_N \rangle$ that maximizes $p(c_1, \ldots, c_N, s_1, \ldots, s_N)$ or $p(c_1, \ldots, c_N \mid s_1, \ldots, s_N)$. Among these models, HMMs and MEMMs are directed graphic models while CRFs and our model are undirected graphical models. These models can be found in Figure 2.5. Comparing with these two undirected graphic models, each c_i links to c_{i-1} and s_i in CRFs. Therefore, c_i is dependent on c_{i-1} and s_i. In contrast, in our model, each c_i links to s_i, s_{i-1}, and s_{i+1}, not the previous category c_{i-1}. Therefore, c_i is not dependent on c_{i-1}. This makes it possible for the sequence $\langle c_1, \ldots, c_N \rangle$ with the maximum value of $p(c_1, \ldots, c_N \mid s_1, \ldots, s_N)$ be determined by finding each c_i that satisfies Equation (2.6). In this way, the time complexity for recognizing a new incoming sequence with N symbols at the worst case is $M * N$, where M is the number of categories. For example, if $C = \{C_1, C_2, C_3\}$, then $M = 3$. Therefore, the time complexity is $O(N)$. The memory space also will be reduced compared the other graphic models because we do not need to store all the previous category chains into the memory. Moreover, our model is more reliable and stable due to the global maximum probability being obtained by the local maximal probabilities.

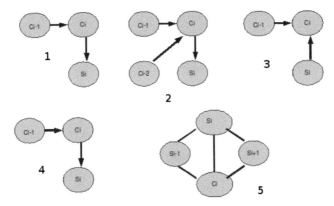

Figure 2.5 (1) A HMM model, (2) a second order HMM $p(s_1, \ldots, s_N, c_1, \ldots, c_n) = \prod_{i=1}^{N} p(s_i \mid c_i) p(c_i \mid c_{i-1}, c_{i-2})$, (3) a MEMM model $p(s_1, \ldots, s_N, c_1, \ldots, c_n) = \prod_{i=1}^{N} p(c_i \mid c_{i-1}, s_i)$, (4) a CRF model $p(c_1, \ldots, c_N \mid s_1, \ldots, s_N) = \prod_{i=1}^{N} p(c_i \mid s_i) p(c_i \mid c_{i-1}) p(c_{i-1})$, and (5) the model presented in this chapter $p(c_1, \ldots, c_N \mid s_1, \ldots, s_N) = \prod_{i=1}^{N} p(s_{i-1} \mid s_i, c_i) p(s_{i+1} \mid s_i, c_i) p(s_i \mid c_i) p(c_i)$

There is no chance to change the previous category chain because of an accidently higher probability at the current state.

A number of NP chunking and WSD methods have been developed over the years. The methods for NP chunking are given in [4, 16–18, 20] while the methods for WSD are given in [5,6,10,11,21]. Our method adopts Ramshaw's idea [17] of assigning different categories to words in a sentence based on whether these words are inside a NP chunk, outside a NP chunk, or start a new NP chunk. For WSD, in contrast with other methods, the polysemous word is represented by a sequences of ordered words with POS tags. Our model assigns a category for a word of the input sentence based on the information of the word, the previous word, and the next word we have met before, which a human often does this in the same way. The experiments in Section 2.3.1 show our model achieves better performance than HMMs and CRFs [18]. For WSD, our method can achieve precision and recall > 90 if number of instances of a sense ≥ 1000.

2.5 Conclusions

Recognizing patterns in a sentence is the first step toward understanding the meaning of the sentence. This chapter presents a new probabilistic graphical model for doing such tasks. Experiments show that our model is effective. We have achieved an average of precision 97.7% and an average of recall 98.7% on WSJ data from the Penn Treebank and an average precision 95.15% and an average recall 96.05% on CoNLL-2000 shared task data set for recognizing NP chunks in a sentence. Moreover, we have achieved an average precision 90.57% and an average of recall 92.35% on recognizing a particular sense of polysemous nouns, an average precision 90.86% and an average recall 91.22% on recognizing a particular sense of a polysemous verb, and an average precision 86.50% and an average recall 91.01% on recognizing a particular sense of a polysemous adjective. From the empirical results, in order to improve the performance of WSD, we need to increase the size of the training set. In the future, we will expand the number of instances for *line_interest_hard_serve* data sets and test our model for other senses of these polysemous words. Moreover, we will test our model on recognizing the semantic roles in a sentence by using CoNLL-2005 shared task data set.

Appendix

Computing $p(c_1, \ldots, c_N \mid s_1, \ldots, s_N)$

Based on Equation (2.4)

$$p(c_1, \ldots, c_N \mid s_1, \ldots, s_N)$$

$$= \frac{\prod_{n=1}^{N-1} p(s_n, s_{n+1}, c_n) p(s_n, s_{n+1}, c_{n+1})}{\prod_{m=1}^{N-1} p(s_m, s_{m+1}) \prod_{m=2}^{N-1} p(s_m, c_m)}$$

$$= \frac{1}{\prod_{m=1}^{N-1} p(s_m, s_{m+1})} \times \frac{\prod_{n=1}^{N-1} p(s_n, s_{n+1}, c_n)}{\prod_{m=2}^{N-1} p(s_m, c_m)} \times \prod_{n=1}^{N-1} p(s_n, s_{n+1}, c_{n+1})$$

$$= \frac{p(s_1, s_2, c_1)}{\prod_{m=1}^{N-1} p(s_m, s_{m+1})} \times \frac{\prod_{n=2}^{N-1} p(s_n, s_{n+1}, c_n)}{\prod_{m=2}^{N-1} p(s_m, c_m)} \prod_{n=1}^{N-1} p(s_n, s_{n+1}, c_{n+1})$$

$$= \frac{p(s_1, s_2, c_1)}{\prod_{m=1}^{N-1} p(s_m, s_{m+1})} \times \prod_{n=2}^{N-1} p(s_{n+1} \mid s_n, c_n) \times \prod_{m=2}^{N} p(s_{m-1}, s_m, c_m)$$

setting $n = m - 1$ in the last product

$$= \frac{p(s_1, s_2, c_1) p(s_{N-1}, s_N, c_N)}{\prod_{m=1}^{N-1} p(s_m, s_{m+1})} \times \prod_{n=2}^{N-1} p(s_{n+1} \mid s_n, c_n)$$

$$\times \prod_{m=2}^{N-1} p(s_{m-1}, s_m, c_m)$$

$$= \frac{p(s_1, s_2, c_1)p(s_{N-1}, s_N, c_N)}{\prod_{m=1}^{N-1} p(s_m, s_{m+1})} \times \prod_{n=2}^{N-1} p(s_{n+1} \mid s_n, c_n)p(s_{n-1}, s_n, c_n)$$

$$= \frac{p(s_1, s_2, c_1)p(s_{N-1}, s_N, c_N)}{\prod_{m=1}^{N-1} p(s_m, s_{m+1})}$$

$$\times \prod_{n=2}^{N-1} p(s_{n+1} \mid s_n, c_n)p(s_{n-1} \mid s_n, c_n)p(s_n \mid c_n)p(c_n)$$

$$= \frac{p(s_2 \mid s_1, c_1)p(s_1 \mid c_1)p(c_1) \times p(s_{N-1} \mid s_N, c_N)p(s_N \mid c_N)p(c_N)}{\prod_{m=1}^{N-1} p(s_m, s_{m+1})}$$

$$\times \prod_{n=2}^{N-1} p(s_{n+1} \mid s_n, c_n)p(s_{n-1} \mid s_n, c_n)p(s_n \mid c_n)p(c_n).$$

Define $p(s_0 \mid s_1, c_1) = 1$ and $p(s_{N+1} \mid s_N, c_N) = 1$, then:

$$p(c_1, \ldots, c_N \mid s_1, \ldots, s_N)$$

$$= \frac{p(s_2 \mid s_1, c_1)p(s_0 \mid s_1, c_1)p(s_1 \mid c_1)p(c_1)}{\prod_{m=1}^{N-1} p(s_m, s_{m+1})}$$

$$\times \frac{p(s_{N+1} \mid s_N, c_N)p(s_{N-1} \mid s_N, c_N)p(s_N \mid c_N)p(c_N)}{\prod_{m=1}^{N-1} p(s_m, s_{m+1})}$$

$$\times \prod_{n=2}^{N-1} p(s_{n+1} \mid s_n, c_n)p(s_{n-1} \mid s_n, c_n)p(s_n \mid c_n)p(c_n)$$

$$= \frac{1}{\prod_{m=1}^{N-1} p(s_m, s_{m+1})}$$

$$\times \prod_{n=1}^{N} p(s_{n-1} \mid s_n, c_n)p(s_{n+1} \mid s_n, c_n)p(s_n \mid c_n)p(c_n).$$

Define $M_{s_1, \ldots, s_N} = \frac{1}{\prod_{m=1}^{N-1} p(s_m, s_{m+1})}$, then:

$$p(c_1, \ldots, c_N \mid s_1, \ldots, s_N)$$

$$= M_{s_1,\dots,s_N} \prod_{n=1}^{N} p(s_{n-1} \mid s_n, c_n) p(s_{n+1} \mid s_n, c_n) p(s_n \mid c_n) p(c_n). \quad (2.12)$$

References

[1] S. Abney and S.P. Abney. Parsing by chunks. In *Principle-Based Parsing*, pages 257–278. Kluwer Academic Publishers, 1991.

[2] C.M Bishop. *Pattern Recognition and Machine Learning*. Springer, 2002.

[3] R. Bruce and J. Wiebe. Word-sense disambiguation using decomposable models. In *Proceedings of the 32nd Annual Meeting of the Association for Computational Linguistics*, pages 139–146, 1994.

[4] K.W. Church. A stochastic parts program and noun phrase parser for unrestricted text. In *Proceedings of the Second Conference on Applied Natural Language Processing*, pages 136–143, 1988.

[5] W. Gale, K. Church, and D. Yarowsky. A method for disambiguating word senses in a large corpus. In *Computers and the Humanities*, pages 415–439, 1992.

[6] M.A. Hearst. Noun homograph disambiguation using local context in large text corpora. In *Proceedings of the Seventh Annual Conference of the UW centre for the New OED and Text Research*, pages 1–22, 1991.

[7] M. Huang and R.M. Haralick. A graphical model for recognizing noun phreases from text. In *Proceedings of the 3rd International Conference on Language and Automata Theory and Applications*, 2009.

[8] D. Jurafsky and J.H Martin. *Speech and Language Processing*. AI Pearson Education, 2006.

[9] J. Lafferty, A. MaCallum, and F. Pereira. Conditional random fields: Probabilistic models for segmenting and labeling sequence data. In *Proceedings of 18th International Conference on Machine Learning*, pages 282–289, 2001.

[10] C. Leacock, G.A. Miller, and M. Chodorow. Using corpus statistics and wordnet relations for sense identification. *Computational Linguistics*, 24:147–165, 1998.

[11] C. Leacock, G. Towell, and E. Voorhees. Corpus based statistical sense resolution. In *Proceedings of the Workshop on Human Language Technology*, pages 260–265, 1993.

[12] E. Levin, M. Sharifi, and J Ball. Evaluation of utility of LSA for word sense discrimination. In *Proceedings of HLT-NAACL*, pages 77–80, 2006.

[13] A. MaCallum, D. Freitag, and F. Pereira. Maximum entropy markov models for information extraction and segmentation. In *Proceedings of 17th International Conference on Machine Learning*, pages 591–598, 2000.

[14] C.D. Manning and H. Schutze. *Foundations of Statistical Natural Language Processing*. The MIT Press Cambridge, 2003.

[15] M.P. Marcus, B. Santorini, and M.A. Marcinkiewicz. Building a large annotated corpus of English: The Penn Treebank. *Computational Linguistics*, 19(2):313–330, 1994.

[16] A. Molina and F. Pla. Shallow parsing using specialized HMMs. *Journal of Machine Learning Research*, 2:595–613, 2002.

[17] L.A. Ramshaw and M.P. Marcus. Text chunking using transformation-based learning. In *Proceedings of the Third Workshop on Very Large Corpora*, pages 82–94, 1995.

[18] F. Sha and F. Fereira. Shallow parsing with conditional random fields. In *Proceedings of HLT-NAACL*, pages 213–220, 2003.

[19] E.F. Tjong and K. Sang. Introduction to the CoNLL-2000 shared task: Chunking. In *Proceedings of CoNLL-2000*, pages 127–132, 2000.

[20] W.-C. Wu, Y.-S. Lee, and J.-C. Yang. Robust and efficient multiclass svm models for phrase pattern recognition. *Pattern Recognition*, 41:2874–2889, 2008.

[21] D. Yarowsky. Decision lists for lexical ambiguity resolution: Application to accent restoration in spanish and frech. In *Preceedings of the 32nd Annual Meeting*, 1994.

[22] D. Yarowsky and R. Florian. Decision lists for lexical ambiguity resolution: Application to accent restoration in Spanish and French. *Natural Language Engineering*, pages 293–310, 2002.

3

Wavelet-Based Computational Method in Image Transformation by Singular Integral Equations with Logarithmic Kernel

Limin Cui[1], Yuan Yan Tang[1], Patrick S.P. Wang[2] and Fucheng Liao[3]

[1]*Department of Computer Science, Hong Kong Baptist University, Hong Kong;*
E-mail: lmcui@comp.hkbu.edu.hk
[2]*College of Computer and Information Science, Northeastern University,*
Boston, MA 02115, USA
[3]*Applied Science School, University of Science and Technology Beijing, China*

Abstract

Image distortion can be characterized by the image geometric transformation model. Once the distorted image is approximated by a certain image geometric transformation model, we can apply its inverse transformation to remove the distortion for the geometric restoration. Consequently, the key in the restoration is finding a mathematical form to approximate the distorted image. Harmonic tranformation is a very important model, which can cover other linear and nonlinear geometric models. In fact, it is represented by partial differential equation (PDE) with boundary conditions. To solve PDE, we need to study the integral equations in depth. The singularity is the difficult problem for integral equations. In this chapter, the Wavelet-Galerkin algorithm for solving the first kind of weak singular integral equations with the logarithmic kernel is presented. Because of the singularity of logarithmic kernel, we use the Tikhonov regularization method to solve the system of stiff equation. Finally, the convergence and numerical result of approximate solutions are discussed.

Patrick Shen-Pei Wang (Ed.), Pattern Recognition and Machine Vision – In Honor and Memory of Professor King-Sun Fu, 37–53.

Keywords: image distortion, geometric transformation, periodic wavelet, Galerkin method, Fourier approximation, Tikhonov regularization, weak singular integral equation.

3.1 Introduction

Image geometric distortion can be characterized by some geometric transformation models, which have been widely used in many areas and can be summarized as follows [1, 3, 5, 6, 8, 12]:

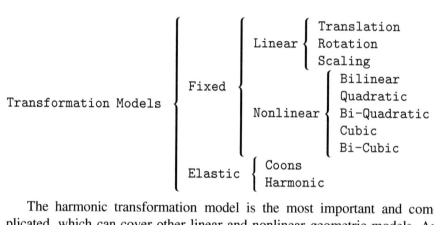

The harmonic transformation model is the most important and complicated, which can cover other linear and nonlinear geometric models. An example can be found in Figure 3.1, where the image of American flag is distorted. The harmonic model can be characterized by partial differential equation, which is described as below:

Let Ω be the region of the elastic plane in the image, and Γ be its boundary. Suppose the functions of the transformation of boundary Γ are $u = f(x_1, x_2)$, and $v = g(x_1, x_2)$. The harmonic transformation

$$T : (x_1, x_2) \rightarrow (u, v)$$

satisfies the partial differential equation:

$$\begin{cases} \Delta u(x_1, x_2) = 0, & (x_1, x_2) \in \Omega, \\ u|_\Gamma = f(x_1, x_2), & (x_1, x_2) \in \Gamma, \end{cases} \qquad \begin{cases} \Delta v(x_1, x_2) = 0, & (x_1, x_2) \in \Omega, \\ v|_\Gamma = g(x_1, x_2), & (x_1, x_2) \in \Gamma, \end{cases}$$

$$(3.1)$$

where

$$\Delta : \frac{\partial^2}{\partial x_1^2} + \frac{\partial^2}{\partial x_2^2}$$

Figure 3.1 Harmonic distortion and restoration

is Laplace's operator, thus the above partial differential equation is called Laplace's equation or harmonic equation.

The distorted image in Figure 3.1 can be approximated by harmonic transformation. As the corresponding harmonic equation is solved, and its inverse transformation is utilized, the restored image can be obtained. Hence, solving the harmonic equation, Equation (3.1) plays a key role in the geometric restoration.

One common method to solve PDE is that PDE is changed into the form of integral equation and integral representation on boundary Γ. There are a lot of advantages to solve integral equations with Wavelet-Galerkin algorithm, such as the numerical sparsity of stiff matrix, the diagonal precondition procedure etc. A great deal of difficulties arises if we only use Wavelet-Galerkin algorithm to solve singular integral equations. This chapter studies the solution to weak singular integral equations with logarithmic kernel based on wavelet. For the purpose of illuminating this question, Laplace's equation is firstly presented.

The first kind weak singular integral equation is as follows:

$$-2\int_0^1 f(s)[\log|\gamma(t) - \gamma(s)|]ds = g(t). \tag{3.2}$$

It will occur when we use the single potential representation.

The interior question about Laplace's equation can be described as follows:

$$\begin{cases} \Delta\omega(x) = 0, & x = (x_1, x_2) \in \Omega, \\ \omega(x) = q(x), & x = (x_1, x_2) \in \Gamma, \end{cases} \tag{3.3}$$

where $q(x)$ is the given boundary function and $q(x) \in C(\Gamma)$, Γ is the smooth boundary of Ω, and $\omega(x)$ is the function to be found. Through the single potential representation, we know that solving (3.3) can be transformed into solving the following integral equations

$$\omega(x) = -\int_\Gamma \rho(y)[\log|x - y|]dS_y, \quad x = (x_1, x_2) \in \Omega.$$

We must solve $\rho(x)$ by the following boundary integral equation:

$$-\int_\Gamma \rho(y)[\log|x-y|]dS_y = q(x), \quad x = (x_1, x_2) \in \Gamma.$$

We can write $x = \gamma(t)$, $t \in [0, 1]$, as the representation of Γ because Γ is smooth. Then

$$-\int_0^1 \rho(\gamma(s))\gamma'(s)[\log|\gamma(t)-\gamma(s)|]ds = q(\gamma(t)). \qquad (3.4)$$

Let

$$f(s) = \frac{\rho(\gamma(s))\gamma'(s)}{2},$$

then (3.4) can be transformed into (3.5)

$$-2\int_0^1 f(s)[\log|\gamma(t)-\gamma(s)|]ds = q(\gamma(t)). \qquad (3.5)$$

Let

$$Mf(t) = -2\int_0^1 f(s)[\log|\gamma(t)-\gamma(s)|]ds, \ g(t) = q(\gamma(t)),$$

then (3.5) can be transformed into (3.2), and we denote

$$Mf = g. \qquad (3.6)$$

Here we propose a novel method to solve PDE in the image transformation using the characteristic of periodic wavelet on $L^2[0, 1]$. At first, we decompose the unknown function on scaling space not only to decrease computational complexity but also to get a sparse matrix. The solutions to the discrete system of equations which are obtained above are ill-posed problems, because of the characteristic of periodic wavelet on $L^2[0, 1]$ and logarithmic kernel. Then we solves this ill-posed system of equations with Tikhonov regularization method. Next by adding some valve value according to requirements, a quick and simple numerical implementation to solve (3.2) is formed. In the end, an example of calculation is presented for the purpose of accounting for the algorithm's validity. Periodic wavelet and multiresolution analysis is introduced in Section 3.2. Wavelet-Galerkin algorithm for solving integral equations is introduced in Section 3.3. Section 3.4 analyses the properties of operator M. The deduction and characteristic of stiff matrix is introduced in

Section 3.5. Section 3.6 solves the system of equations which is ill-posed after discretization with Tikhonov regularization method. An algorithm to solve the first kind of singular integral equation with the logarithmic kernel and the verification convergence of this algorithm are presented in Section 3.7, and an example of computation is given in Section 3.8.

3.2 Periodic Wavelet

Let $\varphi(x)$ and $\psi(x)$ be the normal orthogonal scaling function and the compactly supported wavelet function. Suppose $\varphi(x)$ and $\psi(x)$ adapt to multiresolution analysis respectively. Meanwhile $\varphi(x)$ and $\psi(x)$ satisfy the double-scaling equations as follows:

$$\varphi(x) = \sqrt{2} \sum_n h_n \varphi(2x - n) \quad \text{and} \quad \psi(x) = \sqrt{2} \sum_n g_n \psi(2x - n),$$

where $\{\varphi_{j,k}\}_k$ and $\{\psi_{j,k}\}_k$ are orthonormal basis of V_j and W_j respectively. We periodize scaling function $\varphi(x)$ to $\tilde{\varphi}(x)$ and wavelet function $\psi(x)$ to $\tilde{\psi}(x)$ on $L^2[0, 1]$ [4], namely

$$\tilde{\varphi}(x) := \sum_l \varphi(x + l), \quad \tilde{\psi}(x) := \sum_l \psi(x + l).$$

Periodic scaling and wavelet function space are marked with

$$\tilde{V}_J := \text{closedspan}\{\tilde{\varphi}_{J,k}, k = 0, 1, \ldots, 2^J - 1\},$$

$$\tilde{W}_J := \text{closedspan}\{\tilde{\psi}_{J,k}, k = 0, 1, \ldots, 2^J - 1\}.$$

where $\{\tilde{\varphi}_{J,k}\}_k$ and $\{\tilde{\psi}_{J,k}\}_k$ are orthonormal basis of $\tilde{V}_J$ and $\tilde{W}_J$ respectively. We obtain some equations as the following [9]:

$$\begin{cases} \tilde{\varphi}(x) = \sqrt{2} \sum h_k \tilde{\varphi}(2x - k), & \{h_k\} \in l^2, \\ \tilde{\psi}(x) = \sqrt{2} \sum g_l \tilde{\varphi}(2x - l), & \{g_l\} \in l^2, \\ \tilde{V}_j = \tilde{W}_{j-1} \oplus \tilde{W}_{j-2} \oplus \cdots \oplus \tilde{W}_0 \oplus \tilde{V}_0, & j \in Z^+, \\ \tilde{V}_j \perp \tilde{W}_l, & j \neq l, \\ \bigcup_{j \in Z} \tilde{W}_j = L^2[0, 1], \end{cases}$$

where $\{\tilde{\varphi}_{0,0}\} \bigcup \bigcup_{0 \leq j \leq J-1} \{\tilde{\psi}_{j,k}\}_k$ is a group of orthonormal basis of $\tilde{V}_J$ too, because

$$\tilde{V}_J = \tilde{W}_{J-1} \oplus \tilde{W}_{J-2} \oplus \cdots \oplus \tilde{W}_0 \oplus \tilde{V}_0, \quad J \in Z^+.$$

We know that a transition matrix $W_{\tilde{\varphi},\tilde{\psi}}$ exists between $\{\tilde{\varphi}_{J,k}\}_k$ and $\{\tilde{\varphi}_{0,0}\} \bigcup \bigcup_{0 \leq j \leq J-1} \{\tilde{\psi}_{j,l}\}_l$ and $W_{\tilde{\varphi},\tilde{\psi}}$ is positive definite. In fact $\tilde{\varphi}_{0,0} = 0$.

3.3 Wavelet-Galerkin Algorithms for Solving Integral Equation

The Wavelet-Galerkin Method is a standard Galerkin method with wavelet $\bigcup_{-1 \leq j \leq J-1} \{\tilde{\psi}_{j,k}\}_k$ as the trail basis and the test basis, where

$$\{\tilde{\psi}_{-1,k}\}_k = \{\tilde{\psi}_{0,0}\}. \tag{3.7}$$

Denote f with (3.7) under scale J. Then we have

$$f_J = \sum_{j=-1}^{J-1} \sum_k x_{j,k} \tilde{\psi}_{j,k}. \tag{3.8}$$

Replacing f in (3.6) with (3.8), we obtain

$$M \sum_{j=-1}^{J-1} \sum_k x_{j,k} \tilde{\psi}_{j,k} = g. \tag{3.9}$$

Taking the inner product with $\tilde{\psi}_{j',k'}$ on either hand of (3.9), we have

$$\langle M \sum_{j=-1}^{J-1} \sum_k x_{j,k} \tilde{\psi}_{j,k}, \tilde{\psi}_{j',k'} \rangle = \langle g, \tilde{\psi}_{j',k'} \rangle.$$

So

$$\sum_{j=-1}^{J-1} \sum_k x_{j,k} \langle M \tilde{\psi}_{j,k}, \tilde{\psi}_{j',k'} \rangle = \langle g, \tilde{\psi}_{j',k'} \rangle.$$

Then we solve (3.6) to obtain the approximate solution to (3.10):

$$\mathbf{M}_{\tilde{\psi}} \mathbf{x} = \mathbf{g}, \tag{3.10}$$

where $\mathbf{M}_{\tilde{\psi}}$ is the moment matrix with entries elements $\langle M\tilde{\psi}_{j,k}, \tilde{\psi}_{j',k'} \rangle$, $\mathbf{g}$, $\mathbf{x}$ are vectors with elements $\langle g, \tilde{\psi}_{j',k'} \rangle$ and $x_{j,k}$ respectively, and $\langle \cdot, \cdot \rangle$ is the inner product on $L^2[0, 1]$. But the method is hindered by the computers of the entries of $\mathbf{M}_{\tilde{\psi}}$ and $\mathbf{g}$, which are the integrals of the product of the wavelet basis $\tilde{\psi}$ with the known kernel function in the integral equation (3.6) and the known right hand side $\mathbf{g}$. One method to deal with it is to use the fast transform $\mathbf{W}_{\tilde{\varphi},\tilde{\psi}}$ in which the moment matrix $\mathbf{M}_{\tilde{\varphi}} = [\langle M\tilde{\varphi}_{J,k}, \tilde{\varphi}_{J,k'}\rangle_{k,k'}]$ and the right-hand side vector $\mathbf{g}' = \{\langle g, \tilde{\varphi}_{J,k}\rangle\}_k$ with scaling function basis $\{\tilde{\varphi}_{J,k}\}_k$ are calculated first, and then from relationship, we can find $\mathbf{M}_{\tilde{\psi}}$ and $\mathbf{g}$ by

$$\mathbf{M}_{\tilde{\psi}} = \mathbf{W}_{\tilde{\varphi},\tilde{\psi}} \mathbf{M}_{\tilde{\varphi}} \mathbf{W}_{\tilde{\varphi},\tilde{\psi}}^T, \quad \mathbf{g} = \mathbf{W}_{\tilde{\varphi},\tilde{\psi}} \mathbf{g}'.$$

Solving Equation (3.10) becomes:

$$\text{finding } f_J = \sum_{j=-1}^{J-1} \sum_k x_{j,k} \tilde{\psi}_{j,k} \text{ such that}$$

$$\mathbf{W}_{\tilde{\varphi},\tilde{\psi}} \mathbf{M}_{\tilde{\varphi}} \mathbf{W}_{\tilde{\varphi},\tilde{\psi}}^T \mathbf{x} = \mathbf{W}_{\tilde{\varphi},\tilde{\psi}} \mathbf{g}', \tag{3.11}$$

where $\mathbf{M}_{\tilde{\varphi}}$ is the moment matrix with entries $\langle M\tilde{\varphi}_{J,k}, \tilde{\varphi}_{J,k'}\rangle$ and $\mathbf{g}'$ is a vector with elements $\langle g, \tilde{\varphi}_{J,k}\rangle$.

3.4 Properties of the Operator

Let Γ be a circle whose radius is α, we have $x = (x_1, x_2) = \alpha(\cos(2\pi s), \sin(2\pi s))$. Therefore

$$Mf(t) = -2\int_0^1 f(s)[\log|2\alpha \sin \pi(t-s)|]ds.$$

Using the formula in the theory of generalized function, we obtain

$$\frac{1}{\pi}\log\left|2\sin\left(\frac{\theta}{2}\right)\right| = -\frac{1}{2\pi}\sum_{\substack{n=-\infty \\ n\neq 0}}^{+\infty}\frac{1}{|n|}e^{in\theta} = -\frac{1}{\pi}\sum_{n=1}^{+\infty}\frac{1}{n}\cos n\theta.$$

We obtain the following equalities:

$$-\log|2\alpha \sin \pi(t-s)| = -\log\alpha - \log|2\sin\pi t|$$

$$= -\log\alpha + \frac{1}{2}\sum_{\substack{n=-\infty \\ n\neq 0}}^{+\infty}\frac{1}{|n|}e^{2\pi in(t-s)}.$$

Without loss of generality, let v to be a function with 1-period, the Fourier series expression of v is

$$v(t) = \sqrt{2\pi} \sum_{n=-\infty}^{+\infty} \hat{v}(n) e^{2\pi i n t}.$$

Therefore

$$Mv(t) = -2\sqrt{2\pi}(\log \alpha)\hat{v}(0) + \sqrt{2\pi} \sum_{\substack{n=-\infty \\ n\neq 0}}^{+\infty} \frac{1}{|n|} \hat{v}(n) e^{2\pi i n t} \qquad (3.12)$$

Assume that $\tilde{v}(t)$ is the periodic version of function $v(t)$, namely $\tilde{v}(t) = \sum_{l=-\infty}^{+\infty} v(t+l)$, we have

$$\tilde{v}(t) = \sqrt{2\pi} \sum_{l=-\infty}^{+\infty} \hat{v}(2\pi l) e^{2\pi i l t}.$$

So the periodic scaling function is given by

$$\tilde{\varphi}_{J,k}(t) = \sqrt{2\pi} \sum_{l=-\infty}^{+\infty} \hat{\varphi}_{J,k}(2\pi l) e^{2\pi i l t}.$$

Then

$$\hat{\tilde{\varphi}}_{J,k}(l) = \hat{\varphi}_{J,k}(2\pi l). \qquad (3.13)$$

Using (3.12), we have

$$M\tilde{\varphi}_{J,k}(t) = -2\sqrt{2\pi}(\log \alpha)\hat{\tilde{\varphi}}_{J,k}(0) + \sqrt{2\pi} \sum_{\substack{n=-\infty \\ n\neq 0}}^{+\infty} \frac{1}{|n|} \hat{\tilde{\varphi}}_{J,k}(n) e^{2\pi i n t} \qquad (3.14)$$

3.5 The Computation and Properties of the Moment Matrix

Using (3.13) and (3.14), the elements of $\mathbf{M}_{\tilde{\varphi}}$ can be gained by the deduction:

$$\langle M\tilde{\varphi}_{J,k}, \tilde{\varphi}_{J,k'} \rangle = -4\pi(\log \alpha)\hat{\tilde{\varphi}}_{J,k}(0)\hat{\tilde{\varphi}}_{J,k'}(0)$$

$$+ 2\pi \sum_{\substack{n=-\infty \\ n\neq 0}}^{+\infty} \frac{1}{|n|} \hat{\tilde{\varphi}}_{J,k}(n) e^{2\pi i n t} \hat{\tilde{\varphi}}_{J,k'}(n') e^{2\pi i n' t}.$$

By the orthogonality of $e^{2\pi int}$ and

$$\hat{\varphi}_{J,k}(n) = 2^{-J/2} e^{-in(k/2^J)} \hat{\varphi}\left(\frac{n}{2^J}\right),$$

we have

$$\langle M\tilde{\varphi}_{J,k}, \tilde{\varphi}_{J,k'} \rangle = -4\pi (\log \alpha) \cdot 2^{-J}$$

$$+ 2\pi \cdot 2^{-J} \sum_{\substack{n=-\infty \\ n\neq 0}}^{+\infty} \frac{1}{|n|} \exp\left[-\frac{i(k-k')2\pi n}{2^J}\right] \left|\hat{\varphi}\left(\frac{2\pi n}{2^J}\right)\right|^2.$$

Namely

$$\langle M\tilde{\varphi}_{J,k}, \tilde{\varphi}_{J,k'} \rangle = -4\pi (\log \alpha) \cdot 2^{-J}$$

$$+ 4\pi \cdot 2^{-J} \sum_{n=1}^{+\infty} \frac{1}{n} \cos \frac{(k-k')2\pi n}{2^J} \left|\hat{\varphi}\left(\frac{2\pi n}{2^J}\right)\right|^2. \quad (3.15)$$

Theorem 1. *Equation* (3.15) *satisfies properties as follows:*

1. *Symmetry, namely* $m_{k,k'} = m_{k',k}$;
2. *Circulation, namely* $m_{k,k'} = m_{k+1,k'+1}$;
3. *Repetitiveness, namely* $m_{k,k'} = m_{k,2^J-k'+2}$.

Proof. We only prove (3).

$$m_{k,2^J-k'+2} = \langle M\tilde{\varphi}_{J,k}, \tilde{\varphi}_{J,2^J-k'+2} \rangle$$

$$= -4\pi (\log \alpha) \cdot 2^{-J}$$

$$+ 4\pi \cdot 2^{-J} \sum_{n=1}^{+\infty} \frac{1}{n} \cos \frac{[k-(2^J-k'+2)2\pi n]}{2^J} \left|\hat{\varphi}\left(\frac{2\pi n}{2^J}\right)\right|^2$$

$$= -4\pi (\log \alpha) \cdot 2^{-J}$$

$$+ 4\pi \cdot 2^{-J} \sum_{n=1}^{+\infty} \frac{1}{n} \cos \frac{(k-k')2\pi n}{2^J} \left|\hat{\varphi}\left(\frac{2\pi n}{2^J}\right)\right|^2$$

$$= \langle M\tilde{\varphi}_{J,k}, \tilde{\varphi}_{J,k'} \rangle$$

$$= m_{k,k'}.$$

$\square$

We obtain from Theorem 1 that $\mathbf{M}_{\tilde{\varphi}}$ is a circular symmetrical matrix with rank 2^J, so only $2^{J-1}+1$ elements are need to calculate. In this way the computational complexity and the storage are decreased sharply. The computation of compactly supported scaling function $\varphi(x)$ is as same as the counterpart in Section 3.4. When g is smooth, the computation of $\langle g, \tilde{\varphi}_{J,k} \rangle$ is relatively easy, which results are only normal numerical solutions on the functional integral based on wavelet [11]. For example, Gauss-type quadrature

$$\int_R f(t)\varphi_{J,k}(t)dt \approx 2^{-J/2} f\left(\frac{m_1 + k}{2^J}\right), \tag{3.16}$$

where $m_n = \int_R t^n \varphi(t)dt$, and m_n can be calculated by [2]

$$\begin{cases} m_0 = 1, \\ \\ m_l = \dfrac{1}{2^l - 1} \dfrac{\sqrt{2}}{2} \sum_{j=0}^{l-1} \binom{l}{j} m_j \left(\sum_n h_n n^{l-j}\right). \end{cases}$$

The Fourier transform of compactly supported scaling function $\varphi(x)$ can be calculated by

$$\hat{\varphi}(x) = \frac{1}{\sqrt{2\pi}} \prod_{j=1}^{\infty} H(2^{-j}\omega), \quad H(\omega) = \frac{1}{\sqrt{2}} \sum_{k \in Z} h_k e(-ik\omega).$$

3.6 Wavelet Regularization Method

Rieder [10] concretely described a Schwarz iteration algorithm which is used to rapidly solve linear ill-posed problems, in order to dispose these ill-posed problems. This algorithm adopts the Tikhonov regularization method. It employs spline function or Daubechies wavelet and applies them to the first kind integral equation. On the basis of [10], this chapter further decomposes functions on wavelet space, so it can be organically linked up with the Wavelet-Galerkin method mentioned in Section 3.5 and takes wavelet function as basis.

3.6.1 Regulation Method Using Additive Schwarz Iteration for Ill-Posed Problems

We mainly study how to apply periodic wavelet in the problem of solving the first kind operator equation as follows

$$Mf = g, \tag{3.17}$$

where $M : X \to Y$ is a compact operator, so (3.17) is ill-posed.

Considering problem (3.17), we now assume that noisy data $g^\varepsilon \in Y$ are available satisfying $\|g - g^\varepsilon\|_Y \leq \varepsilon$ for a known error bound $\varepsilon > 0$. According to Tikhonov regularization method, a computable approximation to f^* is then provided by the unique solution $f_l^{\varepsilon,\alpha}$ of the finite dimensional normal equation

$$(M_l^* M_l + \alpha I) f_l = M_l^* g^\varepsilon, \quad \alpha > 0. \tag{3.18}$$

In (3.18), $M_l = M P_l$, where $P_l : X \to \tilde{V}_l$ is the orthogonal projection on the meaning of inner product, and space $\tilde{V}_l$ is the scaling function space of periodic wavelet, and $\tilde{V}_l \subset X$. Under these assumptions the quantity

$$\gamma_l := \|M - M_l\| = \|M(I - P_l)\|. \tag{3.19}$$

Based on the characteristic of multiresolution analysis of periodic wavelet, we have $\gamma_l \leq \gamma_{l+1}$ and $\gamma_l \to 0 \; (l \to 0) \Leftrightarrow M$ is a compact operator [7]. In addition, on the basis of the characteristic of orthogonal decomposition of space $\tilde{V}$, we have

$$\tilde{V}_l = \tilde{V}_{l_{\min}} \bigoplus \bigoplus_{j=l_{\min}}^{l-1} \tilde{W}_j, \quad l_{\min} \leq l - 1. \tag{3.20}$$

Accordingly, the following decomposition of the projection operator can be obtained: $P_l = P_{l_{\min}} + \sum_{j=l_{\min}}^{l-1} Q_j$, where Q_j is the orthogonal projection from X to space $\tilde{W}_j$.

Lemma 1. *Let $\tilde{V}_j$ and $\tilde{W}_j$ be the scaling function space and the wavelet space of periodic wavelet respectively, and let operator $M : X \to Y$ be a compact linear operator. Then $\|M Q_l\| \leq \gamma_l \to 0 \;\; (l \to \infty)$, where γ_l is defined in (3.19).*

Taking the inner product with v_l on either side of (3.18) as $\langle (M_l^* M_l + \alpha I) f_l, v_l \rangle = \langle M_l^* g^\varepsilon, v_l \rangle$, we have $\langle M_l^* M_l f_l, v_l \rangle + \langle f_l, v_l \rangle = \langle M_l^* g^\varepsilon, v_l \rangle$, namely $\langle M_l f_l, M_l v_l \rangle + \langle f_l, v_l \rangle = \langle M_l^* g^\varepsilon, v_l \rangle$.

Define the bilinear form $a : X \times X \to L^2[0, 1]$ as

$$a(u, v) := \langle Mu, Mv \rangle_Y + \alpha \langle u, v \rangle_X. \tag{3.21}$$

Clearly, α is symmetric positive definite.

Therefore, the problem of solving (3.18) can be transformed into solving the problem described as below: finding $f_l^{\varepsilon,\alpha} \in V_l$, such that $f_l^{\varepsilon,\alpha}$ satisfies

$$a(f_l^{\varepsilon,\alpha}, v_l) = \langle K_l^* g^\varepsilon, v_l \rangle_X, \quad \forall v_l \in \tilde{V}_l. \tag{3.22}$$

Define the following operators:

$$A_l = M_l^* M_l + \alpha P_l, \quad B_l = Q_l M_l^* M_l Q_l + \alpha Q_l.$$

Obviously,

$$a(u_l, v_l) = \langle A_l u_l, v_l \rangle_X, \quad \forall u_l, v_l \in \tilde{V}_l$$

and

$$a(w_l, z_l) = \langle B_l w_l, z_l \rangle_X, \quad \forall w_l, z_l \in \tilde{W}_l.$$

Considering the definition of A_l, (3.18) can be transformed into $A_l f_l = K_l^* g^\varepsilon$. Another bilinear form $b_j : \tilde{W}_j \times \tilde{W}_j \to L^2[0, 1]$ is defined by $b_j(w_j, u_j) := \alpha \langle w_j, u_j \rangle_X$. We know from Lemma 1 that b_j is approximating α defined by (3.21) on space $\tilde{W}_j$ when j is increased enough.

By introducing an operator $F_j = \alpha^{-1} Q_j A_l$ for $v_l \in \tilde{V}_l$ and $w_j \in \tilde{W}_j$, we have $b_j(F_j v_l, w_j) = \alpha(v_l, w_j)$. Let $\tilde{V}_{l_{min}}$ be the coarsest approximate space, for $v_l \in \tilde{V}_l$ and $v_{l_{min}} \in \tilde{V}_{l_{min}}$, and we introduce another operator: $R_{l_{min}} = P_{l_{min}} A_{l_{min}}^{-1} P_{l_{min}} A_l$. Then $R_{l_{min}} : \tilde{V}_l \to \tilde{V}_{l_{min}}$ and $a(R_{l_{min}} v_l, v_{l_{min}}) = a(v_l, v_{l_{min}})$. Next, adding all F_j to $R_{l_{min}}$, we have

$$R_{l_{min}} + \sum_{j=l_{min}}^{l-1} F_j = \left(P_{l_{min}} A_{l_{min}}^{-1} P_{l_{min}} + \alpha^{-1} \sum_{j=l_{min}}^{l-1} Q_j \right) A_l.$$

Through the formula about decomposition in space (3.20), we know that

$$C_{l,l_{min}} := P_{l_{min}} A_{l_{min}}^{-1} P_{l_{min}} + \alpha^{-1} \sum_{j=l_{min}}^{l-1} Q_j$$

is the approximation of A_l^{-1}.

For (3.22), we can use additive Schwarz iteration to solve this problem rapidly:

$$u_l^{\mu+1} = u_l^\mu - \omega C_{l,l_{min}} (A_l u_l^\mu - K_l^* g^\varepsilon), \quad \mu = 0, 1, 2, \ldots, \tag{3.23}$$

where an initial guess $\mu_l^0 \in \tilde{V}_l$ and a damping parameter $\omega \in R$.

3.6.2 Deduction of Algebraic Equations

In this section we present a matrix version of the iteration (3.23) given suitable bases in $\tilde{V}_l$ and $\tilde{W}_l$.

Let space $X = L^2[0, 1]$, $\tilde{\varphi}$ and $\tilde{\psi}$ are scaling function and wavelet function of periodic wavelet separately. They satisfy multiresolution analysis of periodic wavelet. Since function $f_l = \sum_k c_k^l \tilde{\varphi}_{l,k} \in V_l$ and function $g_l = \sum_k d_k^l \tilde{\psi}_{l,k} \in W_l$, we have

$$f_l + g_l = \sum_k c_k^{l+1} \tilde{\varphi}_{l+1,k}.$$

Also

$$c_k^{l+1} = \sum_i h_{k-2i} c_k^l + \sum_j h_{k-2j} d_j^l,$$

which we write in matrix notation as $c^{l+1} = H_{l+1}^T c^l + G_{l+1}^T d^l$. Clearly, $H_{l+1} : R^{n_l+1} \to R^{n_l}$ and $G_{l+1} : R^{n_l+1} \to R^{n_l}$.

The solution $f^{\varepsilon,\alpha}$ can be expanded in the basis of $\tilde{W}_l$ as $f_l^{\varepsilon,\alpha} = \sum_k (\xi_l)_k \tilde{\varphi}_{l,k}$, where the vector ξ_l of the expansion coefficients is the unique solution of the linear system $A_l \xi_l = \beta_l$. The concrete forms of the matrix A_l and the vector β_l are given in [10] as follows:

$$(A_l)_{i,j} = \langle M\tilde{\varphi}_{l,i}, M\tilde{\varphi}_{l,j} \rangle_Y + \alpha \langle \tilde{\varphi}_{l,i}, \tilde{\varphi}_{l,j} \rangle_X, \quad (\beta_l)_j = \langle g^\varepsilon, M\tilde{\varphi}_{l,j} \rangle_Y.$$

$H_{l,j}$ and $G_{l,j}$ are defined by

$$\mathbf{H}_{l,j} := H_{j+1} H_j \cdots H_{l-1} H_l : R^{n_l} \to R^{n_j},$$

$$\mathbf{G}_{l,j} := G_{j+1} H_j \cdots H_{l-1} H_l : R^{n_l} \to R^{n_j}.$$

where $j \leq l - 2$. Let $\mathbf{H}_{l,l-1} := H_l$ and $\mathbf{G}_{l,l-1} := G_l$, then the above iteration can be transformed into

$$z_l^{k+1} = z_l^k - \omega C_{l,l_{\min}}(A_l z_l^k) - \beta_l, \quad k = 0, 1, 2, \ldots, \qquad (3.24)$$

where

$$C_{l,l_{\min}} = H_{l,l_{\min}}^T A_{l,l_{\min}}^{-1} H_{l,l_{\min}} + \alpha^{-1} \sum_{j=l_{\min}}^{l-1} G_{l,j}^T B_j^{-1} G_{l,j} \qquad (3.25)$$

$$z_l^0 = H_{l,l_{\min}}^T A_{l,l_{\min}}^{-1} H_{l,l_{\min}}. \qquad (3.26)$$

Different from [10], this chapter defines a matrix A_l and a vector β_l as follows:

$$(A_l)_{i,j} = \langle M\tilde{\psi}_{l,i}, M\tilde{\psi}_{l,j}\rangle_Y + \alpha\langle\tilde{\psi}_{l,i}, \tilde{\psi}_{l,j}\rangle_X, \quad (\beta_l)_j = \langle g^\varepsilon, M\tilde{\psi}_{l,j}\rangle_Y.$$

Then we use the iteration format as similar to (3.24), but (3.25) and (3.26) respectively are changed to

$$C_{l,l_{\min}} = W_{\tilde{\varphi},\tilde{\psi}}(H_{l,l_{\min}}^T A_{l,l_{\min}}^{-1} H_{l,l_{\min}} + \alpha^{-1}\sum_{j=l_{\min}}^{l-1} G_{l,j}^T B_j^{-1} G_{l,j})W_{\tilde{\varphi},\tilde{\psi}}^T, \quad (3.27)$$

$$z_l^0 = W_{\tilde{\varphi},\tilde{\psi}}(H_{l,l_{\min}}^T A_{l,l_{\min}}^{-1} H_{l,l_{\min}})W_{\tilde{\varphi},\tilde{\psi}}^T\beta_l. \quad (3.28)$$

3.7 Algorithm and Convergence

On the basis of the above analysis, we conclude the Wavelet-Galerkin algorithm for solving (3.2) as follows:

Step 1 Compute $g_l, m_{k,l}, l = 0, 1, \cdots, 2^J - 1$ by (3.16), (3.15). We get the matrix $M_{\tilde{\varphi}}^J$ and the vector $\mathbf{g}$.

Step 2 Calculating $W_{\tilde{\varphi},\tilde{\psi}}$ and we get Equation (3.11).

Step 3 Solving regularization equation (3.18) by (3.24), (3.27) and (3.28), we get $x_k, k = 0, 1, \ldots, 2^J - 1$.

Theorem 2. *Let $\tilde{m}_{\tilde{\psi}kl}$, $m_{\tilde{\psi}kl}$ be the elements of $\tilde{M}_{\tilde{\psi}}$ and $M_{\tilde{\psi}}$ respectively. Then we get f_J by (3.8). In the Wavelet-Galerkin algorithm, $\tilde{m}_{\tilde{\psi}kl}$ uniformly converges to $m_{\tilde{\psi}kl}$ as $n \to \infty$, and f_J uniformly converges to the solution $f(x)$ of function (3.2) as $J \to \infty$ and $n \to \infty$.*

Proof. Since $\tilde{M}_{\tilde{\psi}} = W_{\tilde{\varphi},\tilde{\psi}}\tilde{M}_{\tilde{\varphi}}W_{\tilde{\varphi},\tilde{\psi}}^T$, $M_{\tilde{\psi}} = W_{\tilde{\varphi},\tilde{\psi}}M_{\tilde{\varphi}}W_{\tilde{\varphi},\tilde{\psi}}^T$, we can find that uniformly converges to m_{kl} when $n \to \infty$. So $\|\tilde{M}_{\tilde{\varphi}} - M_{\tilde{\varphi}}\| \to 0$ $(n \to \infty)$. Since $W_{\tilde{\varphi},\tilde{\psi}}$ is orthogonal matrix, we have

$$\|\tilde{M}_{\tilde{\psi}} - M_{\tilde{\psi}}\| = \|W_{\tilde{\varphi},\tilde{\psi}}(\tilde{M}_{\tilde{\varphi}} - M_{\tilde{\varphi}})W_{\tilde{\varphi},\tilde{\psi}}^T\| = \|\tilde{M}_{\tilde{\varphi}} - M_{\tilde{\varphi}}\| \to 0 \quad (n \to \infty).$$

Then $\tilde{m}_{\tilde{\psi}kl}$ uniformly converges to $m_{\tilde{\psi}kl}$ as $n \to \infty$. We can also define the operator

$$Q_J : \tilde{V}_J \to \tilde{W}_{J-1} \bigoplus \tilde{W}_{J-2} \bigoplus \cdots \bigoplus \tilde{W}_0 \bigoplus \tilde{V}_0.$$

Obviously, Q_J is orthogonal transformation. Since $f_J = Q_J F_n P_J f(x)$ and following Peng et al. [9], f_J uniformly converges to the solution $f(x)$ of Equation (3.2) as $J \to \infty, n \to \infty$. □

We can prove the convergence of wavelet regularization by [10].

3.8 Numerical Examples

Let $g(x) = e^{x_1} \cos x_2$ in the Laplace interior problem (3.3), and assume $\alpha = 0.5$. Then we solve the boundary integral equation (3.2) by the above algorithm.

3.8.1 Computation of Stiff Matrix

We use $J = 4$ and $J = 7$ as two kinds of scaling discretization for $[0, 1]$, namely dividing $[0, 1]$ into 16 and 128 equal parts. We assign Daubechies scaling function with $N = 6$ as $\varphi_{4,0}(x)$ and $\varphi_{7,0}(x)$, then transform them into periodic function on space $L^2[0, 1]$.

Since elements in $\mathbf{M}_\varphi^J$ are circular symmetry, only $2^{J-1} + 1$ of these elements are needed to compute and to be stored. Let $n = 100$ in (3.14). For scaling discrete $J = 4$, the matrix $\mathbf{M}_\varphi^J$ is 16×16, and for scaling discrete $J = 7$, the matrix $\mathbf{M}_\varphi^J$ is 128×128.

Similarly, we know that $\mathbf{M}_{\tilde{\psi}}^J$ is sparser than $\mathbf{M}_{\tilde{\varphi}}^J$ and increasing J is the key to improve effect.

3.8.2 Analysis of Regularization

1. Figures 3.2 and 3.3 present the situation when $J = 4$. The error is no more than $1.677093352110495e^{-004}$ with Tikhonov regularization using Schwartz iteration, while the error is no more than $5.863185146877874e^{-005}$ with Tikhonov regularization using Newton iteration. The number of times conducting both iterations is large.
2. Figures 3.4 and 3.5 present the situation when $J = 7$. The error is no more than $5.183141329873582e^{-007}$ with Tikhonov regularization using Schwartz iteration, while the error is no more than $5.959144566129969e^{-005}$ with Tikhonov regularization using Newton iteration. The times of Schwartz iteration are 1, while the times of Newton iteration are 11.
3. We make only one decomposition in the regularization method using the Schwartz iteration when $J = 4$ and $J = 7$. Solutions become more precise when the times of decomposition are smaller, and the solution become coarser when the times of decomposition are bigger, though regularization method using Schwartz iteration is better than regularization

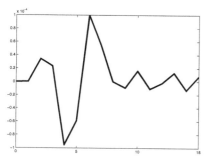

Figure 3.2 Error of Tikhonov regularization using Schwartz iteration when $J = 4$

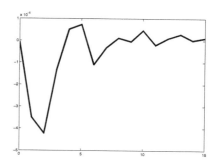

Figure 3.3 Error of Tikhonov regularization using Newton iteration when $J = 4$

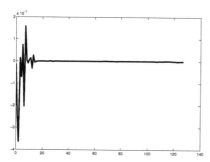

Figure 3.4 Error of Tikhonov regularization using Schwartz iteration when $J = 7$

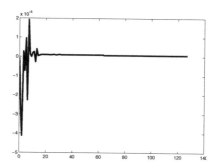

Figure 3.5 Error of Tikhonov regularization using Newton iteration when $J = 7$

method using Newton iteration in precision, regularization method using Schwartz iteration is better than regularization method using Newton iteration in precision in iterations when J is large.

4. The errors in Figures 3.2–3.5 are obtained by comparing the double ends of $\mathbf{M}_{\tilde{\psi}}^{J}\mathbf{x} = \mathbf{g}$ after substituting results for the counterpart of the double ends of this equation.

The difficulty of solving a singular integral equation with the wavelet method corresponds to different singular kernels, we should use different numerical method to gain satisfied effectiveness on stable and rapid algorithms of calculation of singular integral. This paper uses the character of periodic wavelet on $L^2[0, 1]$ and logarithmic kernel to make stiff matrix sparser and uses Tikhonov regularization method to remove singularity for the purpose of approximate calculating this kind of weak singular integral equation rapidly.

The method discussed in this paper can be used to deal with other kinds of singular integral problems with trigonometric function singular kernel.

References

[1] A.C. Berg, T.L. Berg, and J.Malik. Shape matching and object recognition using low distortion correspondences. In *Proceedings of the 2005 IEEE Computer Society Conference on Computer Vision and Pattern Recognition*, 2005.

[2] A. Cohen, I. Daubechies, and P. Vial. Wavelet on the interval and fast wavelet transforms. *Appl. Comp. Harm. Anal.*, 1(1):54–81, 1993.

[3] L. Cui, Y.Y. Tang, F. Liao, and X. Feng. Wavelet approach to image transformation. In *Proceedings of the 4th International Symposium on Information and Communication Technologies*, 2005.

[4] I. Daubechies. The orthonomal bases of compactly supported wavelets. *Comm.Pure Appl. Math*, 41:906–996, 1988.

[5] P. Dong, J.G. Brankov, N.P. Galatsanos, Y. Yang, and F. Davoine. Digital watermarking robust to geometric distortions. *IEEE Trans. on Image Processing*, 14(12):2140–2150, 2005.

[6] T. Ernst, O. Speck, L. Itti, and L. Chang. Simultaneous correction for interscan patient motion and geometric distortions in echoplanar imaging. *Magnetic Resonance in Medicine*, 42:201–205, 1999.

[7] C.W. Groetsch. *The Theory of Tikhonov Regularization for Fredholm Equations of the First Kind*. Pitman, Boston, MA, 1984.

[8] X. Kang, J. Huang, and Y.Q. Shi. *An Image Watermarking Algorithm Robust to Geometric Distortion*. Springer, Berlin/Heidelberg, 2003.

[9] S.L. Peng, D.F. Li, and Q.H. Sheng. *Theory and Applications of the Periodic Wavelets*. Science Press, Beijing, 2003.

[10] A. Rieder. A wavelet multilevel method for ill-posed problems stabilized by Tikhonov regularization. *Numer.Math.*, 75(4):501–522, 1997.

[11] G. Strong and T. Nguyen. *Wavelets and Filter Banks*. Cambridge Press, Wellesley, 1996.

[12] Y.Y. Tang and C.Y. Suen. Image transformation approach to nonlinear shape restoration. *IEEE Trans. System, Man, and Cybernetics*, 23(1):151–172, 1993.

4

Analysis of DNA Microarray Gene Expression Data Based on Pattern Recognition Methods

Alan Wee-Chung Liew[1] and Hong Yan[2,3]

[1]*School of Information and Communication Technology, Griffith University, Gold Coast Campus, Southport, QLD 4222, Australia; E-mail: a.liew@griffith.edu.au*
[2]*Department of Electronic Engineering, City University of Hong Kong, Kowloon, Hong Kong; E-mail: h.yan@cityu.edu.hk*
[3]*School of Electrical and Information Engineering, University of Sydney, Sydney, NSW 2006, Australia*

Abstract

The advancement in DNA microarray technology has generated an unprecedented amount of gene expression data. The challenge now is to analyze and extract useful information from the data. In this chapter, we discuss how gene expression study is performed and how to analyze these data based on pattern recognition methods. The problems we address include microarray image segmentation, missing value imputation, gene expression data clustering and biclustering and spectral analysis of time series data.

Keywords: bioinformatics, DNA microarray technology, gene expression data analysis, clustering, biclustering, spectral analysis, gene regulation.

4.1 Introduction

Recent advancement in molecular biology and genomic research, such as high throughput sequencing methods and DNA microarray technology, has

Patrick Shen-Pei Wang (Ed.), Pattern Recognition and Machine Vision – In Honor and Memory of Professor King-Sun Fu, 55–69.

generated an unprecedented amount of data. With the completion of the Human Genome Project and the availability of the vast number of genomic sequences, the next major challenge is to study gene activities and gene functions. Important insights into gene function can be gained by gene expression profiling.

Gene expressing profiling is the process of determining when and where particular genes are expressed. For example, some genes are turned on (expressed) or turned off (suppressed) when there is a change in external conditions or stimuli. In multi-cellular organisms, gene expressions in different cell types are different during different developmental stages in life. Even within the same cell type, gene expressions are dependent on the cell cycle the cells are in. DNA mutation may alter the expression of certain genes, which causes illness such as cancer. Furthermore, the expression of one gene is often regulated by the expression of another gene. A detail analysis of all these information will provide an understanding about the networking of different genes and their functional roles.

In the past, genes and their expression profiles are studied one at a time. However, this is inadequate for the holistic study of the complete genome of an organism since the expressions of different genes are generally interdependent. Microarray technology, which allows massively parallel, high throughput profiling of gene expression in a single hybridization experiment, has emerged as a powerful tool for genomic research [16, 20]. By using an array containing many DNA samples, scientists can determine, in a single experiment, the expression levels of tens of thousands of genes within a cell. Besides the enormous scientific potential of DNA microarrays in the study of gene expressions, gene regulations and interactions, they also have very important applications in pharmaceutical and clinical research. For example, by comparing the ways in which genes are expressed in a normal and diseased organ, scientists can identify the genes and hence the associated proteins that are part of the disease process. Researchers could then use that information to synthesize drugs that interact with these proteins, thus reducing the disease's effect on the body.

4.2 DNA Microarray Technology

DNA Microarrays are small, solid supports of, i.e. glass microscope slides, silicon chips or nylon membranes, onto which sequences from tens of thousands of different genes known as probes are immobilized at fixed locations. The DNA is spotted or synthesized directly onto the support.

In spotted microarrays, the probes are oligonucleotides, cDNA or small fragments of PCR products that correspond to mRNAs. The probes are synthesized prior to deposition on the array surface and are then spotted onto the glass using an array of fine pins or needles controlled by a robotic arm.

Oligonucleotide arrays, on the other hand, are produced by synthesizing short oligonucleotide sequences (i.e. 25-mer probes in Affymetrix chip) directly onto the array surface using a photolithographic process, where light and light-sensitive masking agents are used to build a sequence one nucleotide at a time [19]. Each applicable probe is selectively unmasked prior to bathing the array in a solution of a single nucleotide, then a masking reaction takes place and the next set of probes are unmasked in preparation for a different nucleotide exposure. This process is repeated until the probes are fully constructed.

In a two-channel microarray experiment, two samples of cRNA, which are reversed transcribed from mRNA purified from cellular contents, are labeled with two different fluorescent dyes (i.e. cyanine3 (Cy3) and cyanine5 (Cy5)) to constitute the cDNA targets. The two cDNA targets are then hybridized onto a cDNA microarray. If a target contains a cDNA whose sequence is complementary to the DNA probe on a given spot, that cDNA will hybridize to the spot, where it will be detected by its fluorescence. Once the cDNA targets have been hybridized onto the array, the array is laser scanned to determine how much of each targets is bound to each spot. The hybridized microarray is scanned for the red wavelength (at approximately 635 nm for Cy5) and the green wavelength (at approximately 530 nm for Cy3), which produces two sets of images in 16 bits TIFF format. The ratio of the two fluorescence intensities at each spot indicates the relative abundance of the corresponding DNA sequence in the two cDNA samples that are hybridized to the DNA probe on the spot. As one sample is the control, the ratio expresses the extent to which the other sample is differentially expressed with respect to the control. By examining the expression ratio of each spot, gene expression study can be performed. Figure 4.1 shows a schematic of the cDNA microarray technique and the steps in performing a cDNA microarray experiment.

4.3 Gene Expression Data Extraction

Given a microarray image, the first task is to extract the gene expression ratios from the image. This requires the accurate segmentation of spots in the microarray image. As the occurrence of missing values due to the omission

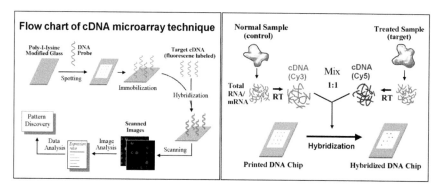

Figure 4.1 Left: a schematic of the cDNA microarray technique. Right: the steps involved in a cDNA microarray experiment

of unreliable expression ratios is common, missing value imputation is often needed.

4.3.1 Microarray Image Analysis

The goal in microarray image analysis is to automatically quantify each spots giving information about the relative extent of hybridization of the two cDNA samples. This generally involves several steps: (i) block segmentation, where each block within a microarray image is delineated, (ii) gridding, where the location of each spot within a block is determined, and, (iii) spot segmentation, where each spot is segmented and its intensity value determined. In [15], we proposed a robust microarray image segmentation algorithm called GeneIcon. Our algorithm starts by generating a single gray level image from the two TIFF images. Then the blocks in a microarray image are segmented by analyzing the vertical and horizontal projection profiles obtained from an adaptively binarized image. For gridding, we first locate the good quality spots which we called guide spots. To account for the variable background and spot intensity, a novel adaptive thresholding and morphological processing algorithm are used to detect the guide spots. The geometry of the grid is then inferred from the guide spots. Finally, spot segmentation is performed in each of the sub-regions defined by the grid. The intensity distribution of the pixels within the subregion is modeled using a 2-class Gaussian-Mixture to find the optimum threshold. Then, a best-fit circle is computed for the final spot segmentation. Once the spots are extracted, the intensity value of each spot can be obtained and the expression ratio which indicates the differential expression of the two samples can be computed. Figure 4.2 shows

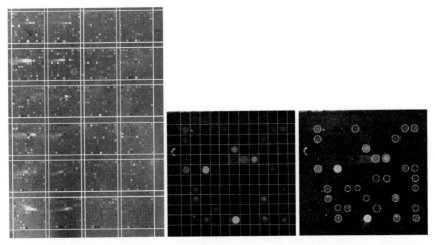

Figure 4.2 Left: the segmentation of a microarray image into blocks. Middle: gridding in a block. Right: cDNA Microarray spot segmentation results

block segmentation (left), gridding (middle), and spot segmentation (right) of a microarray image using GeneIcon.

The raw intensity values of the spots need to undergo preprocessing involving background correction, data normalization, and data filtering before subjected to further analysis. Background correction aims to remove from the spot's measured intensity a contribution not due to the specific hybridization of the target to the probe. This contribution could arise from non-specific hybridization and stray fluorescence on the slide. A simple background correction is by subtracting the local background intensity from the spot intensity [13]. Data normalization adjusts for any bias that arises from variation in the microarray process rather than from biological differences between the RNA samples. One common bias is the red-green bias due to the differences in labeling efficiencies. Loess normalization is often used for data normalization [21,28]. Finally, poor quality ratios are usually filtered out prior to subsequent data analysis, which results in missing values in the data.

By conducting a series of N microarray experiments (at different time points or under different experimental conditions) on the same set of genes M, one obtain a $M \times N$ gene expression matrix, where the rows are the M genes under study and the columns are the N conditions under which the study is performed. Each row in a gene expression matrix gives the expression profile of a gene in response to a set of experimental conditions.

4.3.2 Missing Value Imputation

It is not uncommon to find gene expression data with up to 90% of genes affected by missing values [18]. Instead of simply discarding gene expression profiles containing missing values, or replacing missing values by zeros or by the average of the expression profile, they can be estimated by exploiting the correlation in the data [25].

Microarray experiments that study cyclic systems, e.g. cell cycle [22], are usually carried out by synchronizing a population of cells. Synchronization is achieved by first arresting cells at a specific biological life point and then releasing cells from the arrest so that all cells are at the same point when the experiment begins [11, 22]. However, even if cells are synchronized perfectly at the beginning of the experiment, they do not remain synchronized forever [3]. Due to the loss of synchronization, the peaks and troughs of the expression profile of most cyclic genes decrease in magnitude with time.

In [9], we proposed a set theoretic framework based on projection onto convex sets (POCS) for missing data imputation called POCSimpute, which takes into consideration the phenomenon of synchronization loss. The main idea of POCS is to formulate every piece of prior knowledge into a corresponding convex set and then use a convergence-guaranteed iterative procedure to obtain a solution in the intersection of all these sets. Our POCS imputation method captures localized gene-wise correlation in the gene expression data by constructing a convex set based on local least square regression using the K-most correlated genes. The array-wise correlation is captured by using the PCA method. Finally, to account for synchronization loss in microarray experiments, we constrain the vector length of the missing values to be bounded by the vector length of the observed values within the same cycle.

Our method provides a flexible framework to incorporate all a priori information in the solution. Regardless of whether it is a consistent or inconsistent problem, the convergence of the POCS algorithm is guaranteed as long as the constraint sets are convex. For noisy gene expression data with imprecise prior information, this tolerance to imprecision is very important. Another useful feature of our POCSimpute algorithm is its adaptivity in finding a good solution. In POCS, when the information is more reliable, the corresponding convex set will be smaller. Since POCS always converge to the intersection, the final solution will always be dominated by the smaller set, while still satisfying the constraint imposed by the less reliable set.

4.4 Gene Expression Data Analysis

A useful tool in gene expression data analysis is cluster analysis. Since genes with related functions are expected to have similar expression patterns, clustering of genes may suggest possible roles for genes with unknown functions based on the known functions of some other genes that are placed in the same cluster. Time series gene expression data are useful in the study of dynamic cellular process. By studying the regulatory relationship between genes, one can infer the underlying gene networks that control these processes.

4.4.1 Detection of Coherently Expressed Genes

Many clustering algorithms such as k-means, self-organizing maps, hierarchical clustering, and principal component analysis, have been applied to the study of high-dimension gene expression data [1, 7, 24, 26, 29]. In [23], we proposed a clustering algorithm called Binary Hierarchical Clustering (BHC). BHC combines the features of hierarchical and partition-based clustering by performing a successive binary subdivision of the data in a hierarchical manner, until further splitting of a larger partition into two smaller partitions is insignificant anymore. The hierarchical structure is manifested in the binary tree structure of the clustering result, where a parent node gives rise to two children nodes if the projection onto the discriminant axis is greater than a certain threshold. The tree structure allows the relationship between adjacent clusters and the variation within each cluster to be observed easily. In BHC, a variant of the fuzzy C-means clustering algorithm that can handle non-ellipsoidal clusters is used to split a parent cluster into two children clusters.

In partition-based clustering, if the number of prototypes is less than that of the natural clusters in the dataset, a prototype could win patterns from more than one cluster. In addition, a natural cluster might be erroneously divided into two or more classes, or several natural clusters or part of them are erroneously grouped into one class. To uncover the natural clusters in the gene expression dataset, we proposed a novel partition-based clustering framework called self-splitting and merging competitive learning clustering (SSMCL) [27]. SSMCL uses a competitive learning paradigm that allows a cluster prototype to focus on just one natural cluster, while minimizing the competitions from other natural clusters. To find all the natural clusters in the dataset, an over-clustering and merging strategy is used in SSMCL. SSMCL was used to cluster the yeast cell cycle data and the resulting 22 clusters are all visually distinct from each other.

x	y	z	w
1.2	1.2	1.2	1.2
1.2	1.2	1.2	1.2
1.2	1.2	1.2	1.2
1.2	1.2	1.2	1.2

(a)

x	y	z	w
1.2	1.2	1.2	1.2
2.0	2.0	2.0	2.0
1.5	1.5	1.5	1.5
3.0	3.0	3.0	3.0

(b)

x	y	z	w
1.2	2.0	1.5	3.0
1.2	2.0	1.5	3.0
1.2	2.0	1.5	3.0
1.2	2.0	1.5	3.0

(c)

x	y	z	w
1.2	2.2	0.2	3.2
2.0	3.0	1.0	4.0
1.4	2.4	0.4	3.4
2.4	3.4	1.4	4.4

(d)

x	y	z	w
1.0	2.0	0.5	1.5
2.0	4.0	1.0	3.0
1.4	2.8	0.7	2.1
2.4	4.8	1.2	3.6

(e)

x	y	z	w
1.0	2.1	0.6	1.7
2.0	4.1	1.1	3.2
1.4	2.9	0.8	2.3
2.4	4.9	1.3	3.8

(f)

Figure 4.3 Different linear bicluster patterns: (a) constant values, (b) constant rows, (c) constant columns, (d) additive coherent values, (e) multiplicative coherent values, and (f) linear coherent values

In traditional clustering, the expression patterns are grouped either along the row direction or the column direction. However, in many situations, an interesting cellular process is active only under a subset of conditions, or a single gene may participate in multiple pathways that may or may not be co-active under all conditions. In addition, the data to be analyzed often include many heterogeneous conditions from many experiments. In these instances, it is often unrealistic to require that related genes behave similarly across all measured conditions as in conventional clustering. This requires simultaneous clustering along both the row and column directions, and is called biclustering [4,10,17]. Biclustering allows us to consider only a subset of conditions when looking for similarity across a subset of genes. In biclustering, the aim is to find coherent submatrices within the data matrix.

In [10], we formulate the biclustering problem in terms of identifying linear geometric structures in a high dimensional data space. This perspective views biclusters of linear patterns as hyperplanes in a high dimensional space, and allows a unified treatment in detecting these linear biclusters simultaneously. Figure 4.3 shows different linear bicluster patterns: (a) constant values, (b) constant rows, (c) constant columns, (d) additive coherent values, where each row/column is obtained by adding a constant to another row/column, (e) multiplicative coherent values, where each row/column is obtained by multiplying another row/column by a constant value, and (f) linear coherent values, where each column is obtained by multiplying another column by a constant value and then adding a constant. The linear coherent pattern of (f) actually subsumes all previous five patterns.

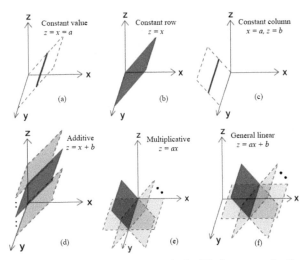

Figure 4.4 Different geometries (lines or planes) in the 3D data space for the corresponding bicluster patterns

If we treat each measurement (column) as a variable in the 4D space $[x, y, z, w]$ and each object (row) as a point in the 4D space, the six pattern in Figures 4.3(a–f) would correspond to the following six geometric structures respectively: (a) a cluster at a single point with coordinate $[x, y, z, w] = [1.2, 1.2, 1.2, 1.2]$, (b) a cluster defined by the lines $x = y = z = w$, (c) a cluster at a single point with coordinate $[x, y, z, w] = [1.2, 2.0, 1.5, 3.0]$, (d) a cluster defined by the lines $x = y-1 = z+1 = w2$, (e) a cluster defined by the lines $x = 0.5y = 2z = 2w/3$, and (f) a cluster defined by the lines $x = 0.5(y-0.1) = 2(z-0.1) = 2(w-0.2)/3$. Each object (row) in a cluster is a point lying on one of these points or lines. When a pattern is embedded in a larger data matrix with extra measurements, i.e., a bicluster that covers only part of the measurements in the data, the points or lines defined by the bicluster would sweep out a hyperplane in the high dimensional data space. In a 3D space, if we denote the three measurements as x, y and z respectively, and assume a bicluster covers x and z only, we can generate 3D geometric views for the different patterns as shown in Figure 4.4.

We use the Hough transform (HT) to detect the biclusters in the multidimensional data space [10]. Statistical properties of the HT, such as robustness, consistency and convergence, as well as its ability to identify geometric patterns in noisy data, make it highly attractive for bicluster analysis of noisy microarray data. Our geometric biclustering algorithm is applied to the human lymphoma dataset and we are able to discover biologically meaningful biclusters.

4.4.2 Uncovering Gene Regulatory Relationships

Time-series expression data are a particularly valuable source of information because they can describe a dynamic biological process such as the cell cycle or metabolic process [5,22]. They allow the determination of regulatory relationships between the expressions of different genes. Such relationship could lead to a better understanding of the gene networking process and gene regulation within a cell.

Several methods were proposed to extract the significant modes of variation from the time series expression data. Holter et al. [12] used SVD to extract the "characteristic modes" of gene expression. They showed that the behavior of the widely disparate gene systems analyzed in their work is dominated by a small subset of the characteristic modes and that a linear combination of just a few modes provides a good approximation to the behavior of the entire system in most cases. Alter et al. [2] used a similar analysis on two cell cycle time series and found that the first two modes for the cell cycle time series are approximately sinusoidal and 90° out of phase. The temporal nature of time series gene expression data was explicitly modeled by Dewey and Galas [6] using a dynamic model. They modeled the entire set of gene expression data using a first order Markov model which is equivalent to a first order autoregressive (AR) model. The construction of a gene network consisting of dynamic classes based on the transition matrix was also demonstrated.

Other methods attempt to perform pairwise comparison of gene expressions to identify pairs of genes that have direct regulatory relationships from the set of gene expression profiles [7, 8, 14, 30]. Among the various pairwise comparison methods, correlation-based method is perhaps the most popular. This method determines whether or not two genes have a regulatory relationship by checking the global similarity between their expression profiles using the Pearson correlation measure. However, correlation method does not take into account the fact that there is often a time delay before the regulator gene product can exert its influence on the target gene. Such time delay can significantly degrade the performance of the method. Correlation method also strong favors global similarity over more localized similarities arising from conditional regulatory relationships.

If the expression of gene A varies periodically at a constant frequency, we expect the expression of gene B to be varying more or less at the same frequency. This frequency of variation, however, may not be easily seen from the two time-series expression profiles due to noise and other factors. In addition, if gene B is under the influence of both gene A and gene C (a "two-regulating-one" situation), and the expression profiles of these influencing genes are varying at different frequencies, then the relationship between gene A and gene B may not be easily seen from their time-series profiles. This would cause problem for correlation-based similarity comparison.

In [30], we propose a spectral component correlation approach for measuring the correlation between time-series expression data, and use the results to infer the potential regulatory relationships between genes. Our technique summarizes the essential features of an expression pattern by means of a frequency spectrum estimated by autoregressive modeling. Specifically, the pattern $x[n]$ is decomposed into a set of damped sinusoids of different frequencies,

$$x[n] = \sum_{i=1}^{M} x_i[n] = \sum_{i=1}^{M} \alpha_i \exp(\sigma_i n) \cos(\omega_i n + \phi_i) \qquad (4.1)$$

so that each sinusoid $x_i[n]$ can be considered separately during the analysis. The parameters α_i, σ_i, ω_i, and φ_i are the amplitude, damping factor, normalized frequency and phase angle, respectively, of spectral component i. The correlation of $x[n]$ with $y[n]$ can then be reformulated as a sum of component-wise correlations between each spectral component,

$$x[n] \circ y[n] = \sum_{i} \sum_{j} \sqrt{E_{x_i} E_{y_i} / E_x E_y} x_i[n] \circ y_j[n], \qquad (4.2)$$

where the symbol ∘ denotes correlation operation and each term with letter E represents either total energy of a sequence or energy of a particular component. Such component-wise correlation provides more insights into the regulatory relationship. For instance, for the "two-regulating-one" situation, correlation between the expression profiles of gene A and gene B may not be strong enough to suggest their relationship due to the presence of spectral components in gene B induced by gene C. However, the spectral components of gene B due to gene A would exhibit strong correlations to gene A's expression profile.

Transcriptional regulation can involve activation or inhibition. In the activation process, the product of gene A affects the transcription process of

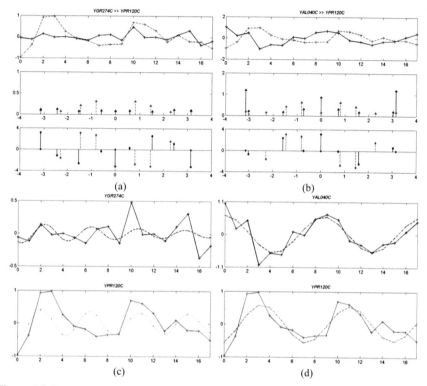

Figure 4.5 Two activation regulations with gene YPR120C as an activatee. (a) Activation regulation with gene YGR274C as an activator. (b) Activation regulation with gene YAL040C as an activator. (c) Correlated frequency components for the first pair. (d) Correlated frequency components for the second pair

gene B such that the production rate for gene B increases. On the other hand, the inhibition process involves gene A's product decreasing the production of gene B. In [30], we used the spectral component correlation algorithm to analyze the alpha-synchronized yeast cell-cycle dataset [22]. We were able to detect many regulatory pairs that were missed by the traditional correlation method due to weak correlation value.

The spectral component correlation method allows us to identify strongly oscillatory but time-shifted expression pairs by using only the spectral magnitude information. It also allows us to detect regulatory relationships involving multiple genes by checking for the existence of regulators' frequencies from the expression profile of the gene being regulated. Figure 4.5 shows two known activation regulations with a common gene YPR120C as an activatee. It reveals that the first regulation has its expression profiles correlated at frequency of around 1.48 rad/s, whereas the second regulation has its profiles correlated at around 0.76 rad/s. In [30], we have identified many known activation sets involving multiple genes using the spectral component correlation method.

When the component-wise correlation analysis is applied to all 439 known regulations, the results indicated that 223 out of 343 activations and 55 out of 96 inhibitions have their component-wise correlations score greater than 0.5 [30]. We found that a large number of visually dissimilar expression pairs do have very similar dominant frequency components. For example, among those 307 pairs having traditional correlation coefficients of less than 0.5, 196 of them have greater than 0.5 component-wise correlation coefficients. Furthermore, 60 out of this 196 pairs have their component-wise correlation coefficients greater than 0.9 and the expression patterns in each of these pairs strongly oscillate at almost identical frequencies. The spectral component correlation method therefore allows the hidden component-wise relationships between two expression profiles to be revealed, which are otherwise hidden in the traditional correlation method.

4.5 Conclusions

DNA microarray technology, which allows massively parallel, high throughput profiling of gene expression, has emerged as a powerful tool for genomic research. In this chapter, we present an overview of DNA microarray technology and gene expression data analysis. We described the extraction of gene expression data from microarray images, which involves image analysis, data preprocessing, and missing value imputation. We presented some of our recent work on cluster and bicluster analysis of gene expression data. Finally, we described our work for gene regulation study using the technique of spectral component correlation. The technique allows many weak but biologically significant correlations between time series gene expression profiles to be detected, and facilitates the inference of complex gene regulatory relationships between multiple genes. To conclude, we remark that many of the problems in microarray data analysis involve identifying and analyzing features and patterns in the data, the pattern recognition community could certainly make significant contribution to this fascinating field.

Acknowledgement

This work is supported by the Hong Kong Research Grant Council (Project CityU122607).

References

[1] U. Alon, N. Barkai, D.A. Notterman, K. Gish, S. Ybarra, D. Mack, and A.J. Levine. Broad patterns of gene expression revealed by clustering analysis of tumor and normal colon tissues probed by oligonucleotide arrays. *PNAS*, 96(12):6745–6750, 1999.

[2] O. Alter, P.O. Brown, and D. Botstein. Singular value decomposition for genome-wide expression data processing and modeling. *PNAS*, 97(18):10101–10106, 2000.

[3] Z. Bar-Joseph, S. Farkash, D.K. Gifford, I. Simon, and R. Rosenfeld. Deconvolving cell cycle expression data with complementary information. *Bioinformatics*, 20:i23–i30, 2004.

[4] K.-O. Cheng, N.-F. Law, W.-C. Siu, and A.W.C. Liew. Identification of coherent patterns in gene expression data using an efficient biclustering algorithm and parallel coordinate visualization. *BMC Bioinformatics*, 9:210, 2008.

[5] J.L. DeRisi, V.R. Lyer, and P.O. Brown. Exploring the metabolic and genetic control of gene expression on a genomic scale. *Science*, 278:680–686, 1997.

[6] T.G. Dewey and D.J. Galas. Dynamic models of gene expression and classification. *Functional & Integrative Genomics*, 1(4):69–278, 2001.

[7] M.B. Eisen, P.T. Spellman, P.O. Brown, and D. Botstein. Cluster analysis and display of genome-wide expression patterns. *PNAS*, 95(25):14863–14868, 1998.

[8] V. Filkov, S. Skiena, and J. Zhi. Analysis techniques for microarray time series data. *Journal of Computational Biology*, 9(2):317–330, 2002.

[9] X. Gan, A.W.C. Liew, and H. Yan. Microarray missing data imputation based on a set theoretic framework and biological consideration. *Nucleic Acids Research*, 34(5):1608–1619, 2006.

[10] X. Gan, A.W.C. Liew, and H. Yan. Discovering biclusters in gene expression data based on high-dimensional linear geometries. *BMC Bioinformatics*, 9:209, 2008.

[11] A.P. Gasch, P.T. Spellman, C.M. Kao, O. Carmel-Harel, M.B. Eisen, G. Storz, D. Botstein, and P.O. Brown. Genomic expression programs in the response of yeast cells to environmental changes. *Molecular Biology of the Cell*, 11:4241–4257, 2000.

[12] N.S. Holter, M. Mitra, A. Maritan, M. Cieplak, J.R. Banavar, and N.V. Fedoroff. Fundamental patterns underlying gene expression profiles: Simplicity from complexity. *PNAS*, 97(15):8409–8414, 2000.

[13] C.L. Kooperberg, T.G. Fazzio, J.J. Delrow, and T. Tsukiyama. Improved background correction for spotted DNA microarrays. *Journal of Computational Biology*, 9(1): 55–66, 2002.

[14] A.T. Kwon, H.H. Hoos, and R. Ng. Inference of transcriptional regulation relationships from gene expression data. *Bioinformatics*, 19(8):905–912, 2003.

[15] A.W.C. Liew, H. Yan, and M. Yang. Robust adaptive spot segmentation of DNA microarray images. *Pattern Recognition*, 36(5):1251–1254, 2003.

[16] D.J. Lockhart and E.A. Winzeler. Genomics, gene expression and DNA arrays. *Nature*, 405:827–846, 2000.

[17] S.C. Madeira and A.L. Oliveira. Biclustering algorithms for biological data analysis: A survey. *IEEE/ACM Transactions on Computational Biology and Bioinformatics*, 1(1):24–45, 2004.

[18] M. Ouyang, W.J. Welsh, and P. Georgopoulos. Gaussian mixture clustering and imputation of microarray data. *Bioinformatics*, 20:917–923, 2004.

[19] A.C. Pease, D. Solas, E.J. Sullivan, M.T. Cronin, C.P. Holmes, and S.P.A. Fodor. Light-generated oligonucleotide arrays for rapid DNA sequence analysis. *PNAS*, 91(11):5022–5026, 1994.

[20] M. Schena, D. Shalon, R.W. Davis, and P.O. Brown. Quantitative monitoring of gene expression patterns with a complementary DNA microarray. *Science*, 270:467–470, 1995.

[21] G.K. Smyth and T.P. Speed. Normalization of cDNA microarray data. *Methods*, 31:265–273, 2003.

[22] P.T. Spellman, G. Sherlock, M.Q. Zhang, V.R. Iyer, K. Anders, M.B. Eisen, P.O. Brown, D. Botstein, and B. Futcher. Comprehensive identification of cell cycle-regulated genes of the yeast Saccharomyces cerevisiae by microarray hybridization. *Molecular Biology of the Cell*, 9:3273–3297, 1998.

[23] L.K. Szeto, A.W.C. Liew, H. Yan, and S.-S. Tang. Gene expression data clustering and visualization based on a binary hierarchical clustering framework. *Journal of Visual Languages and Computing*, 14(4):341–362, 2003.

[24] P. Tamayo, D. Slonim, J. Mesirov, Q. Zhu, S. Kitareewan, E. Dmitrovsky, E.S. Lander, and T.R. Golub. Interpreting patterns of gene expression with self-organizing maps: Methods and application to hematopoietic differentiation. *PNAS*, 96(6):2907–2912, 1999.

[25] O. Troyanskaya, M. Cantor, G. Sherlock, P. Brown, T. Hastie, R. Tibshirani, D. Botstein, and R.B. Altman. Missing values estimation methods for DNA microarrays. *Bioinformatics*, 17:520–525, 2001.

[26] K.P. White, S.A. Rifkin, P. Hurban, and D.S. Hogness. Microarray analysis of Drosophila development during metamorphosis. *Science*, 286:2179–2184, 1999.

[27] S. Wu, A.W.C. Liew, and H. Yan. Cluster analysis of gene expression data based on self-splitting and merging competitive learning. *IEEE Transactions on Information Technology in Biomedicine*, 8(1):5–15, 2004.

[28] Y.H. Yang, S. Dudoit, P. Luu, D.M. Lin, V. Peng, J. Ngai, and T.P. Speed. Normalization for cDNA microarray data: a robust composite method addressing single and multiple slide systematic variation. *Nucleic Acids Research*, 30(4):e15, 2002.

[29] K.Y. Yeung and W.L. Ruzzo. Principal component analysis for clustering gene expression data. *Bioinformatics*, 17(9):763–774, 2001.

[30] L.K. Yeung, L.K. Szeto, A.W.C. Liew, and H. Yan. Dominant spectral component analysis for transcriptional regulations using microarray time-series data. *Bioinformatics*, 20(5):742–749, 2004.

5

Invariants and Covariants in Pattern Recognition and Image Analysis

Marco Reisert, Nikos Canterakis and Hans Burkhardt

Computer Science Department, Institute for Pattern Recognition and Image Processing, Albert-Ludwigs University of Freiburg, Georges-Koehler-Allee 052, D-79110 Freiburg, Germany; E-mail: marco.reisert@uniklinik-freiburg.de, {canterakis, burkhardt}@informatik.uni-freiburg.de

Abstract

Invariance and covariance against group transformations play a crucial role for a variety of applications in pattern recognition and image processing. Invariant features are able to map equivalence classes based on the group action onto one point of a corresponding feature space and thus simplify the design of classifiers. They mostly find their applications in shape and image retrieval and object classification tasks. On the other hand, covariant image transformations inherit the group transformation behavior from the space of input images onto the transformation space. For example, a Gaussian smoothing is probably the most simple motion-covariant image transformation. In general, they play an important role in recognition, image processing and low-level vision tasks, e.g. object detection, image reconstruction and restoration.

This contribution presents the foundations and recent advances in the field, with a focus on the Euclidean Group transformation. Different methodologies how to obtain invariant features and covariant image transformations are discussed. With respect to the computation of group invariant features for pattern recognition, we consider a variety of instances where groups are acting on pattern spaces. Starting from simple finite (commutative and noncommutative) groups up to only locally compact noncommutative Lie groups like 2D and 3D Euclidean motion, we indicate the associated suit-

Patrick Shen-Pei Wang (Ed.), Pattern Recognition and Machine Vision – In Honor and Memory of Professor King-Sun Fu, 71–85.

able system of basis functions to work with in each case. Based on that we develop and discuss techniques for the computation of complete and hierarchically optimal systems of invariants. On the other hand, the group-associated basis system induces covariant differential operators and covariant tensor products. These operations allow the development of a large variety of nonlinear, motion-covariant image transformations in a systematic manner, with applications ranging from classic filtering to recognition specific preprocessing.

Keywords: invariants, covariants, group transformations, motion group, group representations, harmonic filters.

5.1 Invariants for Pattern Recognition

In this part of the contribution we will discuss a number of cases around the problem of forming features for pattern recognition that are invariant with respect to some transformation group. We will be concerned with the group of Euclidean motions as well as with subgroups and specialisations of it.

We consider patterns x being elements of a pattern space Ω and some group G with elements g that are acting in the pattern space by mapping some x to $gx \in \Omega$. The set $\{gx : g \in G\}$ is an orbit being an equivalence class in Ω for which x is a representative. The equivalence relation itself is denoted by $\sim$. A central problem of pattern recognition is to find a number of features F_n for the patterns x that are invariant w.r.t. the group action, i.e.

$$F_n(gx) = F_n(x) \, \forall \, n, \, x \text{ and } g$$

If, in addition, a certain set of such invariant features allows to separate all orbits, i.e. allows to conclude from $F_n(x_1) = F_n(x_2) \, \forall n$ that $x_1 \sim x_2$, then this set is said to be complete. In this case there is always for two nonequivalent patterns $x_1 \nsim x_2$ at least one feature F_i with $F_i(x_1) \neq F_i(x_2)$.

In a very general sense, it can be shown that any invariant feature F can be computed by group averaging of some measurement function m performed upon the pattern [9]. We formally have that if we define

$$F(x) = \int_G m(g^{-1}x) \, dg$$

then F will be an invariant feature with $F(hx) = F(x) \, \forall x \in \Omega$ and $h \in G$. The integral above is a so called Haar integral where integration is performed over the whole group. For continuous groups with some parameterisation, dg has to be a normalised left invariant differential defined in the parameter space of the group [5]. For finite groups the integral simplifies to a summation over all group elements and results to the so called Hurwitz group averaging.

This kind of producing invariants poses the questions of how to choose the measurements m and how to efficiently perform the required integration. Answers to these questions are guided mostly by practical considerations and have been explored in a number of research works [13, 14] yielding techniques showing partially excellent results in their own field of application. On the other hand, especially for the choice of the measurement functions m there are no general analytic tools to apply (other that they necessarily have to be nonlinear, with a single exception) in order to have some guarantee of producing independent, discriminative or even complete invariant features.

A more analytic direction to tackle the problem is to make use of the results of noncommutative harmonic analysis and group representation theory [5] in order to obtain in a preprocessing step a decomposition of some pattern in subspaces as small as possible that are invariant w.r.t. the considered group and where the latter is acting homomorphically via its irreducible representations. Intuitively, this approach facilitates the analysis because, also mathematically, smaller spaces are easier overlooked. We will try to present these ideas in the simplest and as elementary as possible way.

5.1.1 A Very Simple Case

We start with a toy size problem that clarifies the introduced notions and the difference between the two approaches.

Let the patterns be functions x on the ordered set [0 1 2] with real values $x(0) = a$, $x(1) = b$ and $x(2) = c$ and the group be cyclic translations. Trying the (linear) measurement function $m(x) = x(0) = a$ (take the zeroth element) and averaging over the group, we obtain the simple invariant $F_0 = a + b + c$. Proceeding with a second measurement $m(x) = x(0)^2 = a^2$ and again averaging yields the invariant $F_1 = a^2 + b^2 + c^2$. Since this problem seems to be very simple we might develop the ambition to compute here also complete invariants. Now we observe that we have to include also some invariant that is not symmetric in the elements of x because otherwise we will not be able to discriminate between the orbit of x and of its reflection giving together six functions arising from all permutations of the three elements. We therefore now average the nonsymmetric monomial a^2b giving a third invariant which precludes the reflection orbit (if they do not happen to coincide) and it is very tempting to suppose that we are done. However, by computing back from these three invariants the function, e.g. via Groebner elimination [2], we discover that they are again common to two orbits, the one we want to isolate and a second one, not the reflection orbit. So we have to keep trying with averaging new measurements until we can, somehow, show that we have reached completeness.

In contrast to the procedure above, the analytic approach asks first for the minimal, not further reducible in dimension invariant subspaces of the problem at hand. In the present case and due to commutativity of the group, they are all one-dimensional and are spanned by the basis functions (modulo normalizing constant factors), $b_0 = [1\ 1\ 1]$, $b_1 = [1\ \omega\ \omega^2]$ and $b_2 = [1\ \omega^2\ \omega]$ with ω being a primitive third root of unity. With ρ and $r \in \{0, 1, 2\}$ we can write $b_\rho(r) = \omega^{\rho r}$ and the action of a single step cyclic translation g or repetitions g^α thereof on the basis functions is seen not to leave the respective 1D subspace since we have $g^\alpha\, b_\rho(r) = \omega^{\rho\alpha} \cdot b_\rho(r)$. Now, the values of x w.r.t. the new basis are given through projections via inner products onto the basis elements and read: $\hat{x}_0 = a + b + c$, $\hat{x}_1 = a + b\omega^2 + c\omega$ and $\hat{x}_2 = a + b\omega + c\omega^2$. They constitute, of course, the DFT of the original pattern with the property $\hat{x}_0(g\,x) = \hat{x}_0(x)$, $\hat{x}_1(g\,x) = \omega\hat{x}_1(x)$ and $\hat{x}_2(g\,x) = \omega^2\hat{x}_2(x)$. With the exception of the zeroth DFT-component which is already an invariant, the first and the second are 1D *covariants* each, which however make the computation of further invariants especially easy because they exhibit a very simple law of variation upon the action of the group. The resulting procedure we have to follow in order to obtain invariants is immediately seen to be the formation of products of powers of the covariants so as to eliminate the factors ω. For example, $\hat{x}_1^3$ is such an invariant which, in addition, together with the invariant $\hat{x}_0$ turns out to constitute a complete system since it is very easy to show that they allow unique determination of the orbit. Note that by writing out the resulting expressions for the real and the imaginary part of $\hat{x}_1^3$ we can see what measurements we have to average in order to achieve completeness. But, of course, the resulting measurements are by no means unique. If we rebuild this first complete set of invariants by elementary manipulations, we may arrive at the equivalent set $\{a + b + c, a^2b + b^2c + c^2a, ab^2 + bc^2 + ca^2\}$ which is again complete and shows that averaging of the two indicated asymmetric measurements together with the linear one is enough to achieve completeness. The fact that third degree expressions of spectral components, which appeared here in a natural way, suffice, seems to be a general law. The roots for showing this have been put in the work of Kakarala [6].

The very simple case considered above was about a commutative group and it is a general law that commutative groups always lead to 1D minimal invariant subspaces and therefore to simple covariants that make the computation of invariants especially easy like above. For example, 2D discrete images under cyclic translations in both directions are such a case that can be treated with the 2D DFT. Another commutative case are 2D images under pure rotations. The problem here can be solved by starting to use as basis functions the complex polynomials $(x + jy)^p(x - jy)^q$ where $j^2 = -1$ which give 1D invariant subspaces w.r.t. rotations. But two such bivariate polynomials are not orthogonal within the unit disc if the differences $p - q$ coincide. The particularity here, however, is that all such functions with equal differences $p - q$ are also equally transformed by a rotation and that allows us to form specific linear combinations of them which now produce orthogonal functions without destroying the invariance property. The result is the set of 2D Zernike polynomials having some additional useful properties [7].

5.1.2 A Noncommutative Case with Extensions

Things become abruptly more complicated and more interesting if we come across noncommutative groups. In this subsection we will expose the probably simplest case of this kind which are 3×3 discrete images under cyclic translations in both directions and rotations around the middle pixel with angles that are multiples of $\pi/2$. As we will see, although still finite, this group possesses parallels to both, 3D rotations which is also a noncommutative but compact group, and to the full 2D Euclidean motion group which is noncommutative and noncompact, meaning that the group parameters do not always remain within a bounded region. The cardinality of this group is $9 \cdot 4 = 36$ because there are 9 different translational and 4 rotational positions. The nine pixels are described by the set of vectors $S = \{\mathbf{r} = [r \ s]^T\}$ where r and s are $-1, 0$, or 1, and a rotation by $\pi/2$ is described by multiplication with the matrix $\mathbf{R} = \left(\begin{smallmatrix} 0 & -1 \\ 1 & 0 \end{smallmatrix}\right)$. A group element g is described by the two parameters $\lambda \in \{0, 1, 2, 3\}$ and $\mathbf{t} \in S$ and maps grid positions $\mathbf{r}$ to $g\mathbf{r} = \mathbf{R}^\lambda(\mathbf{r} + \mathbf{t})$. After the addition, the coordinates have, of course, to be reduced cyclically back to the allowed set $\{-1, 0, 1\}$. This group is not commutative because $g_1 g_2 \neq g_2 g_1$ in general. Now, what are the suitable basis functions for this group? If we try to start again with the DFT in order to cope with translation we have to ask what happens with the spectrum if we also rotate the image. As is well known, the spectrum rotates by the same rotation thereby singling out two 'circles' consisting of the spectral points $\{[1 \ 0], \ [0 \ 1], \ [-1 \ 0], \ [0 \ -1]\}$

and $\{[1\ 1],\ [-1\ 1],\ [-1\ -1],\ [1\ -1]\}$. We therefore use again DFT within the spectrum and along the two 'circles', now a four-points one, and obtain eventually the basis functions

$$b_\rho^n(\mathbf{r}) = \sum_{\nu=0}^{3} j^{n\nu} \omega^{\mathbf{v}_\rho^T \mathbf{R}^\nu \mathbf{r}}.$$

With $n \in \{0, 1, 2, 3\}$, $\rho \in \{1, 2\}$, $\mathbf{v}_1^T = [1\ 0]$ and $\mathbf{v}_2^T = [1\ 1]$ this gives eight basis functions. Finally, the overall constant function $b_0 \equiv 1$ completes the needed set of nine mutually orthogonal basis functions for this problem. For example, up to some normalizing constant factor, the basis function b_1^1 looks as follows:

$$b_1^1 = \begin{pmatrix} 1-j & 1 & 1+j \\ -j & 0 & j \\ -1-j & -1 & -1+j \end{pmatrix}.$$

Note that these functions could be named discrete Bessel functions as will become apparent later. Also note that they can be interpreted as being DFT-coefficients of the ν-periodic kernel which allows to obtain for the latter

$$\omega^{\mathbf{v}_\rho^T \mathbf{R}^\nu \mathbf{r}} = \frac{1}{4} \sum_{n=0}^{3} j^{-n\nu} b_\rho^n(\mathbf{r})$$

and this, in turn, allows us to give the behavior of the basis functions under group action:

$$b_\rho^n(g\mathbf{r}) = b_\rho^n(\mathbf{R}^\lambda(\mathbf{r}+\mathbf{t})) = \frac{j^{-n\lambda}}{4} \sum_{\alpha=0}^{3} b_\rho^{n-\alpha}(\mathbf{t}) b_\rho^\alpha(\mathbf{r}).$$

This shows that besides b_0, which forms a 1D invariant subspace, we have two four-dimensional invariant subspaces which can be put together in the vectors

$$\mathbf{b}_\rho(\mathbf{r}) = [b_\rho^0,\ b_\rho^1,\ b_\rho^2,\ b_\rho^3]^T(\mathbf{r}) \text{ with } \rho = 1, 2.$$

The group action can be now formulated compactly by the formula $\mathbf{b}_\rho(g\mathbf{r}) = \mathbf{D}_\rho(g)\mathbf{b}_\rho(\mathbf{r})$, where $\mathbf{D}_\rho(g)$ are unitary, irreducible representations of the considered group obeying the homomorphism $\mathbf{D}_\rho(g_1 g_2) = \mathbf{D}_\rho(g_1)\mathbf{D}_\rho(g_2)$ and are given by

$$\mathbf{D}_\rho(g) \sim \mathrm{diag}([1,\ -j,\ -1,\ j])^\lambda \begin{pmatrix} b_\rho^0 & b_\rho^3 & b_\rho^2 & b_\rho^1 \\ b_\rho^1 & b_\rho^0 & b_\rho^3 & b_\rho^2 \\ b_\rho^2 & b_\rho^1 & b_\rho^0 & b_\rho^3 \\ b_\rho^3 & b_\rho^2 & b_\rho^1 & b_\rho^0 \end{pmatrix} \text{(t)}.$$

Now, for any given pattern x we first compute its coefficients w.r.t. the new basis via inner products and organise them in the respective vectors according to $\hat{\mathbf{x}}_\rho = \langle x, \mathbf{b}_\rho \rangle$ which gives the behavior under group action $\hat{\mathbf{x}}_\rho(g) = \mathbf{D}_\rho(g)\hat{\mathbf{x}}_\rho$. These coefficient vectors are therefore *covariants* that facilitate as usual the computation of invariants: Namely, since the representations $\mathbf{D}_\rho$ are unitary, we first obtain the invariants $\hat{\mathbf{x}}_\rho^\dagger \hat{\mathbf{x}}_\rho$ which gives for $\rho = 0, 1$, and 2 already a set of three invariants which are of course not yet that many. However, by forming tensor (Kronecker) products between the above coefficient vectors we obtain new vectors of larger dimension that transform according to the corresponding Kronecker products of the respective irreducible representations: $\hat{\mathbf{x}}_1(g) \otimes \hat{\mathbf{x}}_2(g) = (\mathbf{D}_1(g) \otimes \mathbf{D}_2(g))(\hat{\mathbf{x}}_1 \otimes \hat{\mathbf{x}}_2)$. Now, taking the norm of the resulting product would certainly not give anything really new because this invariant would be dependend upon the previous ones. However, it is a general fact in the representation theory of groups that Kronecker products of irreducible representations are themselves reducible and can be decomposed in a direct sum of a certain subset of the irreducible representations of the given group. This decomposition is obtained through a so called Clebsch Gordan (CG) matrix $\mathbf{C}_{12}$ as follows [8]:

$$\mathbf{D}_1(g) \otimes \mathbf{D}_2(g) = \mathbf{C}_{12}^\dagger(\mathbf{D}_\alpha(g) \oplus \cdots \oplus \mathbf{D}_\beta(g))\mathbf{C}_{12}$$

where the $\mathbf{D}_\alpha, \ldots, \mathbf{D}_\beta$ come from the set of irreducible representations of the considered group. The matrix $\mathbf{C}_{12}$ consists of constants that can be computed beforehand and is also unitary. This fact can be use d to reformulate the behavior of the Kroneker product of the coefficient vectors above as follows:

$$\mathbf{C}_{12}(\hat{\mathbf{x}}_1(g) \otimes \hat{\mathbf{x}}_2(g)) = (\mathbf{D}_\alpha(g) \oplus \cdots \oplus \mathbf{D}_\beta(g))\mathbf{C}_{12}(\hat{\mathbf{x}}_1 \otimes \hat{\mathbf{x}}_2)$$

$$= (\mathbf{D}_\alpha(g) \oplus \cdots \oplus \mathbf{D}_\beta(g))((\hat{\mathbf{x}}_1 \bullet_\alpha \hat{\mathbf{x}}_2) \oplus \cdots \oplus (\hat{\mathbf{x}}_1 \bullet_\beta \hat{\mathbf{x}}_2)).$$

In words, the matrix $\mathbf{C}_{12}$ maps the original Kronecker product of coefficient vectors $\hat{\mathbf{x}}_1 \otimes \hat{\mathbf{x}}_2$ in such a way as to obtain a behavior under group action that goes via a direct sum of irreducible, unitary representations. The induced products are denoted by $\bullet_\alpha, \ldots, \bullet_\beta$ where the subscript refers to the corresponding irreducible subspace. That means that if we take now the norms of all appearing subvectors $\hat{\mathbf{x}}_1 \bullet_\alpha \hat{\mathbf{x}}_2, \ldots, \hat{\mathbf{x}}_1 \bullet_\beta \hat{\mathbf{x}}_2$ separately, we obtain an additional number of invariants. Some of them might be trivial and some dependend upon the old ones but some are really new and it is a current point of investigation to clarify things here. Note that generalisation of this process to three or more factors is straightforward.

Regarding patterns defined as functions on the unit sphere under the 3D rotation group $SO(3)$, the suitable system of basis functions to work with here are the well known spherical harmonics that are organized in invariant subspaces expressed by vectors $\mathbf{Y}^\alpha(\mathbf{r})$ where $\|\mathbf{r}\| = 1$. And the only principal difference with the previous simple case considered above is that we now have an infinite number of irreducible representations that are however themselves, thanks to compactness of the group $SO(3)$, again finite dimensional. Projection of some pattern onto the functions, forming Kronecker products of the resulting coefficient vectors and reduction via CG matrices follow exactly the same lines as above. For patterns defined in the interior of the unit sphere under 3D rotations, 3D Zernike-like polynomials have been developed in [4] that again transform via the irreducible representations of $SO(3)$ and present therefore no new principal difficulties regarding the formation of invariants.

Some formal complications arise if we consider 2D patterns under the group of full Euclidean motions $SE(2)$ since this group, besides being non-commutative, is also noncompact. Nevertheless, the definition of the suitable system of basis functions follows the same lines as above and gives

$$b_\rho^n(\mathbf{r}) = \frac{1}{2\pi} \int_0^{2\pi} e^{jn\theta} e^{j2\pi\rho\mathbf{e}_\theta^T\mathbf{r}} \, d\theta$$

with $\mathbf{e}_\theta^T = [\cos(\theta)\,\sin(\theta)]$, which is nothing else than 2D Bessel functions. Again, the parameter ρ is indexing the invariant subspaces and the functions themselves can be interpreted as Fourier coefficients of the θ-periodic kernel, thus allowing an easy determination of the group action. Now, however, the parameter ρ is traversing the whole continuum $\mathbb{R}^+$ giving a continuum of invariant subspaces and n is an integer running from $-\infty$ to $+\infty$ giving infinite dimensional representations. Nevertheless, by the assumption of having to deal with patterns of finite extent, the first problem can be overcome by sampling ρ at discrete intervals determined by the size of the pattern and the second by introducing a sort of band limitation. And finally, for the problem of 3D patterns under the full 3D motion group $SE(3)$, a setting up of 3D Bessel functions is currently being developed.

5.2 Covariant Image Transformations

The notion of covariance (or relative invariance) appears in lots of different contexts. In this part we focus on covariant image transformations, i.e. images are mapped onto images by preserving their group transformation behavior. In one-dimensional signal processing Volterra Theory [3] explains how arbitrary nonlinear image transformations covariant against time-shifts can be modeled. Here, we want to consider the more difficult multi-dimensional case where the covariance group is the motion group $SE(n)$ for $n = 2, 3$. The probably most simple $SE(n)$-covariant image transformation is a linear convolution with an isotropic Gaussian function, i.e. a blur of the image. But, of course, the usefulness of such a transformation is rather limited. The goal of this section is to give guidelines how to design and implement arbitrary nonlinear $SE(n)$-covariant transformations. Therefor we use two basic building blocks which both itself are necessarily covariant again. On the one hand linear operations are introduced which can be identified to be a convolution or differentiation and on the other hand the product $\bullet_\alpha$ introducing the nonlinearity. The main difficulty to achieve an $SE(n)$-covariant behavior is founded in the rotation behavior. Thus, both operations will be defined on the basis of the above introduced irreducible subspaces of the rotation group $SO(n)$. In this way we will achieve systematically a covariant behavior. Another advantage is the computational efficiency of the irreducible representations. In fact, per mathematical definition performing a group transformation by irreducible representations is the most efficient way under all tensor representations.

For convenience we represent images on the continous $\mathbb{R}^n$, i.e. an image is a mapping $\mathbf{x} : \mathbb{R}^n \mapsto V$, where V is the space of values the image is built of, typically for intensity images one has $V = \mathbb{R}^+$, but in our case it may be an arbitrary vector space. The action of the Euclidean group on such an image is then

$$(g\mathbf{x})(\mathbf{r}) := \mathbf{D}(g)\mathbf{x}(\mathbf{R}_g^T \mathbf{r} - \mathbf{t}_g), \tag{5.1}$$

where $\mathbf{t}_g$ is translation vector and $\mathbf{R}_g$ a rotation matrix defined on the image space. The matrix $\mathbf{D}(g) : V \mapsto V$ is an arbitrary representation of the group $SO(n)$ acting on the associated image values. For example, for vector fields, i.e. $V = \mathbb{R}^3$ we have just $\mathbf{D}(g) = \mathbf{R}_g$.

A covariant image transformation $\mathcal{F}$ is a mapping that preserves the above defined behavior, i.e. it is the same as if an image is translated by g and then the transformation $\mathcal{F}$ is applied or one first applies $\mathcal{F}$ onto the image and then translates the result, formally

$$\mathcal{F}(g\mathbf{x}) = g\mathcal{F}(\mathbf{x})$$

The question in course is, how one can build such transformations systematically? For the translation group a complete theory is established known as Volterra theory. The idea is to generalize from the linear case to the arbitrary non-linear case. But the generalization of this concept to $SE(n)$ is not straightforward. We do not intend to present here a general theory like the Volterra theory but give general guidelines to build such transformations.

5.2.1 Covariant Products

We start with a very simple example. Consider vector images, maybe gradient images or velocity fields. Such vector images obey the transformation law described above. The question is how it is possible to form new images by multiplication while keeping the transformation behavior (5.1). Formally, we search for a product • with the following behavior:

$$(g\mathbf{x}_1) \bullet (g\mathbf{x}_2) = g(\mathbf{x}_1 \bullet \mathbf{x}_2), \tag{5.2}$$

that is, the product is covariant with respect to $SE(n)$. For example, in 3D we can use the cross-product (or vector-product). One just connects two images point-by-point with the cross-product multiplication and obtains another image which again obeys the law in question. The main reason for that is the rotation behavior of the cross-product

$$(\mathbf{R}_g \mathbf{v}_1) \times (\mathbf{R}_g \mathbf{v}_2) = \mathbf{R}_g (\mathbf{v}_1 \times \mathbf{v}_2).$$

The translation-covariance is ensured by applying the operation point-by-point. Another simple idea is to use the scalar-product in the associated vector space to, for example, compute the squared norm of the vector image. Doing so the result is not a vector image anymore but an intensity image, but it still obeys the corresponding transformation law given in Equation (5.1) with the trivial representation $\mathbf{D}(g) = 1$.

The goal is now to generalize this idea. We want to keep the idea of a point-by-point multiplication, because this ensures the translation-covariance. So, we have to bother with the rotation behavior. Therefore, consider again the subspaces V^α corresponding to the irreducible representations of $SO(2)$ or $SO(3)$. We denote the space of images of type $\mathbb{R}^n \mapsto V^\alpha$ by T^α. We say the images in T^α are of degree α. It is the space of V^α-valued images. In the above example we considered $V^1 = \mathbb{R}^n$-valued images. Now, the goal is to find 'products' that are of type

$$\bullet_\alpha : V^{\alpha_1} \times V^{\alpha_2} \mapsto V^\alpha$$

and simultaneously obey the covariance property for the rotation group. By the term 'product' we mean that $\bullet_\alpha$ is a bilinear-form. Thus, distributivity is ensured, while symmetry and associativity do not have to hold. Explicitly, the resulting vector has to be a weighted sum of products of the components of the incoming vectors:

$$(\mathbf{v}_1 \bullet_\alpha \mathbf{v}_2)_k = \sum_{i,j} b_{kij}^\alpha (\mathbf{v}_1)_i (\mathbf{v}_2)_j .$$

Here the round brackets access the components of the vectors $\mathbf{v} \in V^\alpha$. In fact it is possible to find such weights b_{kij}^α that accomplish the above described ideas. For $SO(2)$ they look rather simple but for $SO(3)$ they involve rather complicated expressions of combinatorial nature. We do not want to go into detail here and refer to [10–12]. The weights are depending on the group and on the degrees of the incoming and outgoing images. Using these products point-by-point we can easily obtain an $SE(n)$-covariant product (obeying Equation (5.2)) of type $\bullet_\alpha : T^{\alpha_1} \times T^{\alpha_2} \mapsto T^\alpha$. That is, the product takes two images of certain degrees and gives an image of degree α. We cannot expect that the output degree is arbitrary. For example, in 3D we have already pictured two types of products, one of type $T^1 \times T^1 \mapsto T^1$ and one of type $T^1 \times T^1 \mapsto T^0$. In fact, there is only one other output degree namely $T^1 \times T^1 \mapsto T^2$. It can be imagined as the dyadic product of the two vectors. All other covariant products that take two vectors are built upon those.

5.2.2 Covariant Differentials

As already depicted a convolution with a rotation symmetric kernel (e.g. a Gaussian) is a covariant image transformation on intensity-valued images. And indeed, this is the only linear covariant image transformation which maps an intensity-valued image onto another intensity-valued image. To gain more richness one can consider, for example vector-valued images. Just consider the gradient operator: it maps an intensity-valued image onto a vector-valued image, where the components are partial derivatives with respect to the coordinate axis. In fact, the gradient operator $\nabla_\mathbf{r}$ is an $SE(n)$ covariant image transformation because

$$\nabla_\mathbf{r} \mathbf{x}(\mathbf{R}_g^T \mathbf{r} - \mathbf{t}_g) = \mathbf{R}_g (\nabla_{\mathbf{r}'} \mathbf{x})(\mathbf{R}_g^T \mathbf{r} - \mathbf{t}_g),$$

where $\nabla_{\mathbf{r}'}$ denotes differentation with respect to the transformed coordinate $\mathbf{r}' = \mathbf{R}_g^T \mathbf{r} - \mathbf{t}_g$. Due to the convolution theorem we know that differentiation and convolution commute. We can use this property to produce new covariant transformations. Assume a rotational symmetric convolution kernel k, then $k * (\nabla \mathbf{x}) = (\nabla k) * \mathbf{x}$. Thus, any convolution with ∇k is a covariant transformation mapping intensities onto vectors. And in fact, one can show that any transformation of this type can be written in this way.

Similarly as for the products we want to generalize this idea for the subspaces V^α corresponding to the irreducible representations of $SO(2)$ or $SO(3)$. Here V^1 is just equivalent to the vector space $\mathbb{R}^n$ we just considered together with the gradient operator. We are now looking for first-order differential operators similar as ∇ connecting the different spaces V^α. In fact, we can use the above introduced products and the original ∇ operator to achieve such operators. Interpret the ∇-operator formally as an ordinary vector living in $\mathbb{R}^n$. Then, we can use the above defined product to let the differential operate on the image. More precisely we take a product of type $V^1 \times V^{\alpha_1} \mapsto V^\alpha$ and use ∇ as the first argument. Using the explicit form of the product this means:

$$(\nabla \bullet_\alpha \mathbf{x})_k = \sum_{i,j} b^\alpha_{kij} \frac{\partial}{\partial r_i} (\mathbf{x})_j,$$

Again the output degree is not arbitrary. For the considered groups it is only possible to lower or raise the degree of the input by 1. So, we define

$$\nabla^1 \mathbf{x} := (\nabla \bullet_{\alpha+1} \mathbf{x}) \quad \text{and} \quad \nabla_1 \mathbf{x} := (\nabla \bullet_{\alpha-1} \mathbf{x})$$

for raising and lowering the degree, respectively. Here the type of the raising operator is $\nabla^1 : \mathcal{T}^\alpha \mapsto \mathcal{T}^{\alpha+1}$ and for the lowering one $\nabla_1 : \mathcal{T}^\alpha \mapsto \mathcal{T}^{\alpha-1}$. For intuition consider another simple example: the ordinary divergence operator known from usual differential calculus is just ∇_1 applied on a vector image. For more detailed description of these operators in 3D, see [11]. In 2D the raising and lowering operators coincide with usual complex derivative appearing in complex calculus.

5.3 Building a Covariant Transformation

Now we want to stick the pieces together. Of course, there are a lot of possibilities to apply convolution, differentiations and products repeatedly to achieve a complex covariant transformation. Here one way is pointed out which was applied successfully to several image processing tasks. The idea behind is the general concept of extracting local, fine grained information from the image (by covariant differentiation), map or filter this information in a meaningful

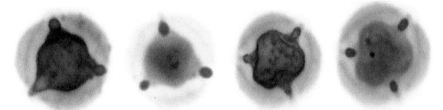

Figure 5.1 Mugwort pollen (green in e-print) with overlayed filter response (red in e-print) for two examples. The filter detects the three porates, but there are also some spurious responses within the pollen, because the pollen has also strong inner structures

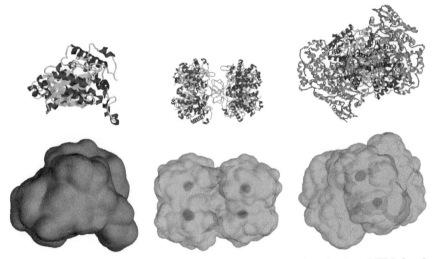

Figure 5.2 Three different protein in cartoon representation (top) and EM-density representation (bottom). In the cartoon representation the motif of interest is highlighted (red/orange in e-print). The EM-density on the left (1AFW) was used for training. For the middle (1M3Z) and the right (1WDL) the filter responses are overlayed in red (in e-print).

and application dependent way (by covariant products), and then spread this information back into the image (again by covariant differentiation).

Let us apply this idea in practice. Our goal is to detect a certain class of visually similar objects in a scene independent of their pose. We do this by constructing an $SE(n)$ covariant image transformation $\mathcal{F}$ that maps the incoming image onto a kind of attention or voting map reflecting the probability that an object is observed at a certain position in space. In a first step local information is gathered by applying repeatedly covariant derivatives. This results in a sequence of 'feature images'

$$\mathbf{p}^\alpha = \underbrace{\nabla^1 \ldots \nabla^1}_{\alpha \text{ times}} x_s = \nabla^\alpha x_s,$$

where x_s is the appropriately smoothed input image $x_s = k_{in} * x$, where k_{in} is for example an isotropic Gaussian. For one position $\mathbf{r} \in \mathbb{R}^n$ the collection of features $[\mathbf{p}^0(\mathbf{r}), \ldots, \mathbf{p}^L(\mathbf{r})]$ describes the local neighborhood of $\mathbf{r}$. This description is very similar to a Taylor expansion around $\mathbf{r}$ while our represent-ation has much nicer properties with respect to rotations as discussed earlier. Now, the goal is to let each point $\mathbf{r}$ on the object cast votes for its center. This approach may be seen as a version of the so called generalized Hough transform [1]. In order to do so we use the covariant products to map the local description onto a new local description representing a kind of probability density for the center of the object of interest. For example, we can construct this mapping by second order products as follows:

$$\mathbf{V}^\alpha[\mathbf{p}^0(\mathbf{r}), \ldots, \mathbf{p}^L(\mathbf{r})] = \sum_{j_1, j_2} a_{j_1, j_2}^\alpha \, \mathbf{p}^{j_1}(\mathbf{r}) \bullet_\alpha \mathbf{p}^{j_2}(\mathbf{r}),$$

where the a_{j_1, j_2}^α depend on the application and have still to be determined. After the determination of the $[\mathbf{V}^0, \ldots, \mathbf{V}^L]$ for each point in the image the contributions are collected in an additive manner. In fact, this can be realized by a kind of inverse procedure as used for the extraction of the $\mathbf{p}^\alpha$. One just has to apply the lowering differential operator multiple times according to the degree of the $\mathbf{V}^\alpha$, smooth the result and add everything up, so

$$\mathcal{F}(x) = k_{out} * \sum_{\alpha=0}^{L} \nabla_\alpha \mathbf{V}^\alpha[\mathbf{p}^0, \ldots, \mathbf{p}^L],$$

where k_{out} is again a rotationally symmetric function. Its width determines how far the voting information is communicated. For a rigorous derivation of the voting picture, see [12].

We applied this scheme to biological volume data for the detection of pollen porates which are small blobs on the surface of airborne pollen, and for the detection of protein motifs (or subdomains) in EM-cryo density data. We adapted the coefficients a_{j_1, j_2}^α by a simple least-square approach on a small set of well-chosen training samples. In Figures 5.1 and 5.2 examples are shown. The data is shown in green while the filter response is overlayed in red.

References

[1] D. Ballard. Generalizing the Hough transform to detect arbitrary shapes. *Pattern Recognition*, 13(2):111–122, 1981.

[2] T. Becker and V. Weispfenning. *Groebner Bases*, Springer Graduate Texts in Mathematics, Vol. 141. Springer, 1998.

[3] S. Boyd, L.O. Chua, and C.A. Desoer. Analytical foundations of Volterra series. *IMA Journal of Mathematical Control and Information*, 1:243–282, 1984.

[4] N. Canterakis. 3D Zernike moments and Zernike affine invariants for 3D image analysis and recognition. In *Proceedings of the 11th Scandinavian Conference on Image Analysis*, Kangerlussuaq, Greenland, Vol. 1, pages 399–406, 1999.

[5] G.S. Chirikjian and A.B. Kyatkin. *Engineering Applications of Noncommutative Harmonic Analysis with Emphasis on Rotation and Motion Groups*. CRC Press, 2001.

[6] R. Kakarala. Triple correlation on groups. PhD Thesis, University of California, Irvine, 1992.

[7] A. Khotanzad and Y.H. Hong. Invariant image recognition by Zernike moments. In *IEEE Trans. Pattern Analysis and Machine Intelligence*, 12(5):489–497, 1990.

[8] R. Kondor. The skew spectrum of functions on finite groups and their homogeneous spaces, available online: http://arxiv.org/abs/0712.4259v1.

[9] T.G. Newman. A group theoretic approach to invariance in pattern recognition. In *Proceedings IEEE Conference on Pattern Recognition and Image Processing (PRIP)*, Chicago, IL, pages 407–412, August 1979.

[10] M. Reisert and H. Burkhardt. Complex derivative filters. *IEEE Trans. Image Processing*, 17(12):2265–2274, December 2008.

[11] M. Reisert and H. Burkhardt. Spherical tensor calculus for local adaptive filtering. In *Tensors in Image Processing and Computer Vision*, S. Aja-Fernández, R. de Luis García, D. Tao, and X. Li, (Eds.), Advances in Pattern Recognition. Springer, 2009.

[12] M. Reisert and H. Burkhardt. Harmonic filters for generic feature detection in 3D. In *Pattern Recognition, Proceedings of the 31st DAGM Symposium*, Jena, Germany, September 2009, Lecture Notes in Computer Science, Vol. 5748. Springer, Berlin, 2009.

[13] O. Ronneberger, H. Burkhardt, and E. Schultz. General-purpose object recognition in 3D volume data sets using gray-scale invariants. In *Proceedings of the International Conference on Pattern Recognition*. IEEE Computer Society, Quebec, Canada, 2002.

[14] H. Schulz-Mirbach. Anwendung von Invarianzprinzipien zur Merkmalgewinnung in der Mustererkennung. PhD Thesis, TU Hamburg Harburg, Reihe 10, Nr. 372. VDI-Verlag, 1995.

6

Scalable Identity Inference from Difficult Face Images via Probabilistic Multi-Region Histograms of Visual Words

Conrad Sanderson[1] and Brian C. Lovell[2]

[1]*NICTA, P.O. Box 6020, St Lucia, Brisbane QLD 4067, Australia*
[2]*School of Information Technology and Electrical Engineering,*
The University of Queensland, St. Lucia, Brisbane, QLD 4072, Australia

Abstract

We propose a fast, scalable identity inference algorithm that provides good levels of recognition on faces that are partially aligned (as typically obtained by current face detection algorithms), and exhibit variations in pose, expression, illumination, resolution and scale. Such variations are typically present in face images obtained in video surveillance contexts (e.g. CCTV), where conventional face recognition techniques tend to perform very poorly. The proposed algorithm is inspired by methods used in text processing as well as 2D Hidden Markov Models. Each face is described in terms of probabilistic multi-region histograms of visual words, with the visual words representing prototype local features (patches). Matching corresponds to a normalised distance calculation between the histograms of two faces, with the normalisation involving a comparison against a set of "cohort" faces. Experiments on the recent "Labeled Faces in the Wild" dataset (faces gathered automatically from the Internet, exhibiting various concurrent and uncontrolled variations) as well as FERET (where each variation was controlled) show that the proposed algorithm obtains performance on par with a more complex method, and exhibits clear advantages over predecessor systems.

Patrick Shen-Pei Wang (Ed.), Pattern Recognition and Machine Vision – In Honor
and Memory of Professor King-Sun Fu, 87–102.

Keywords: visual words, surveillance, face recognition, scalability.

6.1 Introduction

When compared to face images obtained well in controlled conditions (e.g. immigration checkpoints), automatic identity inference based on images obtained in surveillance contexts (e.g. via CCTV) is considerably more difficult. This arises due to several concurrent and uncontrolled factors affecting the face images: pose (this includes both in-plane and out-of-plane rotations), expression, illumination and resolution (due to variable distances to cameras). Furthermore, an automatic face locator (detector) must be used which can induce further problems. As there are no guarantees that the localisation is perfect, faces can be at the wrong scale and/or misaligned [16].

A surveillance system may have further constraints: only one gallery image per person, as well as real-time operation requirements in order to handle large volumes of people (e.g. peak hour at a railway station). In this context the computational complexity of an identity inference system is necessarily limited, suggesting that time-expensive approaches, such as the deduction of 3D shape from 2D images [3] (to compensate for pose variations), may not be applicable.

In this work we describe a Multi-Region Histogram (MRH) based approach, with the aim of concurrently addressing the above-mentioned problems. MRH can be considered to be an evolution of local feature techniques based on Gaussian Mixture Models (GMMs) and 2D Hidden Markov Models (HMMs) presented in [4, 10, 12]. MRH also takes inspiration from recent image categorisation approaches, which in turn were adapted from text classification methods from the fields of natural language processing and information retrieval [5, 14, 18]. The first point of departure from GMMs and 2D HMMs is that instead of directly calculating the likelihood of a given image, probabilistic histograms of occurrences of "visual words" are first built, followed by histogram comparison. The second difference is that only one probabilistic model is used for all persons, instead of one per person – this directly leads to much better scalability.

We continue as follows. Section 6.2 describes the proposed MRH approach in detail, along with associated histogram approximation and distance normalisation methods. In Section 6.3 the MRH approach is briefly contrasted to related methods, taking scalability into account. Results from evaluations and comparisons on the Labeled Faces in the Wild [7] and FERET [15] datasets are given in Section 6.4. A prototype face search application is briefly presented in Section 6.5. The main findings and an outlook are presented in Section 6.6.

6.2 Multi-Region Histograms of Visual Words

Each face is divided into several fixed and adjacent regions, with each region comprising a relatively large part of the face (see Figure 6.1). For region r a set of feature vectors is obtained, $X_r = \{\mathbf{x}_{r,1}, \mathbf{x}_{r,2}, \ldots, \mathbf{x}_{r,N}\}$, which are in turn attained by dividing the region into small blocks (or patches) and extracting descriptive features from each block via 2D DCT [6] decomposition.[1] Each block has a size of 8×8 pixels and overlaps neighbouring blocks by 75%. To account for varying contrast, each block is normalised to have zero mean and unit variance. Based on preliminary experiments we elected to retain 15 of the 64 DCT coefficients, by taking the top-left 4×4 submatrix of the 8×8 coefficient matrix and disregarding the first coefficient (as it carries no information due to the above normalisation).

For each vector $\mathbf{x}_{r,i}$ obtained from region r, a probabilistic histogram is computed:

$$\mathbf{h}_{r,i} = \left[\frac{w_1 p_1\left(\mathbf{x}_{r,i}\right)}{\sum_{g=1}^{G} w_g p_g\left(\mathbf{x}_{r,i}\right)}, \frac{w_2 p_2\left(\mathbf{x}_{r,i}\right)}{\sum_{g=1}^{G} w_g p_g\left(\mathbf{x}_{r,i}\right)}, \ldots, \frac{w_G p_G\left(\mathbf{x}_{r,i}\right)}{\sum_{g=1}^{G} w_g p_g\left(\mathbf{x}_{r,i}\right)} \right]^T \tag{6.1}$$

where the g-th element in $\mathbf{h}_{r,i}$ is the posterior probability of $\mathbf{x}_{r,i}$ according to the g-th component of a visual dictionary model.

The visual dictionary model employed here is a convex mixture of Gaussians [2], parameterised by $\lambda = \{w_g, \mu_g, \mathbf{C}_g\}_{g=1}^{G}$, where G is the number of Gaussians, while w_g, μ_g and $\mathbf{C}_g$ are, respectively, the weight, mean vector and covariance matrix for Gaussian g. The mean of each Gaussian can be thought of as a particular "visual word". The dictionary is obtained by pooling a large number of feature vectors from training faces, followed by employing the Expectation Maximisation algorithm [2] to optimise the dictionary's parameters (i.e. λ).

---

[1] It is possible to use other local feature extraction methods, e.g. Gabor wavelets [9].

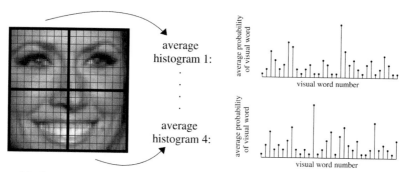

Figure 6.1 Conceptual example of MRH face analysis using 2×2 regions, where each region is divided into small blocks. For each block descriptive features are placed into a vector. The posterior probability of the vector is then calculated using each Gaussian in a visual dictionary, resulting in a histogram of probabilities. For each region the histograms of the underlying blocks are then averaged

Once the histograms are computed for each feature vector from region r, an average histogram for the region is built:

$$\mathbf{h}_{r,\text{avg}} = \frac{1}{N} \sum_{i=1}^{N} \mathbf{h}_{r,i} \tag{6.2}$$

The DCT decomposition acts like a low-pass filter, with the information retained from each block being robust to small alterations (e.g. due to in-plane rotations, expression changes or smoothing due to upsampling from low resolution images). The overlapping during feature extraction, as well as the loss of spatial relations within each region (due to averaging), results in robustness to translations of the face which are caused by imperfect face localisation. We note that in the 1×1 region configuration the overall topology of the face is effectively lost, while in configurations such as 3×3 it is largely retained (while still allowing for deformations in each region).

6.2.1 Normalised Distance

Comparison of two faces can be accomplished by comparing their corresponding average histograms. Based on [8] we define an L_1-norm based distance measure between faces A and B:

$$d_{\text{raw}}(A, B) = \frac{1}{R} \sum_{r=1}^{R} \left\| \mathbf{h}_{r,\text{avg}}^{[A]} - \mathbf{h}_{r,\text{avg}}^{[B]} \right\|_1 \tag{6.3}$$

where R is the number of regions. To reach a decision as to whether faces A and B come from the same person or from two different people, $d_{\text{raw}}(A, B)$ can be compared to a threshold. However, the optimal threshold might be dependent on the image conditions of face A and/or B, which are not known a-priori. Inspired by cohort normalisation [17], we propose a normalised distance in order to reduce the sensitivity of threshold selection:

$$d_{\text{normalised}}(A, B) = \frac{d_{\text{raw}}(A, B)}{\frac{1}{2}\left(\frac{1}{M}\sum_{i=1}^{M} d_{\text{raw}}(A, C_i) + \frac{1}{M}\sum_{i=1}^{M} d_{\text{raw}}(B, C_i)\right)} \quad (6.4)$$

where C_i is the i-th cohort face and M is the number of cohorts. In the above equation cohort faces are assumed to be reference faces that are known not to be of persons depicted in A or B. As such, the terms $\frac{1}{M}\sum_{i=1}^{M} d_{\text{raw}}(A, C_i)$ and $\frac{1}{M}\sum_{i=1}^{M} d_{\text{raw}}(B, C_i)$ estimate how far away, on average, faces A and B are from the face of an impostor. This typically results in Equation (6.4) being approximately 1 when A and B represent faces from two different people, and less than 1 when A and B represent two instances of the same person. If the conditions of given images cause their raw distance to increase, the average raw distances to the cohorts will also increase. As such, the division in Equation (6.4) attempts to cancel out the effect of varying image conditions.

6.2.2 Fast Histogram Approximation

As will be shown in Section 6.4, the size of the visual dictionary needs to be relatively large in order to obtain good performance. Typically about 1000 components (Gaussians) are required, which results in the calculation of histograms via Equation (6.1) to be time consuming. Based on empirical observations that for each vector only a subset of the Gaussians is dominant, we propose a dedicated algorithm that adaptively calculates only a part of the histogram. The algorithm is comprised of two parts, with the first part done during training.

In the first part, the Gaussians from the visual dictionary model are placed into K clusters via the k-means algorithm [2]. Euclidean distance between the means of the Gaussians is used in determining cluster memberships. For each cluster, the closest member to the cluster mean is labelled as a *principal Gaussian*, while the remaining members are labelled as *support Gaussians*.

For a feature vector $\mathbf{x}$ an approximate histogram is then built as follows. Each of the K principal Gaussians is evaluated. The clusters are then ranked according to the likelihood obtained by each cluster's principal Gaussian (highest likelihood at the top). Additional likelihoods are produced cluster by cluster, with the production of likelihoods stopped as soon as the total number of Gaussians used (principal and support) exceeds a threshold. The histogram is then constructed as per Equation (6.1), with the likelihoods of the omitted Gaussians set to zero.

6.3 Related Methods and Scalability

The use of probabilistic histograms in MRH differs to the histograms used in [14] (for image retrieval/categorisation purposes), where a Vector Quantiser (VQ) based strategy is typically used. In the VQ strategy each vector is forcefully assigned to the closest matching visual word, instead of the probabilistic assignment done here.

For the purposes of face classification, MRH is related to, but distinct from, the following approaches: Partial Shape Collapse (PSC) [10], pseudo-2D Hidden Markov Models (HMMs) [4, 12] and Probabilistic Local PCA (PLPCA) [11].

MRH is also somewhat related to the recently proposed and more complex Randomised Binary Trees (RBT) method [13], aimed for more general object classification. While both MRH and RBT use image patches for analysis, RBT also uses: (i) quantised differences, via "extremely-randomised trees", between corresponding patches, (ii) a cross-correlation based search to determine patch correspondence, and (iii) an SVM classifier [2] for final classification.

The differences between MRH and PSC include: (i) the use of fixed regions for all persons instead of manually marked regions for each person, (ii) each region is modelled as a histogram rather than being directly described by a Gaussian Mixture Model (GMM), leading to (iii) MRH using only one GMM (the visual dictionary), common to all regions and all persons, instead of multiple GMMs per person in PSC. The use of only one GMM directly leads to much better scalability, as the number of Gaussians requiring evaluation for a given probe face is fixed, rather than growing with the size of the face gallery. In the latter case the computational burden can quickly become prohibitive [17].

The MRH approach has similar advantages over PLPCA and HMM in terms of scalability and histogram based description. However, there are additional differences. In PLPCA each region is analysed via PCA instead of being split into small blocks. While the probabilistic treatment in PLPCA affords some robustness to translations, the use of relatively large face areas is likely to have negative impact on performance when dealing with other image transformations (e.g. rotations and scale changes). In HMM approaches the region boundaries are in effect found via an automatic alignment procedure (according to the model of each person the face is evaluated against) while in the MRH approach the regions are fixed, allowing straightforward parallel processing.

6.4 Experiments

The experiments were done on two datasets: Labeled Faces in the Wild (LFW) [7], and subsets of FERET [15]. We will show results of LFW first (Section 6.4.1), where the number of face variations (as well as their severity) is uncontrolled, followed by a more detailed study on FERET (Section 6.4.2), where each variation (e.g. pose) is studied separately.

6.4.1 Performance on Labeled Faces in the Wild (LFW)

The recent LFW dataset contains 13,233 face images which have several compound problems – e.g. in-plane rotations, non-frontal poses, low resolution, non-frontal illumination, varying expressions as well as imperfect localisation, resulting in scale and/or translation issues. The images were obtained by trawling the Internet followed by face centering, scaling and cropping based on bounding boxes provided by an automatic face locator. The original bounding boxes were expanded to include context. In our experiments we extracted closely cropped faces[2] using a fixed bounding box placed in the same location in each LFW image.[3] The extracted faces were size normalised to 64×64 pixels, with an average distance between the eyes of 32 pixels. Examples are shown in Figure 6.2.

LFW experiments follow a prescribed protocol [7], where the task is to classify a pair of previously unseen faces as either belonging to the same person (matched pair) or two different persons (mismatched pair). The protocol specifies two views of the dataset: *view 1*, aimed at algorithm development & model selection, and *view 2*, aimed at final performance reporting (to be

---

[2] By *closely cropped* we mean face images that generally do not contain components such as hair and clothing. This is deliberately unlike other algorithms evaluated on the LFW dataset, which typically use the entire image and hence may rely on components that are easily changeable.

[3] The upper-left and lower-right corners of the bounding box were: (83,92) and (166,175), respectively. The box location was determined based on 40 randomly selected LFW faces.

Figure 6.2 Examples from the LFW dataset: (i) master images (resulting from centering, scaling and cropping based on bounding boxes provided by an automatic face locator); (ii) processed versions used in experiments, extracted using a fixed bounding box placed on the master images. The faces typically have at least one of the following issues: in-plane rotations, non-frontal poses, low resolution, non-frontal illumination, varying expressions as well as imperfect localisation, resulting in scale and/or translation issues

used sparingly). In view 1 the images are split into two sets: the training set (1100 matched and 1100 mismatched pairs) and the testing set (500 matched and 500 mismatched pairs). The training set is used for constructing the visual dictionary as well as selecting the decision threshold. The threshold was optimised to obtain the highest *average accuracy* (averaged over the classification accuracies for matched and mismatched pairs). In view 2 the images are split into 10 sets, each with 300 matched and 300 mismatched pairs. Performance is reported using the mean and standard error of the average accuracies from 10 folds of the sets, in a leave-one-out cross-validation scheme (i.e., in each fold nine sets are used training and one set for testing). The standard error is useful for assessing the significance of performance differences across algorithms [7].

In experiment 1 we studied the effect of increasing the size of the visual dictionary (from 2 to 4096 components) and number of regions (from 1×1 to 4×4) on the LFW dataset. As these variations constitute model selection, view 1 was used. The system used probabilistic histograms and normalised distances. Based on preliminary experiments, 32 randomly chosen cohort faces from the training set were used for the distance normalisation. The results, in Figure 6.3(i), suggest that performance steadily increases up to about 1024 components, beyond which performance changes are mostly minor. Dramatic improvements are obtained by increasing the number of regions

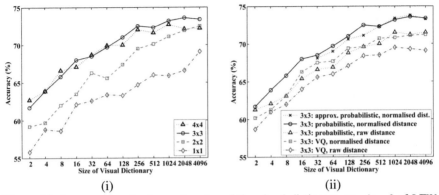

Figure 6.3 Accuracy rate for increasing size of the visual dictionary, on *view 1* of LFW (compound face variations). (i): MRH (probabilistic, normalised distance), with the number of regions varying from 1×1 to 4×4. (ii): 3×3 MRH, with either probabilistic or VQ based histogram generation, as well with and without distance normalisation

from 1×1 to 3×3. Using more regions (i.e. 4×4) shows no appreciable further performance gains.

In experiment 2 we fixed the number of regions at 3×3 and varied the size of the visual dictionary. The performance of systems using exact probabilistic histograms, approximate probabilistic and VQ based was compared. We also evaluated the performance of raw and normalised distances on both probabilistic and VQ based systems. Based on preliminary experiments, approximate histograms used $K = G/10$ clusters and a maximum of $G/4$ Gaussians, where G is the size of the visual dictionary. The results, in Figure 6.3(ii), point to the distance normalisation being helpful, with a consistent advantage of about two percentage points over raw distances (e.g. 72% vs 70%). The results further suggest that probabilistic histograms outperform VQ based histograms, also with an advantage of about two points. Finally, the performance of the computationally less expensive approximate probabilistic histograms is on par with exact probabilistic histograms.

In experiment 3 we used view 2 of LFW, allowing comparison with previously published as well as future results. Several configurations of MRH were evaluated as well as a baseline PCA system. Based on preliminary experiments, the baseline PCA based system used the euclidean distance as its raw distance and 61 eigenfaces (eigenfaces 4 to 64 of the training images, following the recommendation in [1] to skip the first three eigenfaces). The results, presented in Table 6.1, indicate that the performance of MRH based systems is consistent with experiments 1 and 2. Furthermore, the probabil-

Table 6.1 Results on *view 2* of LFW. Results for RBT obtained from http://vis-www.cs.umass.edu/lfw (accessed 2008-09-01), using the method published in [13]. MRH approaches used a 1024 component visual dictionary and closely cropped faces

Method	Mean Accuracy	Standard Error
3×3 MRH (approx probabilistic, normalised distance)	72.35	0.54
3×3 MRH (probabilistic, normalised distance)	**72.95**	0.55
3×3 MRH (probabilistic, raw distance)	70.38	0.48
3×3 MRH (VQ, normalised distance)	69.35	0.72
3×3 MRH (VQ, raw distance)	68.38	0.61
1×1 MRH (probabilistic, normalised distance)	67.85	0.42
PCA (normalised distance)	59.82	0.68
PCA (raw distance)	57.23	0.68
Randomised Binary Trees (RBT)	72.45	0.40

istic 3×3 MRH method is on par with the more complex RBT method. The performance of PCA considerably lags behind all other approaches.

6.4.2 Performance on FERET

For consistency, experiments on FERET were designed to follow a similar pair classification strategy as in the previous section, albeit with manually found eye locations. The "fa" and "fb" subsets (frontal images) were used for training – constructing the visual dictionary as well as selecting the decision threshold. Using persons which had images in both subsets, there were 628 matched and 628 randomly assigned mismatched pairs. The "b" subsets were used for testing, which contain controlled pose, expression and illumination variations for 200 unique persons.

For each image condition there were 200 matched and 4000 mismatched pairs, with the latter obtained by randomly assigning 20 persons to each of the 200 available persons. Image transformations were applied separately to the frontal source images ("ba" series), obtaining the following versions: in-plane rotated (20°), scaled (bounding box expanded by 20% in x and y directions, resulting in shrunk faces), translated (shifted in x and y directions by 6 pixels, or 20% of the distance between the eyes), upsampled from a low resolution version (with the low resolution version obtained by shrinking the original image to 30% of its size, resulting in an average eye distance of ~10 pixels). Example images are shown in Figure 6.4.

The performances of probabilistic MRH with 3×3 and 1×1 configurations, as well as the baseline PCA based system, were compared. Both raw

Figure 6.4 Top row: examples of cropped images from FERET (neutral, followed by expression, illumination and pose change). Bottom row: transformed and cropped versions of the neutral source image (in-plane rotation, scale change, translation and upsampled low-res version)

and normalised distances were evaluated. For testing, each image condition was evaluated separately. Moreover, for each image pair to be classified, the first image was always from the "ba" series (normal frontal image).

The results, presented in Figure 6.5, indicate that increasing the number of regions from 1×1 to 3×3 improves accuracy in most cases, especially when dealing with illumination changes and low resolution images. The notable exceptions are faces with pose changes and in-plane rotations. We conjecture that the histogram for each region (3×3 case) is highly specific and that it has simply altered too much due to the pose change; in the 1×1 case, the single overall histogram is more general and parts of it are likely to be describing face sections which have changed relatively little. For the in-plane rotation, we conjecture that the performance drop is at least partially due to face components (e.g. eyes) moving between regions, causing a mismatch between the corresponding histograms of two faces; in the 1×1 case there is only one histogram, hence the movement has a reduced effect.

The use of normalised distances improved the average performance of all approaches. This is especially noticeable for MRH when dealing with pose changes and in-plane rotations. In all cases the 3×3 MRH system considerably outperformed the baseline PCA system, most notably for faces subject to scale changes and translations.

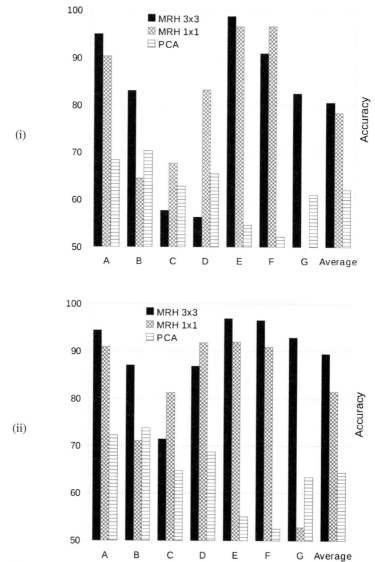

Figure 6.5 Performance on FERET (separate face variations), using: (i) raw distances, (ii) normalised distances. A: expression change, B: illumination change, C: pose change, D: in-plane rotation, E: scale change, F: translation (shift), G: upsampled low resolution image

6.5 Prototype Application

As part of an ongoing research and development effort at NICTA, we have employed the MRH algorithm in a prototype face search application. Here, a query face is compared against a large set of gallery faces, with the quality and conditions of the query and gallery faces being uncontrolled. The matching faces are presented in a rank order fashion.

The web-based interface for the prototype application (built on the Linux, Apache, MySQL and Python (LAMP) software stack) is shown in Figure 6.6. Reliable identity searches over thousands of images are performed in milliseconds on a standard PC, with no explicit hardware acceleration. In this case a photograph of a television image of President George H. Bush is matched to a news photograph. It is interesting to note that his son, President George W. Bush, is also a strong match.

6.6 Main Findings and Outlook

In this chapter we proposed a face matching algorithm that describes each face in terms of multi-region probabilistic histograms of visual words, followed by a normalised distance calculation between the corresponding histograms of two faces. We have also proposed a fast histogram approximation method which dramatically reduces the computational burden with minimal impact on discrimination performance. The matching algorithm was targeted to be scalable and deal with faces subject to several concurrent and uncontrolled factors, such as variations in pose, expression, illumination, as well as misalignment and resolution issues. These factors are consistent with face images obtained in surveillance contexts.

Experiments on the recent and difficult LFW dataset (unconstrained environments) show that the proposed algorithm obtains performance on par with the recently proposed and more complex Randomised Binary Trees method [13]. Further experiments on FERET (controlled variations) indicate that the use of multiple adjacent histograms (as opposed to a single overall histogram) on one hand reduces robustness specific to in-plane rotations and pose changes, while on the other hand results in better average performance. The experiments also show that use of normalised distances can considerably improve the robustness of both multiple- and single-histogram systems.

The robustness differences between multiple- and single-histogram systems suggest that combining the two systems (e.g. by a linear combination of distances) could be beneficial. We note that the MRH approach is eas-

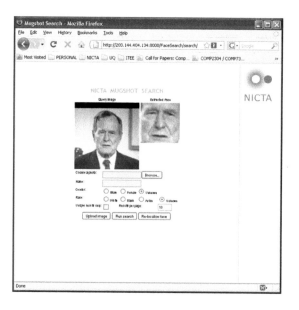

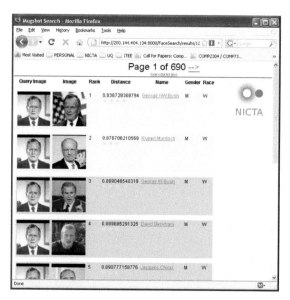

Figure 6.6 Screenshots of the scalable identity inference application prototype. Top: a face image being selected and cropped to extract face features only (hairstyle and background are thus excluded from the matching process). Bottom: the virtually instantaneous results of the search over approximately 7000 face images and 4000 identities

ily amenable to parallelisation: a multi-CPU machine can process regions concurrently, thereby providing a significant speed-up.

Lastly, we have briefly explored the scalability of the approach by presenting our prototype face search application, which is capable of searching through large databases of varying quality faces in just milliseconds.

Acknowledgements

NICTA is funded by the Australian Government via the Department of Broadband, Communications and the Digital Economy, as well as the Australian Research Council through the ICT Centre of Excellence program.

References

[1] P.N. Belhumeur, J.P. Hespanha, and D.J. Kriegman. Eigenfaces vs. fisherfaces: Recognition using class specific linear projection. *IEEE Trans. Pattern Analysis and Machine Intelligence*, 19(7):711–720, 1997.

[2] C.M. Bishop. *Pattern Recognition and Machine Learning*. Springer, 2006.

[3] V. Blanz and T. Vetter. Face recognition based on fitting a 3D morphable model. *IEEE Trans. Pattern Analysis and Machine Intelligence*, 25(9):1063–1074, 2003.

[4] F. Cardinaux, C. Sanderson, and S. Bengio. User authentication via adapted statistical models of face images. *IEEE Trans. Signal Processing*, 54(1):361–373, 2006.

[5] G. Csurka, C.R. Dance, L. Fan, J. Willamowski, and C. Bray. Visual cetegorization with bags of keypoints. In *Proceedings of Workshop on Statistical Learning in Computer Vision (in conjunction with ECCV'04)*, 2004.

[6] R. Gonzales and R. Woods. *Digital Image Processing*, 3rd edition. Prentice Hall, 2007.

[7] G. Huang, M. Ramesh, T. Berg, and E. Learned-Miller. Labeled Faces in the Wild: A database for studying face recognition in unconstrained environments. Technical Report 07-49, University of Massachusetts, Amherst, October 2007.

[8] T. Kadir and M. Brady. Saliency, scale and image description. *International Journal of Computer Vision*, 45(2):83–105, 2001.

[9] T. S. Lee. Image representation using 2D Gabor wavelets. *IEEE Trans. Pattern Analysis and Machine Intelligence*, 18(10):959–971, 1996.

[10] S. Lucey and T. Chen. A GMM parts based face representation for improved verification through relevance adaptation. In *Proceedings of Computer Vision and Pattern Recognition (CVPR)*, volume 2, pages 855–861, 2004.

[11] A.M. Martínez. Recognizing imprecisely localized, partially occluded, and expression variant faces from a single sample per class. *IEEE Trans. Pattern Analysis and Machine Intelligence*, 24(6):748–763, 2002.

[12] A. Nefian and M. Hayes. Face recognition using an embedded HMM. In *Proceedings of Audio Video-Based Biometric Person Authentication (AVBPA)*, pages 19–24, 1999.

[13] E. Nowak and F. Jurie. Learning visual similarity measures for comparing never seen objects. In *Proceedings of Computer Vision and Pattern Recognition (CVPR)*, pages 1–8, 2007.

[14] E. Nowak, F. Jurie, and B. Triggs. Sampling strategies for bag-of-features image classification. In *Proceedings of European Conference on Computer Vision (ECCV), Part IV*, Lecture Notes in Computer Science (LNCS), Vol. 3954, pages 490–503. Springer, 2006.

[15] P.J. Phillips, H. Moon, S.A. Rizvi, and P.J. Rauss. The FERET evaluation methodology for face-recognition algorithms. *IEEE Trans. Pattern Analysis and Machine Intelligence*, 22(10):1090–1104, 2000.

[16] Y. Rodriguez, F. Cardinaux, S. Bengio, and J. Mariethoz. Measuring the performance of face localization systems. *Image and Vision Comput.*, 24:882–893, 2006.

[17] C. Sanderson. *Biometric Person Recognition – Face, Speech and Fusion*. VDM Verlag, 2008.

[18] J. Sivic and A. Zisserman. Video google: A text retrieval approach to object matching in videos. In *Proceedings of 9th International Conference on Computer Vision (ICCV)*, Vol. 2, pages 1470–1477, 2003.

PART II

FUNDAMENTAL PRINCIPLES
AND METHODS IN
MACHINE VISION

7

Pattern Recognition for Genomics Context

Feng Chen[1] and Yi-Ping Phoebe Chen[1,2]

[1]*School of Information Technology, Deakin University, Burwood, Victoria 3125, Australia*
[2]*Australia Research Council Centre of Excellence in Bioinformatics;*
E-mail: phoebe.chen@deakin.edu.au

Abstract

Genomics represents a research of the whole genome, which includes the investigation of DNA, RNA, protein, relationships between them and so on. Pattern recognition (PR) focuses on classifying or clustering data according to either prior knowledge or statistical information in the multi-dimensional data space. When the traditional laboratory techniques can not extract useful information from the increasing biological databases, introducing PR, as well as data mining and machine learning, is a good way to analyze genomics data. This chapter reviews important pattern recognition techniques and their applications in genomics.

Keywords: pattern recognition, genomics, bioinformatics, supervised methodology, unsupervised methodology.

7.1 Introduction

Genomics [1, 2] is concerned about the organism genomes. It includes three objects: (1) the understanding of the genome's structure, function and evolution; (2) the exploration of the genetic information and the reaction among the information in the genome; (3) the analysis and therapy of the diseases. As the development of genomics, there are many branches emerged, as shown

Patrick Shen-Pei Wang (Ed.), Pattern Recognition and Machine Vision – In Honor and Memory of Professor King-Sun Fu, 105–118.

Table 7.1 Genomics branch description

Genomics branches	Descriptions
Comparative genomics	Understanding the biological issues by comparing genomes
Functional genomics	Exploring the gene functions by gene chips
Proteomics	Illustrating protein-protein interaction based on protein level
Disease genomics	Understanding the interactions between genome and diseases
Pharmacogenomics	Researching how genome is used to drug discovery
Bioinformatics	Investigating biological knowledge by computer science

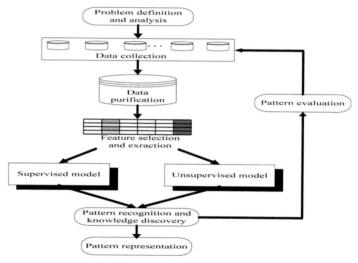

Figure 7.1 The stochastic view of phases in pattern recognition process

in Table 7.1. At the same time, the traditional methods can not obtain new and enough information. For example, a "gene chip", which can integrate a variety of gene expression in diverse samples, costs from about $2000 to $10000 [5]. But the knowledge in the chips such as the relationships between genes is hard to find out by laboratory techniques. Therefore, new techniques such as pattern recognition are provided for this issue.

Pattern recognition [3, 4] focuses on classifying or clustering data by either prior knowledge or statistical information in the multi-dimensional data space. Most of the research in this field is about supervised methodology and unsupervised methodology. In Figure 7.1, after collecting data from different data sources, all the data will be transferred into a data warehouse after data purification, including noise filtering and error correction. Then we choose supervised or unsupervised model to analyse the problem. The supervised

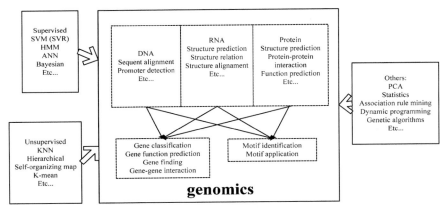

Figure 7.2 The application of PR in genomics

techniques include K-nearest neighbors (KNN) [32, 49], Bayesian classifiers [38, 42], neural networks [39], support vector machine (SVM) [12, 13, 41–43], decision tree [33, 50]. The unsupervised techniques have hierarchical clustering [35] and mixture models [6]. The next step is to evaluate the found patterns. The evaluation results will be taken to determine if the model or data are good enough for dealing with the problem. The last step is to choose a good visualized method to represent the results.

As shown in Figure 7.2, PR can be used for exploring DNA, RNA and protein. In addition, gene discovery and the relationships between genes are involved too. PR can improve the accuracy of genomics research and extract novel information from the datasets. The patterns that are obtained can provide more correct and comprehensive analysis preparation for the following research and applications such as drug discovery [8, 9]. At the same time, PR is closely related with other computer science techniques, including data mining and machine learning. Therefore when we introduce the application of RP in genomics, we also analyse several data mining algorithms for offering the readers a comprehensive understanding of computer science in genomics.

7.2 Promoter and Sequence Alignment in DNA Detection

Promoters can control genes' activation because they are close to transcription start sites (TSSs). Promoter prediction algorithms fall into two categories: classification and coordinate location [10, 11]. PR in genomics is involved with SVM [12, 13], clustering [14], feature extraction [15, 17], ANN [16, 18, 19] and Markov model and its variants [18–20].

SVM suits multiple dimensional issues of genomics data [12]. Through proper kernel function, SVM can reflect low dimension data into high dimension data so that classifier can be constructed easily and correctly. SVM can generate a matrix for measuring the similarity of two sequences to reduce the probability of detecting false positives [12]. Promoters can identify protein-encoding genes by detecting TSSs and transcription regulation networks. ARTS uses SVM with string kernels to identify TSSs, which can improve prediction accuracy because of solving the problem of overfitting [13]. In addition, the self-organizing map can also be used to predict promoters because it can reduce the discovery of false positives [14].

Because genomics data is hard to extract useful information, feature extraction is useful because they can find significant feature set to obtain knowledge [15, 17]. Large-scale structure features in DNA sequences are constructed to detect eukaryotic core promoter regions [10], which can find the common promoters in diverse species. Promoter Explorer [15] considers every feature separately so that more proper feature set is chosen to generate classifier, which builds a cascade structure to predict promoters. In addition, ANN based feature extraction [16] can recognize *E. coli* promoters in di-nucleotide feature space, which can predict promoters and analyse the linear relationship between promoter regions and non-promoter regions.

ANN and HMM are always combined for pattern recognition because ANN can deal with complex classification while HMM is good at tackling sequence data [18–20]. A stochastic example about the combination of ANN and HMM for promoters is shown in Figure 7.3. Interpolation Markov Chains (IMCs) are introduced to recognize promoters, with Simulated annealing (SA) and genetic algorithm [20], which reduces complexity and the possibility of detecting local minimum points. ANN and HMM are also used to detect prokaryote promoters on large sequences in *Drosophia* genome [18, 19].

Sequence alignment involves the detection in DNA, RNA and protein [21–26]. Here we focus on DNA sequence alignment because through this, the readers can obtain a good understanding of sequence alignment on RNA and protein. DNA sequence alignment is to align two or more sequences to identify similar regions which probably represent similar function, structure, or genes. Sequence alignment includes local, global and multiple sequence alignment [21]. Backward search [24] is to implement Burrows-Wheeler Alignment tool for detecting short alignment. Dynamic programming [25, 26]

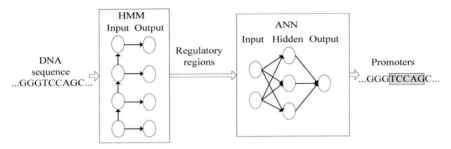

Figure 7.3 The stochastic review of the combination of ANN and HMM

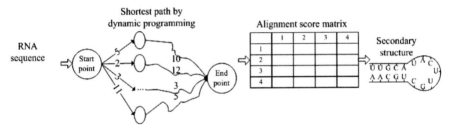

Figure 7.4 Dynamic programming based secondary structure from secondary

is also used to find global and local alignment. Other supervised methods, such as SVM and ANN are also used to recognize PR in genome [23].

7.3 RNA Structure Analysis

The current research of RNA focuses on the understanding of RNA sequence [27–30], secondary [31–33] and tertiary structure [34, 35]. PR is used to predict structure, identify genes and gene regulation networks by understanding the structures of RNAs and RNA regulation networks.

Due to the evolutionary conservation, sequence is always used to predict structure [27–29, 31–33, 36]. Compared with [36] which are time-consuming, dynamic programming illustrated in Figure 7.4, is more suitable for discovering the relationship between sequence and secondary structure [27]. It can detect global similarity for finding common substructures in two RNAs. Dynamic programming is also used in [29] to detect pairwise local structural alignment. Because of the introduction of multiple sequence and bifurcated structures, this method does not require the sequences with high similarity, which means it can be applied widely. There are two algorithms based on unaligned sequences [28, 32]. Do et al. [28] use a method which is similar with

SVM while [32] is based on probability sampling method. trCYK [33] also focuses on detecting local RNA structure alignment. Its merit is to use binary tree to predict the combination of sequence conservation so that it works well with incomplete sequences. Homologous sequence [31] is taken into account to predict secondary structure by modelling the probability distribution between RNA and homologous sequence.

In RNA tertiary structure, sarcin-ricin motifs are detected to recognize tertiary structure by graph-grammars algorithm [34]. It is based on sequence because RNA sequence has a very close relationship with RNA tertiary. Another method for identifying motifs in tertiary structure is to use hierarchical clustering and the nearest neighbour [35]. The cyclic motifs can be used to classify tertiary structure. This method has the same merits as the one in [36].

7.4 Protein Structure Exploration

We introduce PR in proteins from four categories: primary structure, secondary structure, tertiary structure and quaternary structure [37–39]. Primary structure is an amino acids sequence in a polypeptide chain linked by peptide bonds. Understanding primary structure is the basis of understanding of protein because it determines the tertiary structure [38, 40–43]. Secondary structures are the patterns linked by hydrogen bonds between backbone amide and carboxyl groups [44–47]. Tertiary structure is a three-dimensional structure determined by the atomic coordinates [39, 46, 47]. Quaternary structure involves multiple folded protein molecules. Many proteins are combined by a multiple polypeptide chain [37, 48–60].

7.4.1 Protein Primary and Secondary Structure

Primary structure is explored by HMM [40], Bayesian [38, 42] and SVM [41–43]. HMM [40] is to analyse remote homology in probabilistic protein family modelling. It suits little training data in molecular biological analysis. Based on Bayesian, MSpresso [38] explores the relationship between mRNA and protein presence probability for detecting primary structure in large-scale shotgun proteomics data. PROSO [42] combines Bayesian and SVM in *E. coli*, which categories proteins as solubility and insolubility. SVM [41] and its variant [43] are also used for the research related with primary structure. An extending standard kernel based on protein sequence similarity is used to classify protein with noise [41]. Support vector regression (SVR) [43]

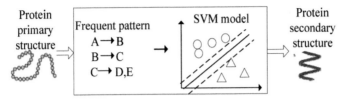

Figure 7.5 Protein secondary structure prediction based on frequent pattern mining and SVM

uses multiple sequence features to predict primary structure, which is used to predict disulfide connectivity.

SVM [44, 45] is also suitable for exploring secondary structure because: (1) the data is not linear separately, and (2) the data is in the high-dimensional space. SVM is taken to construct prediction model from frequent patterns [45], as shown in Figure 7.5. It can provide a new and clear representation of secondary structure and improve the accuracy of consensus prediction. Two SVM algorithms with different hybrid kernels are taken to find the best kernel functions for distinct data types and problem definition [44]. The chosen function is to predict secondary structure with higher accuracy.

7.4.2 Protein Tertiary and Quaternary Structure

Tertiary structure related algorithms can be divided as unsupervised [46, 47] and supervised [39]. Protein classification can be determined by tertiary structure, which is explored by matching and clustering surface graphs [46]. EM graph clustering is taken to cluster protein surface graph data, which is used to classify protein. The method constructs a topological structure to find the general features in many different applications. Tertiary structure classification can also be used to predict protein circular dichroism spectra [47], which can use most of the clustering algorithms to cluster tertiary structure. Then the classification of tertiary structure is taken to generate the tertiary-class specific reference protein set. After being evaluated by cross-validation, neural networks [39] are employed to generate a prediction model for secondary and tertiary structures.

Quaternary structure can also be predicted by supervised algorithms [37, 50]. These two algorithms are both based on primary structure. SVM and covariant discriminant algorithms [37] consider amino acid composition and correlation function. Rule based algorithm is used to construct the decision tree which can predict quaternary structure by focusing on distinguishing homodimers from non-homodimers [50]. In addition, principal component

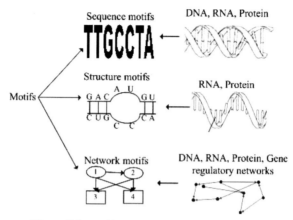

Figure 7.6 Motif categories and their applications

analysis (PCA) can also be used [48] to extract key features and enhance prediction accuracy because it can deal with the issue of high-dimensional space. Nearest neighbours [49] is used to classify quaternary structures based on functional domain composition, which can explore the protein's biological function and understand the molecule relationships.

7.5 Motifs Finding

Motifs can indicate functional regions in genome, which are important to detect unknown genes or explore the gene functions [51–53]. Based on the location, they can be divided into distinct categories (Figure 7.6). A sequence motif indicates a widespread nucleotide or amino acid pattern. A structure motif is a three-dimensional structure in the chain. Network motifs in gene regulation networks can indicate biological signals, which can illustrate how gene expressions change in cells.

DNA motif detection involves HMM [51], dynamic programming [55], Bayesian [52, 55] and so on. Referring to HMM model, GLAM2 and GLAM2SCAN are proposed for sequence motifs by detecting similarities between multiple sequences [51]. They are good at dealing with subtle and linear motifs for considering insertions and deletions in motifs. Dynamic programming [54] is used to calculate P-values of the motifs, which can improve detected motifs' statistical significance. According to Bayesian, stochastic search [55] is combined to find out transcription factor biding sites. Another method enhances prediction accuracy by taking motif related factors into consideration [52]. In addition, Tomtom [53] is employed to quantify similarities between motifs, which use Pearson distribution to measure motif-motif similarity. It is suitable for DNA, RNA and protein.

Bayesian, dynamic programming and constraint networks are included in RNA motif related research [50, 56, 57]. CMfinder [56] integrates Bayesian and statistics to construct an EM framework using multiple unaligned sequences. So it does not need the sequences with high similarity, which extends the application of this method. Comparatively, Locomotif [57] takes dynamic programming for prediction. A graphical tool is to visualize motif detection and search. Its advantage is that it is a client- server method, which is much easier for users to learn. By constraint networks, MilPat [50] pays attention not only to search motifs, but also to explore the interaction between motifs.

Motif detection in protein can be applied widely [58–61]. SVM is taken in [58,60]. MultiLoc [58] employs SVM for protein motifs. It can also be used for subcellular localization and amino acid composition by taking biological information into account. It contributes to detect novel proteins. SVM in [60] is used to extract motif set, in which a string kernel is used to construct an association with every sequence and sequence profile. Protein- protein interaction can be explored by motif too [59, 61]. Both of them can work for homologous sequence. By considering sequence similarity to the query sequence, sequence filters can get rid of the regions which are not closely related motifs in advance [59]. At the same time, HMM is used to identify motifs. Frequent pattern mining is to find out related motif pairs [61]. Then all the pairs can be connected to become a network, which represent the interaction between proteins. This method simplifies the complex issue so that it can be dealt with effective and efficient.

7.6 Conclusion

Genomics includes the research of DNA, RNA and protein, which contributes to explore the structure, function of genome. The obtained gene information can be used to construct gene regulation networks, which will be used to analyze disease causes from gene angel for improving therapy design and drug discovery.

Pattern recognition is for classifying or clustering data in the multi-dimensional space. It mainly includes supervised and unsupervised techniques, which takes advantage of statistics or prior knowledge to extract

Table 7.2 The summary of PR techniques in genomics

Categories	PR techniques	Applications
DNA	SVM [12, 13]	Promoter detection
	Self-organizing map [14]	
	Feature extraction [15–17]	
	ANN & HMM [16, 18–20]	
	Dynamic programming [24–26]	DNA alignment
	SVM or ANN [23]	
	HMM [51]	DNA motifs
	Dynamic programming [54]	
	Bayesian [52, 55]	
RNA	Dynamic programming [27, 29]	Sequence-secondary structure match
	SVM [28]	Secondary structure prediction
	Decision tree [33]	
	Probability distribution [31, 32]	
	Graph-grammars [34]	Tertiary structure analysis
	KNN [45]	
	Bayesian [55]	RNA motifs
	Dynamic programming [57]	
	Constraint networks [50]	
Protein	SVM [40, 43]	Primary-secondary structure match
	HMM [40]	Protein family modeling
	Bayesian [38, 42]	Primary structure identification
	SVM [44, 45]	Secondary structure identification
	EM graph clustering [46]	Protein surface graph clustering
	ANN [39]	Secondary & tertiary prediction
	SVM [37]	Quaternary identification
	Decision tree [50]	
	PCA [48]	
	KNN [49]	
	SVM [58, 60]	Protein motifs
	HMM [61]	

information and predict unknown knowledge. The important techniques of PR involve SVM, HMM, ANN, KNN and so on.

Consequently, these techniques of pattern recognition have been introduced to solve the problem of extracting knowledge from genomics data. SVM has the advantage of classifying high-dimensional data so it is widely used in classification and prediction of DNA, RNA and protein. The combination of ANN and HMM makes that complex pattern can be extracted from genomics sequence data. KNN is good at clustering genomics data for exploring common structure and detecting functions. In addition, other techniques, such as dynamic programming and Bayesian, can also be used here for genomics research. This chapter summarizes these techniques and their application in genomics, which is listed in Table 7.2. We demonstrate that the introduction of PR techniques contributes a lot to explore genomics research. PR accelerates genomics knowledge discovery process, improves classification (clustering) accuracy and enhances prediction accuracy.

References

[1] S. McGrath and D. Sinderen. *Bacteriophage: Genetics and Molecular Biology*, 1st edn. Caister Academic Press, 2007.

[2] N. Cristianini and M.W. Hahn. *Introduction to Computational Genomics*. Cambridge University Press, 2006.

[3] K. Fukunaga. *Introduction to Statistical Pattern Recognition*, 2nd edn. Academic Press, 1990.

[4] A.K. Jain, J. Mao, R.P.W. Duin, and J. Mao. Statistical pattern recognition: A review. *IEEE Transactions on Pattern Analysis and Machine Intelligence*, 22(1):4–37, 2000.

[5] J.C. Lindon, E. Holmes, and J.K. Nicholsom. Pattern recognition methods and applications in biomedical magnetic resonance. *Progress in Nuclear Magnetic Resonance Spectroscopy*, 39:1–40, 2001.

[6] G. Borries and H. Wang. Partition clustering of high dimensional low sample size data based on p-values. *Computational Statistics & Data Analysis*, 53(12):3987–3998, 2009.

[7] H. Permuter, J. Francos, and J.H. Jermyn. Gaussian mixture models of texture and color for image database retrieval. *Proceedings of Acoustics, Speech and Signal Processing*, 3:569–572, 2003.

[8] Y.P.P. Chen and F. Chen. Using bioinformatics techniques for gene identification in drug discovery and development. *Current Drug Metabolism*, 9(6):567–573, 2008.

[9] Y.P.P. Chen and F. Chen. Targets for drug discovery using bioinformatics. *Expert Opinion on Therapeutic Targets*, 12(4):383–389, 2008.

[10] V.B. Bajie, S.L. Tan, Y. Suzuki, and S. Sugano. Promoter prediction analysis on the whole human genome. *Nat. Biotechnol.*, 22(11):1467–1473, 2004.

[11] T. Abeel, Y.V. Peer, and Y. Saeys. Toward a gold standard for promoter prediction evaluation. *Bioinformatics*, 25(12), i313–i320, 2009.

[12] L. Gordon, A.Y. Chervonenkis, A.J. Gammerman, I.A. Shahmuradov, and V.V. Solovyev. Sequence alignment kernel for recognition of promoter regions. *Bioinformatics*, 19(15):1964–1971, 2003.

[13] S. Sonnenburg, A. Zien, and G. Ratsch. ARTS: Accurate recognition of transcription starts in human. *Bioinformatics*, 22(14):e472–e480, 2006.

[14] T. Abeel, Y. Saeys, P. Rouze, and Y.V. Peer. ProSOM: Core promoter prediction based on unsupervised clustering of DNA physical profiles. *Bioinformatics*, 24(13):i24–i31, 2008.

[15] X. Xie, S. Wu, K. Lam, and H. Yan, PromoterExplorer: An effective promoter identification method based on the AdaBosst algorithm. *Bioinformatics*, 22(22):2722–2728, 2006.

[16] T.S. Rani, S.D. Bhavani, and R.S. Bapi. Analysis of E. coli promoter recognition problem in dinucleotide feature space. *Bioinformatics*, 23(5):582–588, 2007.

[17] T. Abeel, Y. Saeys, E. Bonnet, P. Rouze, and Y.V. Peer. Generic eukaryotic core promoter prediction using structural features of DNA. *Genome Research*, 18(2):310–323, 2008.

[18] S. Mann, J. Li, and Y.P.P. Chen. A phMM-ANN based discriminative approach to promoter identification in prokaryote genomic contexts. *Nucleic Acids Research*, 35(2), e12, 2006.

[19] U. Ohler, G. Liao, H. Niemann, and G.M. Rubin. Computational analysis of core promoters in the Drosophila genome. *Genome Biology*, 3(12):0087.1–0087.12, 2002.

[20] Q. Luo, W. Yang, and P. Liu. Promoter recognition based on the Interpolated Markov Chains optimized via simulated annealing and genetic algorithm. *Pattern Recognition Letters*, 27(9):1031–1036, 2006.

[21] S. Batzoglou. The many faces of sequence alignment. *Briefing in Bioinformatics*, 6(1):6–22, 2005.

[22] S.R. Eddy. A model of the statistical power of comparative genome sequence analysis. *PLoS Biology*, 3(1):95–101, 2005.

[23] C. Romualdi, S. Campanaro, D. Campagna, B. Celegato, N. Cannata, S. Toppo, G. Valle, and G. Lanfranchi. Pattern recognition in gene expression profiling using DNA array: A comparative study of different statistical methods applied to cancer classification. *Human Molecular Genetics*, 12(8):823–836, 2003.

[24] H. Li and R. Durbin. Fast and accurate short read alignment with Burrow-Wheeler transform. *Bioinformatics*, 25(14):1754–1760, 2009.

[25] W. Huang, D.M. Umbach, and L. Li, Accurate anchoring alignment of divergent sequences. *Bioinformatics*, 22(1):29–34, 2006.

[26] T.W. Lam, W.K. Sung, S.L. Tam, C.K. Wong, and S.M. Yiu, Compressed indexing and local alignment of DNA. *Bioinformatics*, 24(6):791–797, 2008.

[27] S. Heyne, S. Will, M. Beckstette, and R. Backofen. Lightweight comparison of RNAs based on exact sequence-structure matches. *Bioinformatics*, 25(16):2095–2102, 2009.

[28] C.B. Do, C. Foo, and S. Batzoglou. A max-margin model for efficient simultaneous alignment and folding of RNA sequences. *Bioinformatics*, 24(13):i68–i76, 2008.

[29] J.H. Havgaard, R.B. Lyngso, G.D. Stormo, and J. Gorodkin. Pairwise local structural alignment of RNA sequences with sequence similarity less than 40%. *Bioinformatics*, 21(9):1815–1824, 2005.

[30] U. Schulze, B. Hepp, and C.S. Ong. PALMA: mRNA to genome alignments using large margin algorithms. *Bioinformatics*, 23(15):1892–1900, 2007.

[31] M. Hamada, K. Sato, H. Kiryu, T. Mituyama, and K. Asai. Predictions of RNA secondary structure by combining homologous sequence information. *Bioinformatics*, 25(12):i330–i338, 2009.

[32] X. Xu, Y. Ji, and G.D. Stormo. RNA sampler: A new sampling based algorithm for common RNA secondary structure prediction and structural alignment. *Bioinformatics*, 23(15):1883–1891, 2007.

[33] D.L. Kolbe and S.R. Eddy. Local RNA structure alignment with incomplete sequence. *Bioinformatics*, 25(10):1236–1243, 2009.

[34] K. St-Onge, P. Thibault, S. Hamel, and F. Major. Modeling RNA tertiary structure motifs by graph-grammars. *Nucleic Acids Research*, 35(5):1726–1736, 2007.

[35] S. Lemieux and F. Major. Automated extraction and classification of RNA tertiary structure cyclic motifs. *Nucleic Acids Research*, 34(8):2340–2346, 2006.

[36] T. Jiang, G. Lin, B. Ma, and K. Zhang. A general edit distance between RNA structures. *J. Comput. Biol.*, 9(2):371–388, 2002.

[37] S. Zhang, Q. Pan, H. Zhang, Y. Zhang, and H. Wang. Classification of protein quaternary structure with support vector machine. *Bioinformatics*, 19(18):2390–2396, 2003.

[38] S.R. Ramakrishnan, C. Vogel, J.T. Prince, Z. Li, L.O. Penalva, M. Myers, E.M. Marcotte, D.P. Miranker, and R. Wang. Integrating shotgun proteomics and mRNA expression data to improve protein identification. *Bioinformatics*, 25(11):1397–1403, 2009.

[39] A. Randall, J. Cheng, M. Sweredoski, and P. Baldi. TMBpro: Secondary structure, β-contact and tertiary structure prediction of transmembrane β-barrel proteins. *Bioinformatics*, 24(4):513–520, 2008.

[40] T. Plotz and G.A. Fink. Pattern recognition methods for advanced stochastic protein sequence analysis using HMMs. *Pattern Recognition*, 39(12):2267–2280, 2006.

[41] C. Yu, N. Zavaljevski, F.J. Stevens, K. Yackovich, and J. Reifman. Classifying noisy protein sequence data: a case study of immunoglobulin light chains. *Bioinformatics*, 21(Suppl. 1):i495–i501, 2005.

[42] P. Smialowski, A.J. Martin-Galiano, A. Mikolajka, T. Girschick, T.A. Holak, and D. Frishman. Protein solubility: Sequence based prediction and experimental verification. *Bioinformatics*, 23(19):2536–2542, 2007.

[43] J. Song, Z. Yuan, H. Tan, T. Huber, and K. Burrage. Predicting disulfide connectivity from protein sequence using multiple sequence feature vectors and secondary structure. *Bioinformatics*, 23(23):3147–3154, 2007.

[44] G. Altun, H. Hu, D. Brinza, R.W. Harrison, A. Zelikovsky, and Y. Pan. Hybrid SVM kernels for protein secondary prediction. In *Proceedings of IEEE International Conference on Granular Computing*, pages 762–765, 2006.

[45] F. Birzele and S. Kramer. A new representation for protein secondary structure prediction based on frequent patterns. *Bioinformatics*, 22(21):2628–2634, 2006.

[46] M.A. Lozano and F. Escolano. Protein classification by matching and clustering surface graphs. *Pattern Recognition*, 39(4):539–551, 2006.

[47] N. Sreerama, S.Y. Venyaminov, and R.W. Woody. Analysis of protein circular dichroism spectra based on the tertiary structure classification. *Analytical Biochemistry*, 299(2):271–274, 2001.

[48] T. Wang, H. Shen, L. Yao, J. Yang, and K. Chou. PCA for predicting quaternary structure of protein. *Front. Electr. Electron. Eng. China*, 3(4):376–380, 2008.

[49] X. Yu, C. Wang, and Y. Li. Classification of protein quaternary structure by functional domain composition. *BMC Bioinformatics*, 7:187, 2006.

[50] P. Thebault, S. Givry, T. Schiex, and C. Gaspin. Searching RNA motifs and their intermolecular contacts with constraint networks. *Bioinformatics*, 22(17):2074–2080, 2006.

[51] M.C. Frith, N.F.W. Saunders, B. Kobe, and T.L. Bailey. Discoverying sequence motifs with arbitrary insertions and deletions. *PLoS Computational Biology*, 4(5):e1000071, 2008.

[52] S.M. Li, J. Wakefield, and S. Self. A transdimensional Bayesian model for pattern recognition in DNA sequences. *Biostatistics*, 9(4):668–685, 2008.

[53] S. Gupta, J. Stamatoyannopoulos, T. Bailey, and W. Noble. Quantifying similarity between motifs. *Genome Biology*, 8(2):R24, 2007.

[54] J. Zhang, B. Jiang, M. Li, J. Tromp, X. Zhang, and M.Q. Zhang. Computing exact P-value for DNA motifs. *Bioinformatics*, 23(5):531–537, 2007.

[55] M.G. Tadesse, M. Vannucci, and P. Lio. Identification of DNA regulatory motifs using Bayesian variable selection. *Bioinformatics*, 20(16):2553–2561, 2004.

[56] Z. Yao, Z. Weinberg, and W.L. Ruzzo. CMfinder – A covariance model based RNA motif finding algorithm. *Bioinformatics*, 22(4):445–452, 2006.

[57] J. Reeder, J. Reeder, and R. Giegerich. Locomotif: From graphical motif description to RNA motif search. *Bioinformatics*, 23(13):i392–i409, 2007.

[58] A. Hoglund, P. Donnes, T. Blum, H. Adolph, and O. Kohlbacher. MultiLoc: Prediction of protein subcellular localization using N-terminal targeting sequences, sequence motifs and amino acid composition. *Bioinformatics*, 22(10):1158–1165, 2006.

[59] H. Dinkel and H. Sticht. A computational strategy for the prediction of functional linear peptide motifs in proteins. *Bioinformatics*, 23(24):3297–3303, 2007.

[60] R. Kuang, E. Le, K. Wang, K. Wang, M. Siddiqi, Y. Freund, and C. Leslie. Profile-based string kernels for remote homology detection and motif extraction. *J. Bio. Comp. Biol.*, 3(3):527–550, 2005.

[61] H. Li, J. Li, and L. Wong. Discovering motif pairs at interaction sites from protein sequences on a proteome-wide scale. *Bioinformatics*, 22(8):989–996, 2006.

8

3D Body Joints-Based Human Action Recognition*

Xiaoqing Ding and Junxia Gu

Department of Electronic Engineering, Tsinghua University, Beijing 100084, China; E-mail: {dxq, gujx}@ocrserv.ee.tsinghua.edu.cn

Abstract

Human action recognition is receiving increasing attention. This chapter reviews the action representation and recognition methods, and presents a new 3D action recognition framework. 3D body joints are employed to represent action. Two classifiers are used to recognize action. One is exemplar-based HMM (EHMM) classifier, and the other is learned by the exemplar-based embedding (EE) method without modeling dynamics. Multiple features are fused to further improve the performance. Our results show that EE method is a little better than EHMM and fusion of multiple features can greatly improve the recognition performance.

Keywords: action recognition, viewpoint difficulty, 3D body joints, exemplar-based HMM, exemplar-based embedding.

8.1 Introduction

Recently, human action recognition is receiving increasing attention in the field of pattern recognition and computer vision. This has been motivated by

*Supported by the National Basic Research Program of China (973 Program) (2007CB311004) and National Natural Science Foundation of China (2006AA01Z115).

Patrick Shen-Pei Wang (Ed.), Pattern Recognition and Machine Vision – In Honor and Memory of Professor King-Sun Fu, 119–133.

the desire in applications of intelligent visual surveillance, human-computer interaction, and computer based animation in games and movie industry.

Many terms like 'action', 'activity', 'behavior', 'event' and 'movement' are used to define action. Action hierarchy, such as this three layers definition: motor primitive, action and activity [10], is an appropriate definition method. In this chapter, Motor primitive means the pose in one frame and is used to build action. Action, meaningful pose sequence, is composed into activity. Action can be classified into some categories, such as 'walk', 'point', 'sit', and so on. The activity should be analyzed with the whole situation, for example, 'someone walks across the street', 'someone points to the old building', and 'someone sits down on the chair'. We focus on the second layer, action recognition, which is the hot and basal research of this field.

Research of human action is originated about 30 years ago. During the 1970s and 1980s, the main contribution of action research was how human understand motion information. In 1975, Johansson reported that a simple point-based model of the human body contained sufficient information to recognize the actions [7]. In 1988, Nagel proposed the hierarchy definition of action [11]. During the 1990s, some important methods of action representation and recognition were introduced. Hidden Markov model (HMM) was firstly used in action recognition [20]. Davis et al. presented two templates, motion-energy image (MEI) and motion-history image (MHI), to represent the action sequences [5]. In this stage, the actions are always recognized in the fixed viewpoint. After 2000, the rapid development was motivated by the desire in applications. Many new action representation and recognition methods were presented and some free-viewpoint approaches introduced.

8.2 Key Issues of Action Recognition

The effective representation and recognition are two key issues. Various interesting approaches have been presented, up to now, but, there is very little clear information about best ways to proceed. In the following we give a short overview of these methods and discuss the viewpoint difficulty which is one of the most challenging problems in action recognition.

8.2.1 Action Representation

There are two action representation methods: appearance-based methods and human model-based methods.

Appearance-based approaches directly represent action using image information, such as silhouette, edge, optical flow and trajectory. The motion mesh feature used silhouette information [20]. Li recognized the action using oriented histograms of optical flow field [9].

In contrast, human model-based approaches represent actions in a joint space or parameter space. Human joints model is one of the most effective models for action representation. These methods based on human joints model can obtain some viewpoint invariance quantities and adapt to the different camera configurations [13, 21]. Compared with the appearance-based representation methods, body joints contain much richer and finer information. Recently the improvement of the marker-less full body tracking algorithm makes the human joints model-based approaches available.

At present, the challenging problem is viewpoint difficulty. 3D reconstruction is one of the effective solutions to this problem. Weinland et al. extracted the motion history volumes (MHV) template from the 3D reconstructed sequence [16]. 3D body joints sequence obtained from 3D reconstructed sequence was also used to represent actions [6, 8].

In this chapter, we focus on the human joint model-based representation method. 3D body joints sequence is extracted from the 3D reconstructed sequence by a human model-based marker-less full body tracker.

8.2.2 Action Recognition

Some methods obtained several 2D static templates from the action sequence, such as MEI and MHI [5]. Conventional pattern recognition methods, such as Mahalanobis distance classifier [5], the nearest-neighbor classifier [15], and Fisher classifier [16], are adopted to recognize these actions. Some other methods extracted feature from each frame and got a feature sequence. And HMM, dynamic time warping, or dynamic Bayesian network is used to model dynamics. Several improved HMM, such as exemplar-based HMM [17] and coupled HMM [3] are also used to model actions. The above-mentioned methods explicitly or implicitly model dynamics of the action. In particular, humans are able to recognize many actions from a single static image. Weinland presented an exemplar-based embedding recognition method without modeling dynamics [18].

In this chapter, two classifiers are adopted to recognize the actions. The one is exemplar-based HMM (EHMM) classifier, and the other is learned by the exemplar-based embedding (EE) method without modeling dynamics.

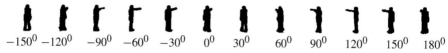

-150^0 -120^0 -90^0 -60^0 -30^0 0^0 30^0 60^0 90^0 120^0 150^0 180^0

Figure 8.1 Projection images of the same pose ('point') in different viewpoints

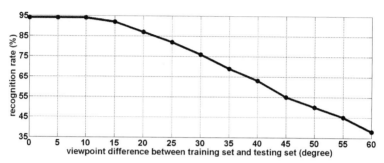

Figure 8.2 Recognition rate vs. viewpoint differences between training set and testing set

8.2.3 Viewpoint Difficulty of 2D Action Recognition

Various viewpoints make the 2D action recognition very challenging. Figure 8.1 shows 12 projection images of the same pose in 12 viewpoints. The projection images change a lot between different viewpoints.

If the same actions are observed in different viewpoints, the recognition performance will become bad. One experiment is performed to show it. The experimental data are form IXMAS dataset [19]. Six actions, including 'kick', 'point', 'stand', 'walk in cycle', 'check watch' and 'wave hand', are used. The EE classifier is adopted. The viewpoints of training set are fixed and those of testing set are changing. Figure 8.2 illuminates that when the viewpoint difference is larger than 10, the recognition rate of testing set is nearly linear decrease.

To conquer the viewpoint difficulty, three methods have been brought forward: viewpoint invariance methods [13,21], multiple viewpoints methods [1, 14], and three dimension methods [16].

The viewpoint invariance methods try to obtain some viewpoint invariance quantities with the corresponding points of 2D images which are captured from different viewpoints. These corresponding points are always the body joints. However, it is very difficult to automatically obtain the body joints from 2D images. So these methods always supposes the body joints can be obtained by hand or a motion capture system.

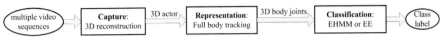

Figure 8.3 Flow chart of the action recognition algorithm

Multiple viewpoints method is a direct solution to the single viewpoint difficulty. For each class of action, action models are learned in all of the viewpoints. $\lambda_{c,v}$ denotes the action model for action c in viewpoint v. The observation action y is classified by $c^* = \arg\max_c\{\max_v P(\lambda_{c,v}|y)\}$. The calculation is linear increase with the number of viewpoints.

By 3D reconstruction, we can observe action sequences as same as the real world. In 3D space, the viewpoints difficulty and self-occluding problems can be conquered easily. The hard part of this method is the hardware setting. Multiple video streams need to be simultaneously captured from several static calibrated cameras.

8.3 Overview of 3D Body Joints-Based Action Recognition

In this chapter, we propose a 3D action recognition framework that represents the action with 3D body joints and employs two classifiers to characterize them. The one classifier is modeled by EHMM, in which the dynamics of actions are encoded by the sate transition probabilities. The other classifier is learned by the EE method without modeling dynamics. 3D body joints-based action recognition includes three basic problems: capture, representation, and classification, as shown in Figure 8.3.

Capture: Multiple video streams are simultaneously captured from static calibrated cameras. And foreground/background segmentation is performed on each through background subtraction method. Then, volumetric represent-ation sequence of actor is created with the visual hull method [16].

Representation: Human joints model-based full body tracking algorithm is adopted to track every body parts and obtain their body joints. However, the human body's large degrees of freedom and the nonlinearity of the dy-namic system bring many difficult problems. To overcome these problems, we propose a tracking approach which fuses the body parts segmentation and adaptive particle filter.

Classification: EHMM and EE method are respectively employed to model the action sequences and classify them. This chapter focuses on the parts of representation and classification.

8.4 Feature Extraction: 3D Body Joints Capture

The 3D reconstructed action sequence is modeled as 3D occupancy grids sequence $A_0, \ldots, A_k, \ldots, A_K$, where $A_k = \{a_n\}_{n=1}^{N_A}$ is the 3D point set of actor in the k_{th} frame and $a_n = (a_n^{(x)}, a_n^{(y)}, a_n^{(z)})$ is the 3D grid coordinates of the point. A human joint model-based full body tracking algorithm is employed to track the movement of every body part of the 3D action sequence and obtain its body joints sequence.

Several 3D full body tracking approaches have been published lately. Kehl et al. [8] tracked the full body with stochastic meta descent method. Cheung et al. [4] used EM approach to segment and track body parts. But the data segmentation would fail when the body parts were too close together. Particle filter provides a suitable framework for state tracking in a nonlinear, non-Gaussian system [2]. However particle filter requires an impractically large number of particles to sample the high dimensional state space effectively. To overcome this problem, we propose an approach which fuses the body parts segmentation and adaptive particle filter. Body parts segmentation reduces the space dimensions and the particle number adaptively changes frame by frame according to the variance.

8.4.1 Human Joints Model

An articulated human joints model $H = \{P, V\}$ is defined and adopted in the tracking method. As shown in Figure 8.4, the human joint model $H = \{P, V\}$ consists of 11 body parts, denoted by $P = \{p_m\}_{m=1}^{11}$, and 15 body joints, denoted by $V = \{v_m\}_{m=1}^{15}$.

We design 23 pose parameters for the human joints model's skeletal structure. There is 1 parameter for the orientation angle (rotation around Z axis), 3 for the location, and the other 19 for the rotation of body joints.

Let u_m be the center (skeleton) axis of body part p_m. We adopt the point-line distance $d(a_n, u_m)$ as the distance between point a_n of actor and body part p_m of human joints model. We normalize it by $\bar{d}(a_n, u_m) = d(a_n, u_m)/r_m$, where r_m is the radius of this body part. Then the distance between actor A and human joints model H is defined as follows:

$$D(A, H) = \sum_{a_n \in A} \min_{u_m}(\bar{d}(a_n, u_m))/|A| \qquad (8.1)$$

Body joints are hierarchical. For example, the waist joint can drive the arm, and some joints, such as left shoulder joint and right shoulder joint, can

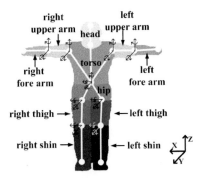

Figure 8.4 Human joints model

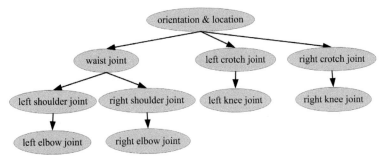

Figure 8.5 Hierarchical human joints model's skeletal structure

be assumed to be independent. As shown in Figure 8.5, the human joints model's skeletal structure is represented by a four layers tree, where the brother nodes are independent. According to this hierarchical search tree we can decompose the search space and search the segmented parameters space in the top-down order.

8.4.2 Human Joints Model-Based Full Body Tracking

Figure 8.6 shows flowchart of the human joints model-based full body tracking algorithm, where Ψ is the pose parameters set and H_Ψ is the human joints model under pose Ψ. The input is 3D reconstructed action sequence. The output is the 3D body joints sequence with pose parameters. The initialization of pose parameters can be completed automatically with the actor under 'standing upright' pose. Then in iteration the body parts segmentation and tracking are performed alternately.

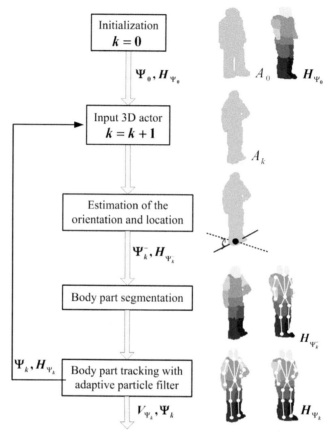

Figure 8.6 Human joint model-based full body tracking algorithm

8.4.2.1 Body Parts Segmentation

Orientation and location parameters of the kth frame should be obtained before body parts segmentation. The location parameters are directly obtained from the center of the 3D point set of actor and orientation parameter is obtained by full search method. Then, we get Ψ_k^- and $H_{\Psi_k^-}$, where orientation and location parameters are updated and the other parameters remain the same as those of the previous frame.

The fusion of top-down strategy and bottom-up strategy is adopted to segment the body parts. The detail process is shown in Figure 8.7.

In the top-down segmentation stage with human joints model, we initially assign the point a_n of actor to the body part p_{m*} if $\bar{d}(a_n, u_{m*})$ has the

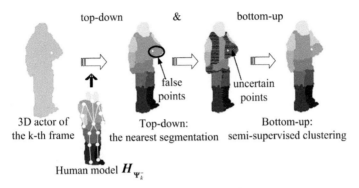

Figure 8.7 Body parts segmentation

smallest value among all of the body parts. But some points may be falsely segmented due to the movement. Then the bottom-up clustering strategy is used to correct the initial segmentation error. These initially assigned points in the body part's interior belong to this body part always correct, and those points around the body part's edge maybe either belong to this body part or others. Finally, semi-supervised clustering method is used to segment these uncertain points. The fusion algorithm of top-down and bottom-up strategy can effectively segment those body parts that are close to each other or move fast.

8.4.2.2 Body Parts Tracking with Adaptive Particle Filter

In order to get the body joints sequence, a body parts tracking algorithm with adaptive particle filter is adopted to conquer the nonlinear of the movement.

Define the particle's state s as the pose parameters and $(s^{(i)}, \omega^{(i)})_{i=1}^{N}$ as the particle set with weight $\omega^{(i)}$. Z_k, the point set tracked of the segmented body part in the k_{th} frame, is the measurement and $T_{S^{(i)}}$ is the corresponding body parts of human joints model under the pose of $S^{(i)}$. $D(Z_k, T_{S^{(i)}})$ indicates the distance between Z_k and $T_{S^{(i)}}$. In prediction stage, the state is propagated through the first order dynamic model. And we use the sigmoid function to approximate measurement likelihood:

$$p(Z_k|s^{(i)}) = 1/(1 + e^{D(Z_k, T_{S^{(i)}})}) \tag{8.2}$$

The particle number varies with the noise variance. If the noise variance is large, we need more particles. The adaptive particle number is embedded into the traditional sampling importance re-sampling filter (SIR) [2]. With

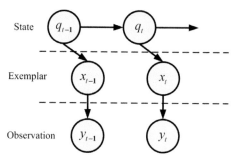

State

Exemplar

Observation

Figure 8.8 The structure of EHMM

this tracker, we track these segmented body parts in the top-down order and obtain their body joints.

On standard PC (Intel Pentium IV 3.0GHz CPU and 1G RAM), the proposed tracker runs at about 530 ms/frame. Compared to the ground truth data labeled by hand, the average error of joint position is 0.83 voxel.

8.5 Action Recognition

8.5.1 EHMM-Based Action Recognition

We model an action as a Markov sequence over a set of key-poses, the exemplars [17]. The structure of EHMM is shown in Figure 8.8. q_t is the hidden state. y_t is the observation. $x_t \in X$ is the exemplar and $X = \{X_1, \ldots, X_e, \ldots, X_E\}$ denotes the exemplar set, which is selected from the training feature set. Both observation and exemplar are represented with 3D body joints vector, which are computed in the above section. Close relationship exits between observation y_t and exemplar x_t.

The EHMM is different from the HMM in the definition of observation densities. For HMM, the general representation of the observation densities is Gaussian mixture model (GMM). While, in EHMM, the definition of observation probability is as follows:

$$p(y_t/q_t) = \sum_{e=1}^{E} p(x_t = X_e/q_t)p(y_t/x_t = X_e) = \sum_{e=1}^{E} c_e N(y_t, X_e, \Sigma_e) \quad (8.3)$$

where $c_e = p(x_t = X_e/q_t)$, the probability of $x_t = X_e$ in state q_t, is the mixture coefficient for the e_{th} mixture, and $N(y_t, X_e, \Sigma_e)$ is a Gaussian density whose mean is fixed at X_e.

Table 8.1 Recognition rates of EHMM action recognition

Data set	Kick	Point	Stand	Turn	Walk	Watch	Wave	Average
training	100.0%	95.6%	97.2%	95.0%	100.0%	98.9%	95.0%	97.4%
testing	100.0%	83.3%	97.2%	77.8%	97.2%	97.2%	83.3%	90.9%

The exemplar selection and the parameters learning are performed alternately with the wrapper forward selection method [17]. Let ζ denote the training feature set, which contains 3D body joints vectors of all frames in the training set. The detail algorithm is shown as follows:

Step 1: Let exemplar set $X = \emptyset$;

Step 2: Find $y^* \in \zeta$ and $y^* \notin X$, where a classifier using exemplar set $\{X \cup \{y^*\}\}$ has best recognition performance in training set;

Step 3: Repeat step 2 until E exemplars are selected.

In EHMM training, each model is learned with the Baum–Welch algorithm [12], except that the means of the Gaussians are not updated.

A testing action sequence $Y = \{y_1, \ldots, y_t\}$ is classified with maximum a posteriori (MAP) estimation:

$$c^* = \arg \max_{c} \{p(Y/\lambda_c)p(c)\} \tag{8.4}$$

where λ_c is the model parameters of EHMM for class c and $p(c)$, the prior of class c, without loss of generality, is assumed to be uniform.

The effectiveness of the proposed algorithm is demonstrated with experiments on 7 actions by 12 actors from IXMAS dataset. All recognition rates are computed with the cross-validation. 10 actors are used to train the classifier, and the remaining 2 actors are used for the evaluation. This is repeated 6 times and the rates are averaged. Table 8.1 shows the results. The average recognition rates are respectively 97.4 and 90.9% for the training set and testing set. The main confusion occurs between 'turn around (turn)', 'stand', and 'walk in cycle (walk)'.

8.5.2 Exemplar Embedding-Based Action Recognition

The other classifier is learned by the exemplar-based embedding (EE) method. Weinland firstly adopted EE method to action recognition with 2D action images in fixed viewpoints [18]. It is a time-invariant representation that does not require a time warping step and is insensitive to variations in speed and length of an action. We apply it to the 3D action recognition.

Exemplars are selected, as in our previous work, using a wrapper forward feature selection technique [17]. We denote $X = \{x_1, \ldots, x_n\}$ as the exemplar

Table 8.2 Recognition rates of EE action recognition

Data set	Kick	Point	Stand	Turn	Walk	Watch	Wave	Average
training	100.0%	100.0%	100.0%	100.0%	100.0%	100.0%	100.0%	100.0%
testing	97.2%	77.8%	97.2%	97.2%	100.0%	91.7%	83.3%	92.1%

set. $Y = y_1, \ldots, y_t$ is the observation action sequence. The feature vector of Y is defined by $H(Y) = (h^*(Y, x_1), \ldots, h^*(Y, x_n))^T$, where $h^*(Y, x_i) = \min\limits_{j=1}^{t} h(y_j, x_i)$ is the smallest distance between exemplar and all of the frames in sequence Y. Both observation y_j and exemplar x_i are represented with 3D body joints vector, and $h(y_j, x_i)$ is defined as the Euclidean distance between y_j and x_i.

In the embedding space, classification of action sequences reduces to classify the vectors $H(Y)$. Each class is modeled with a single Gaussian distribution. And MAP estimation is used to classify them.

Table 8.2 shows the experimental results. The average recognition rates are respectively 100.0 and 92.1% for the training set and testing set. The main confusion occurs between arm actions, such as 'point', 'wave hand(wave)' and 'check watch(watch)'. Our results show that such a representation can effectively model actions and yield recognition rates that are a little better than those of EHMM, with the virtues of simplicity. But no dynamics also brings some problems. For example, 'sitting down' and 'standing up' have the same poses pool but in the opposite dynamic order. In this case, EE method fails, and dynamics must be modeled.

8.6 Multiple-Features Fusion-Based Action Recognition

Both of the above classifiers use parts of information contained in 3D body joints to represent actions. The recognition performances are not as good as we wish. 3D body joints contain much richer and finer information, so we adopt the multiple features fused to further improve the performance. For simplicity, we only use the EHMM classifier to describe this fusion method.

The holistic motion features G_m, body configuration features G_s and arm configuration features A_s are fused to recognize actions. G_m are the actor's location and orientation sequences. G_s are the normalized 3D body joints, which are used in the two above classifiers. A_s are the normalized 3D arms' joints and they help to the problem of 'left-hander'. These features are separately modeled and then fused to classify actions.

Table 8.3 Recognition rates of fusion classifier

Data set	Kick	Point	Stand	Turn	Walk	Watch	Wave	Average
			$\lambda_{G_s}^{(c)}$	+	$\lambda_{G_m}^{(c)}$			
training	100.0%	95.6%	100.0%	100.0%	100.0%	98.9%	95.0%	98.5%
testing	100.0%	86.1%	100.0%	100.0%	97.2%	97.2%	83.3%	94.8%
			$(\lambda_{G_s}^{(c)}$	+	$\lambda_{G_m}^{(c)})$	+	$\lambda_{A_s}^{(c)}$	
training	100.0%	99.4%	100.0%	100.0%	100.0%	100.0%	100.0%	99.9%
testing	100.0%	97.2%	100.0%	100.0%	97.2%	100.0%	97.2%	98.8%

$\lambda_{G_m}^{(c)}$, $\lambda_{G_s}^{(c)}$ and $\lambda_{A_s}^{(c)}$ are respectively denoted as the HMM model for G_m, EHMM model for G_s and EHMM model for A_s. The fusion classifier is realized by two layers fusion. The first layer fuses G_m and G_s as formula (8.5). If the decision of the first layer c_1 belongs to one kind of arm actions, like 'wave hands', 'point', and so on, the action sequence will be recognized with the second layer classifier, as formula (8.6).

$$c_1 = \arg\max_{c}\{p(\lambda_{G_m}^{(c)}\lambda_{G_s}^{(c)}|G_m G_s)\} = \arg\max_{c}\{p(\lambda_{G_m}^{(c)}|G_m)p(\lambda_{G_s}^{(c)}|G_s)\} \quad (8.5)$$

$$c_z = \arg\max_{c}\{p(\lambda_{A_s}^{(c)}/A_s)\} \quad (8.6)$$

Table 8.3 shows the experimental results. With the first layer fusion classifier, $\lambda_{G_s}^{(c)} + \lambda_{G_m}^{(c)}$, 'turn around (turn)' can be effectively distinguish from the 'stand' and 'walk in cycle (walk)' actions. And the two layers fusion classifier, $(\lambda_{G_s}^{(c)} + \lambda_{G_m}^{(c)}) + \lambda_{A_s}^{(c)}$ conquer the problem of 'left-hander' and reduce the confusion between arm actions. The experimental results prove that the fusion of features make the best use of 3D body joints representation and can effectively classify the actions.

We compare our human action recognition approach with other previous researches as given in Table 8.4. It is difficult to compare these approaches, since the data sets and environments are different. However, the results can give a general overview and comparison of some approaches in action recognition. Compared to the mentioned researches, our approach, especially the fusion classifier, yields comparable results.

8.7 Conclusion

This chapter reviewed the existing action representation and recognition methods. Then we addressed a framework for 3D action recognition. 3D body joints sequence captured by a marker-less full body tracker was employed to

Table 8.4 Comparison of the proposed approach with previous researches

Algorithms	Action count	Actor count	Viewpoints	Recognition rate
Wang et al. [15]	10	9	fixed single view	96.7%
Weinland et al. [18]	10	9	fixed single view	100.0%
Davis et al. [5]	18	1	multi-view	83.3%
Ahmad et al. [1]	5		multi-view	87.5%
Weinland et al. [16]	11	10	3D	93.3%
EHMM classifier	7	12	3D	90.9%
EE classifier	7	12	3D	92.1%
$\lambda_{G_s}^{(c)} + \lambda_{G_m}^{(c)}$	7	12	3D	94.8%
$(\lambda_{G_s}^{(c)} + \lambda_{G_m}^{(c)}) + \lambda_{A_s}^{(c)}$	7	12	3D	98.8%

represent action. Two classifiers were used to recognize the actions. One is EHMM classifier with modeling dynamics, and the other is learned by the EE method without modeling dynamics. Our results show that EE method can yield recognition rates that is a little better than EHMM method, with the virtue of simplicity. 3D body joints contain much richer and finer information. The multiple features were fused to improve the performance. The experimental results prove that fusion of features make the best use of 3D body joints representation and effectively classify the actions.

The applications of action recognition require the development of systems is real-time and robust to environment variation and sometimes can not meet the conditions of 3D reconstruction. However, the extraction of 3D body joints requires 3D reconstruction data and is time-consuming. The future work includes the acceleration of full body tracking approach and 3D classifier based 2D action recognition, which we have done a part of.

References

[1] M. Ahmad and S.-W. Lee. HMM-based human action recognition using multiview image sequences. *Proceedings of International Conference on Pattern Recognition*, 2006.
[2] M.S. Arulampalam, S. Maskell, N. Gordon, et al. A tutorial on particle filter for online nonlinear/non-Gaussian Bayesian tracking. *IEEE Trans. on Signal Processing*, 50(2):174–188, 2002.
[3] M. Brand, N. Oliver, and A. Pentland. Coupled hidden Markov models for complex action recognition. In *Proceedings of IEEE Conference on Computer Vision and Pattern Recognition*, pages 994–999, 1997.
[4] K.M. Cheung, T. Kanade, J.-Y. Bouguet, and M. Holler. A real time system for robust 3D voxel reconstruction of human motions. In *Proceedings of IEEE Conference on Computer Vision and Pattern Recognition*, pages 714–720, 2000.

[5] J.W. Davis and A.F. Bobick. The representation and recognition of human movement using temporal templates. In *Proceedings of IEEE Conference on Computer Vision and Pattern Recognition*, pages 928–934, 1997.

[6] J. Gu, X. Ding, S. Wang, et al. Adaptive particle filter with body part segmentation for full body tracking. In *Proceedings of IEEE Conference on Automatic Face and Gesture Recognition*, pages 1–6, 2008.

[7] G. Johansson. Visual motion perception. *Scientific American*, 232(2):76–88, 1975.

[8] R. Kehl, M. Bray, and L. VanGool. Full body tracking from multiple views using stochastic sampling. In *Proceedings of IEEE Conference on Computer Vision and Pattern Recognition*, pages 129–136, 2005.

[9] X. Li. HMM based action recognition using oriented histograms of optical flow field. *Electronics Letters*, 43(10):560–561, 2007.

[10] T.B. Moeslund, A. Hilton, and V. Krüger. A survey of advances in vision-based human motion capture and analysis. *Computer Vision and Image Understanding*, 104(3):90–126, 2006.

[11] H.H. Nagel. From image sequences towards conceptual descriptions. *Image and Vision Computing*, 6(2):59–74, 1988.

[12] L.R. Rabiner. A tutorial on hidden Markov model and selected applications in speech recognition. *Proceedings of the IEEE*, 77(2):257–286, 1989.

[13] Y. Shen, N. Ashraf, and H. Foroosh. Action recognition based on homography constraints. In *Proceedings of International Conference on Pattern Recognition*, pages 1–4, 2008.

[14] R. Souvenir and J. Babbs. Learning the viewpoint manifold for action recognition. In *Proceedings of IEEE Conference on Computer Vision and Pattern Recognition*, 2008.

[15] L. Wang and D. Suter. Informative shape representations for human action recognition. In *Proceedings of International Conference on Pattern Recognition*, pages 1266–1269, 2006.

[16] D. Weinland, R. Ronfard, and E. Boyer. Free viewpoint action recognition using motion history volumes. *Computer Vision and Image Understanding*, 104:249–257, 2006.

[17] D. Weinland, E. Boyer, and R. Ronfard. Action recognition from arbitrary views using 3D exemplars. *Proceedings of International Conference on Computer Vision*, pages 1–7, 2007.

[18] D. Weinland and E. Boyer. Action recognition using exemplar-based embedding. In *Proceedings of IEEE Conference on Computer Vision and Pattern Recognition*, pages 1–7, 2008.

[19] D. Weinland. The IXMAS dataset. https://charibdis.inrialpes.fr/html/sequences.php, 2006.

[20] J. Yamato, J. Ohya, and K. Ishii. Recognition human action in time-sequential images using hidden Markov model. In *Proceedings of IEEE Conference on Computer Vision and Pattern Recognition*, pages 379–385, 1992.

[21] A. Yilmaz and M. Shah. Matching actions in presence of camera motion. *Computer Vision and Image Understanding*, 104(2–3):221–231, 2006.

9

The Average-Half-Face in 2D and 3D Face Recognition

J. Harguess and J.K. Aggarwal

Computer & Vision Research Center (CVRC), Department of Electrical & Computer Engineering, The University of Texas at Austin, 1 University Station, Austin, TX 78712, USA; E-mail: {harguess, aggarwaljk}@mail.utexas.edu

Abstract

It is well known that the human face is inherently symmetric about a bilateral symmetry plane or line (in the case of three- or two-dimensional data respectively). This symmetry has been utilized in applications pertaining to face detection and handling illumination of the face, but its use in face recognition is limited. The average-half-face is one such utilization of symmetry and has been shown to be successful in the face recognition task. To construct the average-half-face, the right and left half-face are averaged together, assuming the bilateral symmetry plane has been found. We present the results of applying the average-half-face to two- and three-dimensional (2D and 3D) face databases by utilizing several popular face recognition algorithms. Our results show that the average-half-face may produce an increase in accuracy over using the original full face for face recognition. This finding may have implications on storage and computation time for face recognition systems. Further, since the success of using the average-half-face in face recognition may depend heavily on the computation of the bilateral symmetry plane (or line) of each face, we present an error analysis of choosing a suboptimal this bilateral symmetry plane (or line) and find that the average-half-face is more robust to this choice than the original full face.

Patrick Shen-Pei Wang (Ed.), Pattern Recognition and Machine Vision – In Honor and Memory of Professor King-Sun Fu, 135–148.

Keywords: face recognition, average-half-face, symmetry, 2D, 3D.

9.1 Introduction

Face recognition has been extensively researched for many years. Recent efforts in improving security, such as automatic surveillance and the use of biometrics in identification, are partly responsible for this interest. However, several challenges remain in improving the accuracy of face recognition under illumination changes, variations in pose, occlusions (including self-occlusion), and image resolution. Many face recognition algorithms have been developed and each has its strengths. It is well-known that the face is symmetric about a bilateral symmetry plane. Surprisingly, not much use of this inherent symmetry of the face has been exploited for recognition. The average-half-face exploits facial symmetry by splitting the frontal full face into two halves about the bilateral symmetry plane, mirroring one of the halves horizontally, and then averaging the two resulting images. We demonstrate the effectiveness of using the average-half-face, versus using the original full face, as an input to face recognition algorithms for a potential increase in accuracy and decrease in storage and computation time.

Two previous analyses have been performed using the average-half-face for face recognition [7,8]. The first analysis showed a large increase in accuracy, over the original full face images, when applying the method along with eigenfaces for face recognition to a 3D range image face database. To further prove the effectiveness of the method, a more complete comparison of the accuracy using both the average-half-face and the original full face image is performed. The following six face recognition algorithms were used for this comparison; eigenfaces or principal components analysis (PCA), multilinear principal components analysis (MPCA), MPCA with linear discriminant analysis (MPCA-LDA), Fisherfaces or linear discriminant analysis (LDA), independent component analysis (ICA), and support vector machines (SVM).

Each of the following five algorithms are based on projection techniques and our used along with nearest neighbors (NN) for classification. Eigenfaces is based on principal components analysis (PCA) and is a common face recognition algorithm for benchmarking newer algorithms. MPCA and MPCA-LDA extend PCA to use the entire face image as a tensor object (or multilinear arrays) in order to try to preserve the relationship between neighboring pixels. Fisherfaces, another popular method in the literature used for benchmarking, is based on linear discriminant analysis (LDA) and thus attempts to utilize class information in the data to maximize between class scatter while minimizing within class scatter. ICA models each face image as a linear combination of non-Gaussian random vectors where the weights of the linear combinations of the training and testing images are used for identification.

A newer classification method known as the SVM finds the hyperplanes that maximize the margins between training data classes and has been applied to face data. We use each face image as a feature vector along with the SVM for classification in our experiments.

In addition to trying multiple methods for face recognition, we also utilize three independent datasets for our experiments; a 3D range image face database [6] and two well-known 2D datasets (the Yale Face Database [18] and the AR Face Database [10]). The results of each algorithm on the different datasets for both the average-half-face and full face are shown. Our results show that the average-half-face produces an increase in accuracy over using the full face in the majority of the experiments.

The sections are organized as follows. The concept of the average-half-face is first introduced. Then, each of the algorithms used for comparison in our experiments are briefly explained. Experiments are then performed using the three datasets along with the aforementioned face recognition methods to compare the accuracy of using the average-half-face to using the full face. Additionally, an error analysis is performed to test the sensitivity of face recognition accuracy to the choice of the bilateral symmetry plane of the face image. A discussion of the results and future work conclude the chapter.

9.2 Average-Half-Face

Utilizing the symmetry of the face to improve face detection and face recognition (in a more limited sense) has been attempted by several other researchers. It is noted and often observed that the "face is roughly symmetrical" [16]. Zhao and Chellappa [20] attempt to solve the problem of illumination in face recognition using Symmetric Shape-from-Shading. Ramanathan et al. [13] introduce the notion of 'Half-faces' (in the sense of exactly one half of the face) to assist in computing a similarity measure between faces using images that have non-uniform illumination. In face recognition, the use of the bilateral symmetry of the face has been limited to extracting facial profiles for recognition [11,19]. Chen et al. [5] automatically compute the symmetry axis (or plane) of a frontal face image for use in face recognition.

The original inspiration for the average-half-face [8] is the symmetry preserving singular value decomposition (SPSVD) [15]. The SPSVD is used to

(a) Full Face   (b) Average-Half-Face   (c) Left Half-Face   (d) Right Half-Face

Figure 9.1 (a) 2D full face image; (b) its average-half-face; (c) its left half-face; and (d) its right half-face

reduce the dimensionality of data while preserving the inherent symmetry. When applying the concept of the SPSVD to a 2D image, such as that of a face, there are two steps. First, once the bilateral symmetry plane of the face has been calculated, the image is centered about the bilateral symmetry plane. This step will preserve the symmetry and ensure that the two spatial halves of the data are near similar (mirrored) images. Second, the face image is divided into two halves and they are averaged together by first reversing the columns of one of the halves. In our experiments, we do not desire a dimensionality reduction on the face image or the average-half-face, so the step of performing the (SVD) on the image is skipped.

As an example of the process of forming the average-half-face, Figure 9.1 displays the full face image, the left and right faces (after centering based on the bilateral symmetry plane), and the average-half-face of an image from the Yale Face database [18].

Since these steps are independent of the face recognition algorithm, the average-half-face can be seen as a preprocessing step. This allows for a direct comparison of the use of the average-half-face with the use of the original full face image in face recognition algorithms.

9.3 Face Recognition Algorithms

Each of the six face recognition methods that will be used in our experiments will be briefly introduced in this section.

9.3.1 Projection Methods

The following methods are all subspace projection methods designed to reduce the feature space for face recognition. For consistency, the classifier used in each case is nearest neighbors (NN) along with the Euclidean distance.

9.3.1.1 Eigenfaces

Eigenfaces [17], one of the first successful face recognition algorithms, is based on principal components analysis (PCA). Eigenfaces is a subspace projection face recognition method that first computes the PCA of a training set which returns a set of basis vectors that maximize the variance within the training data. The training and test images are projected onto the new subspace and NN is then used for classification of the test images.

9.3.1.2 Multilinear Principal Components Analysis

Since we are applying PCA to a 2D image, a method known as multilinear principal components analysis (MPCA) [9] may be used. The idea is to determine a multilinear projection for the image, instead of forming a one-dimensional (1D) vector from the face image and finding a linear projection for the vector. It is thought that the multilinear projection will better capture the correlation between neighborhood pixels that is otherwise lost in forming a 1D vector from the image.

9.3.1.3 MPCA + Linear Discriminant Analysis

The use linear discriminant analysis (LDA) to perform feature selection combined with the projected multilinear arrays results in MPCA+LDA. Example code for the MPCA and MPCA+LDA methods was provided in [9].

9.3.1.4 Fisherfaces

Fisherfaces is the direct application of (Fisher) linear discriminant analysis (LDA) to face recognition [2]. In LDA, the class information is explicitly used to form a linear subspace, unlike PCA which ignores class information. The purpose of LDA is to maximize the objective function

$$J(w) = \frac{w^T S_B w}{w^T S_W w},$$

where S_B and S_W are the *between class scatter* and the *within class scatter* matrices respectively and where w is the normal vector to the discriminant hyperplane. The solution can be posed as the eigenvalue problem

$S_B^{1/2} S_W^{-1} S_B^{1/2} v = \lambda v$, by defining $v = S_B^{1/2} w$. The eigenvectors that correspond to the largest eigenvalues form the subspace for classification. Projecting our training and test vectors into this new subspace, we can classify the test images using NN.

9.3.1.5 Independent Components Analysis

PCA maximizes the variance between pixels to separate linear dependencies between pixels. Independent components analysis (ICA) is a generalization of PCA in that it tries to identify high-order statistical relationships between pixels to form a better set of basis vectors. The pixels are treated as random variables and the face images as outcomes, as described in [1]. The code for ICA was provided by the authors for use in face recognition research [1].

9.3.2 Support Vector Machine

Support Vector Machines (SVM) construct a set of hyperplanes that are designed to maximize the *margin* of the decision boundary between positive and negative examples for classification [14]. We represent each training and test image as a 1D vector, therefore, the feature vectors for the SVM are the face images themselves. For implementation of the SVM, we utilized the powerful LIBSVM library [4] that uses the "one vs. one" approach to multiclass problems. We used default parameters for the SVM, including a radial basis function (rbf) kernel.

9.4 Databases

Three face databases were used for our experiments: (A) the Yale Face database, (B) the AR Face database, and (C) 3D face database.

(A) The Yale Face database [3, 18] consists of a total of 165 gray scale, frontal, 2D face images. There are a total of 15 subjects with 11 images per subject representing changes in illumination and facial expressions. For each of the algorithms, we maintained a consistent use of the database by forming the training data from the first 8 images per subject and using the remaining 3 images per subject for testing.

(B) The AR Face database [10] consists of images from 109 subjects (66 men and 43 women), each with 26 configurations. The different configurations consist of expression changes (such as neutral, smile, anger, and scream), lighting changes, and occlusions. Two different sessions, each with 13 different configurations, were taken to form the database. We used the first 21 configurations per subject for training and the remaining images for testing.

(C) The 3D database we have additionally utilized is a 3D face range image database acquired using an MU-2 stereo imaging system manufactured by 3Q Technologies Ltd. (Atlanta, GA) by the former company Advanced Digital Imaging Research, LLC, Friendswood, TX [6]. The database consists of a total of 1126 images of 104 subjects. There are anywhere from 1 to 55 images per subject. We trained the algorithms using a combination of 360 images from 12 subjects and a single neutral expression from 104 different subjects. The test database consisted of the remaining 662 images from all 104 subjects.

9.5 Experiments

In total, 18 experiments were performed using the six face recognition methods and three face databases. In addition to these experiments, an error analysis was done on the choice of the bilateral symmetry plane of the face.

9.5.1 Varying Algorithms and Databases

The parameters (such as number of eigenvectors used for PCA, size of subspace in LDA, etc.) for each algorithm were kept constant between experiments to maintain a fair comparison of each algorithm's performance on the average-half-face and the full face. The images were centered about the bilateral symmetry plane for both the average-half-face and full face recognition results. Table 9.1 summarizes the results of our experiments. Each of the numbers in the table represents the rank-1 accuracy rate for recognition, which is the accuracy of the closest match of the test data to a corresponding

Table 9.1 Rank-1 accuracy results using the full face (Full) and the average-half-face (AHF)

Database	A		B		C	
	Yale		AR		3D	
Algorithms	Full	AHF	Full	AHF	Full	AHF
PCA	77.8	86.7	49.4	52.3	72.8	80.4
MPCA	80.0	93.3	59.4	57.6	81.0	81.3
MPCA-LDA	66.7	68.9	91.9	88.1	91.8	93.8
LDA	91.1	97.8	54.1	78.0	79.8	82.6
ICA	93.3	100	65.3	60.0	76.9	84.3
SVM	91.1	91.1	44.8	36.1	50.8	51.4

Table 9.2 P-values from testing the difference between accuracies using the full face (Full) and the average-half-face (AHF)

Database	A	B	C
	Yale	AR	3D
Algorithms			
PCA	0.269	0.338	0.001
MPCA	0.064	0.547	0.889
MPCA-LDA	0.823	0.037	0.159
LDA	0.165	8e-17	0.192
ICA	0.077	0.071	0.001
SVM	1.0	0.003	0.827

training sample. The purpose of these experiments is to compare the full face to the average-half-face for recognition, not to compare the accuracy of the algorithms themselves.

Figures 9.2, 9.3, and 9.4 display these results more clearly for each of the three databases involved. From the results of Figures 9.2 and 9.4, we can see that the average-half-face outperforms the full face in every method for the Yale Face database and the 3D face database. However, there are mixed results shown in Figure 9.3 when using the AR Face database. The best performing method for the Yale Face database was ICA with the average-half-face at 100% accuracy. For the AR Face database, the MPCA-LDA method was the best with the full faces at 91.9%. The 3D database saw the best result of 93.8% accuracy with the MPCA-LDA method and the average-half-face.

To test if the differences between the results on the average-half-face and the full face are significant, p-values were calculated on the null hypothesis that the accuracies are equal (assuming a priori that $\alpha = 0.05$ for a two-sided test). The p-values are reported in Table 9.2. The difference in the results of the Yale face database are insignificant except for the accuracy of MPCA and ICA, which are nearly significant. In the case of the AR face database, mostly due to the fact that there are more test images, only the difference in the accuracies when using PCA, MPCA and ICA are insignificant. Only the results of PCA and ICA are significant when using the 3D database.

9.5.2 Bilateral Symmetry Plane Error Analysis

Finally, we performed experiments to analyze the robustness of using the average-half-face with eigenfaces to the error associated with choosing the bilateral symmetry plane (or line, in the case of a frontal 2D image). Choosing the optimal plane of symmetry amounts to selecting the best vertical line in

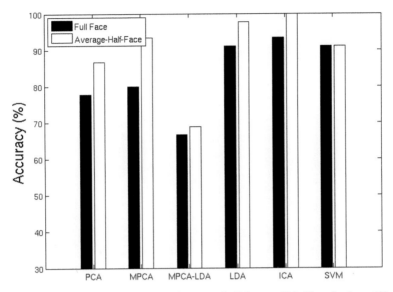

Figure 9.2 Accuracy of full face and average-half-face on Yale Face database (A)

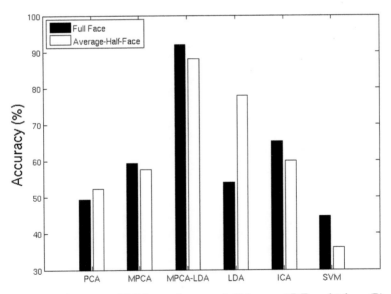

Figure 9.3 Accuracy of full face and average-half-face on AR Face database (B)

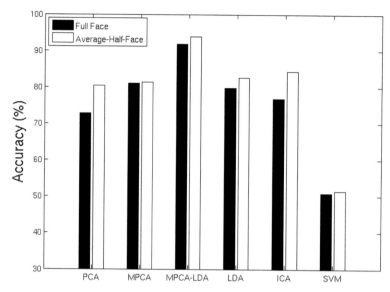

Figure 9.4   Accuracy of full face and average-half-face on 3D Face database (C)

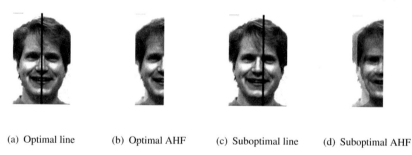

(a) Optimal line   (b) Optimal AHF   (c) Suboptimal line   (d) Suboptimal AHF

Figure 9.5   Examples of (a) optimal bilateral symmetry plane and (b) suboptimal bilateral symmetry plane and the resulting average-half-faces (c) and (d), respectively

the image that divides the face image into left and right halves of the face. A visual example of choosing the bilateral symmetry axis is shown in Figure 9.5 where the suboptimal bilateral symmetry plane is shifted 15 pixels from the optimal plane. We perform the experiments on the 3D face database (where each image is 150×255 pixels) by choosing the plane of symmetry of each image as a random (Gaussian) horizontal offset from the optimal plane of symmetry. We accomplish this by centering a Gaussian distribution at the optimal plane of symmetry with a mean of zero and variance of 5, 10, 15 and 20 pixels and sampling the desired offset from this distribution. Modeling the

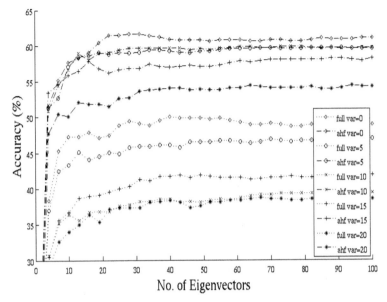

Figure 9.6 Rank 2 accuracy when choosing a suboptimal plane of symmetry

angular error of choosing the optimal plane of symmetry will be considered in future work. We present the results of these experiments in Figure 9.6.

9.6 Discussion

Table 9.1 shows that regardless of the algorithm using average-half-face with the Yale Face database and the 3D database produces an equal or higher accuracy rate than when using the original full face. This is not the case for every method when using the AR Face database. For instance, when using the AR Face database, the rank-1 recognition rates of the MPCA-LDA and SVM methods are significantly better (with p-values of 0.037 and 0.003 respectively) when using the full face versus using the average-half-face. The other methods are very close in accuracy for the AR Face database, except for the Fisherfaces (LDA) method which shows a drastic improvement (p-value near zero) for the average-half-face. For the Yale Face database, the LDA, MPCA and eigenfaces (PCA) methods perform 6–13% better with the average-half-face than with the full face. All other methods with the Yale Face database are comparable, but usually have better results with the average-half-face. The 3D database gives consistently better results when using the average-half-face with a maximum accuracy increase of around 8% with the eigenfaces (PCA) method.

At first glance, the results of using the AR Face database might give evidence that average-half-face is inferior to the full face. However, there are only two instances (from the AR face database experiments with MPCA-LDA and SVM having p-values of 0.037 and 0.003, respectively) out of a total of 18 experiments that give evidence to the full face producing a higher accuracy. Therefore, the average-half-face is clearly of interest, especially since the data stored in the average-half-face is exactly half that of the full face, yet the information stored may be more discriminatory for face identification, especially in the case of the 3D database.

When considering a real world implementation of this algorithm, the error estimation of choosing the optimal plane of symmetry warrants consideration. From the experimental results in Figure 9.6, we can see that method of using the average-half-face (ahf) with eigenfaces is very robust to noise in choosing the optimal plane of symmetry. For most choices of the number of eigenvectors used, the performance of the system using the average-half-face with a Gaussian variance of 5 and 10 pixels is basically equivalent to using the optimal plane of symmetry. This is very promising since choosing the optimal plane of symmetry for newly acquired face images is not a trivial task. Even in the case of a variance of 15 pixels, the performance is acceptable compared to the optimal plane of symmetry. This result can be directly compared to the performance of the same experiments applied to the full face (full), displayed in Figure 9.6. In other words, the error in centering the full face is the same as the error in choosing the optimal plane of symmetry. In contrast, the full face is less robust to error in the choice of the optimal symmetry plane when used with eigenfaces. Each increase in the variance of centering the full faces results in a decrease in the accuracy of the method. This displays yet another advantage of using the average-half-face in 3D face recognition with eigenfaces.

The computation of the average-half-face, given the full face and the position of the middle of the face, is simple. Therefore, with a simple computation step, the accuracy of the majority of the algorithms tested was improved. We believe that this gain in accuracy has its origin in the averaging operation, which produces a new face that contains a set of features that are more discriminatory that those of the full face. More work must be done to verify this claim and to discern the origin of this accuracy gain.

The results in Table 9.1 may not reflect the best accuracy possible for each algorithm because some of the algorithms' parameters may be fine tuned depending on the data set. We utilized each algorithm to compare the accuracy of using the average-half-face and the full face, since we were interested in their relative accuracies, not their absolute accuracies.

9.7 Conclusion

The average-half-face has improved the accuracy, over the use of the full face, in 2D and 3D face recognition in the majority of our experiments. We have also shown that the average-half-face is robust to noise when calculating the bilateral symmetry plane of the face. These results are intriguing, but more results are needed to fully justify its use in the recognition task. We are currently testing the method using the Face Recognition Grand Challenge [12] FRGC database, which is a much larger 2D and 3D face database. Results from the FRGC will help show the efficacy of the average-half-face in face recognition. One direction of future work is to further analyze the source of the accuracy gain of the average-half-face. We would also like to apply the average-half-face to feature extraction methods, such as those using wavelets. As seen from the results on the AR Face Database, further research into the affects of illumination, facial expressions, and occlusions on the average-half-face is needed.

Acknowledgments

We would like to sincerely thank Dr. Patrick Flynn for his valuable comments on this paper. The research was supported in part by Texas Higher Education Coordinating Board award # 003658-0140-2007.

References

[1] M.S. Bartlett, J.R. Movellan, and T.J. Sejnowski. Face recognition by independent component analysis. *IEEE Transactions on Neural Networks*, 13:1450–1464, 2002.

[2] P.N. Belhumeur, J.P. Hespanha, and D.J. Kriegman. Eigenfaces vs. fisherfaces: Recognition using class specific linear projection. In *ECCV '96: Proceedings of the 4th European Conference on Computer Vision-Volume I*, London, UK, pages 45–58. Springer-Verlag, 1996.

[3] P.N. Belhumeur, J.P. Hespanha, and D.J. Kriegman. Eigenfaces vs. fisherfaces: Recognition using class specific linear projection. *IEEE Transactions on Pattern Analysis and Machine Intelligence*, 19(7):711–720, 1997.

[4] C.-C. Chang and C.-J. Lin. *LIBSVM: A library for support vector machines*, 2001. Software available at http://www.csie.ntu.edu.tw/ cjlin/libsvm.

[5] X. Chen, P.J. Flynn, and K.W. Bowyer. Fully automated facial symmetry axis detection in frontal color images. In *Proceedings IEEE Workshop on Automatic Identification Advanced Technologies*, pages 106–111, 2005.

[6] S. Gupta, J.K. Aggarwal, M.K. Markey, and A.C. Bovik. 3D face recognition founded on the structural diversity of human faces. In *Proceedings IEEE Computer Society Conference on Computer Vision and Pattern Recognition*, pages 1–7, 2007.

[7] J. Harguess and J.K. Aggarwal. A case for the average-half-face in 2D and 3D for face recognition. In *Proceedings IEEE Computer Society Workshop on Biometrics (in conjunction with CVPR)*, pages 7–12, 2009.

[8] J. Harguess, S. Gupta, and J.K. Aggarwal. 3D face recognition with the average-half-face. In *Proceedings International Conference on Pattern Recognition ICPR*, pages 1–4, 2008.

[9] H. Lu, K.N. Plataniotis, and A.N. Venetsanopoulos. Mpca: Multilinear principal component analysis of tensor objects. *IEEE Trans. on Neural Networks*, 19(1):18–39, 2008.

[10] A. Martinez and R. Benavente. The AR face database. CVC Technical Report, No. 24, June 1998.

[11] G. Pan and Z. Wu. 3D face recognition from range data. *Int. J. Image Graphics*, 5(3):573–594, 2005.

[12] P.J. Phillips, P.J. Flynn, T. Scruggs, K.W. Bowyer, J. Chang, K. Hoffman, J. Marques, J. Min, and W. Worek. Overview of the face recognition grand challenge. In *CVPR '05: Proceedings of the 2005 IEEE Computer Society Conference on Computer Vision and Pattern Recognition (CVPR'05)*, Washington, DC, USA, Vol. 1, pages 947–954. IEEE Computer Society, 2005.

[13] N. Ramanathan. Facial similarity across age, disguise, illumination and pose. In *Proceedings of International Conference on Image Processing*, 1999.

[14] B. Schölkopf, A.J. Smola, R.C. Williamson, and P.L. Bartlett. New support vector algorithms. *Neural Comput.*, 12(5):1207–1245, 2000.

[15] M.I. Shah and D.C. Sorensen. A symmetry preserving singular value decomposition. *SIAM J. Matrix Anal. Appl.*, 28(3):749–769, 2006.

[16] G. Shakhnarovich and B. Moghaddam. Face recognition in subspaces. In *Handbook of Face Recognition*, S.Z. Li and A.K. Jain (Eds.), pages 141–168. Springer, 2004.

[17] M. Turk and A. Pentland. Eigenfaces for recognition. *J. Cogn. Neurosci.*, 3(1):71–86, 1991.

[18] Yale Univ. Face DB, 2002. http://cvc.yale.edu/projects/yalefaces/yalefaces.html.

[19] L. Zhang, A. Razdan, G. Farin, J. Femiani, M. Bae, and C. Lockwood. 3D face authentication and recognition based on bilateral symmetry analysis. *The Visual Computer*, 22(1):43–55, 2006.

[20] W. Zhao and R. Chellappa. Illumination-insensitive face recognition using symmetric shape-from-shading. In *Proceedings IEEE Computer Society Conference on Computer Vision and Pattern Recognition*, Vol. 1, page 1286, 2000.

10

Model Building in Image Processing by Meta-Learning Based on Case-Based Reasoning

Anja Attig and Petra Perner

Institute of Computer Vision and Applied Computer Sciences, IBaI, 04251 Leipzig, Germany; E-mail: pperner@ibai-institut.de

Abstract

We propose a framework for model building in image processing by meta-learning. The model-building process is seen as a classification process during which the signal characteristics are mapped to the right image-processing parameters to ensure the best image-processing output. The mapping function is realized by case-based reasoning. Case-based reasoning is especially suitable for this kind of process, since it incrementally allows one to learn the model based on the incoming data stream. To find the right signal/image description of the signal/image characteristics that are in relationship to the signal-processing parameters is one important aspect of this work. In connection with this work intensive studies of the theoretical, structural, and syntactical behavior of the chosen image-processing algorithm have to be done. We studied the theoretical and the implementation aspects of the Watershed Transformation and drew conclusions for suitable image descriptions. The central moments is described in this paper as a suitable image description. It can separate well the cases into groups having the same segmentation parameters and it sorts out rotated and rescaled images. Generalization over cases can also be performed over the groups of case. It helps to speed up the retrieval process and to learn incrementally the general model.

Patrick Shen-Pei Wang (Ed.), Pattern Recognition and Machine Vision – In Honor and Memory of Professor King-Sun Fu, 149–164.

Keywords: model building, image segmentation, meta learning, Watershed Transformation, adaptive image segmentation, case-based reasoning.

10.1 Introduction

The aim of image processing is to develop methods for automatically extracting from an image or a video the desired information. The developed system should assist a user in processing or understanding the content of a complex signal, such as an image. Usually, an image consists of thousands of pixels. This information can hardly be quantitatively analyzed by the user. In fact, some problems related to the subjective factor or to the tiredness of the user arise, which may influence reproducibility. Therefore, an automatic procedure for analyzing an image is necessary.

Although in some cases it might make sense to process a single image and to adjust the parameters of the image processing algorithm to this single image manually, mostly the automation of the image analysis makes only sense if the developed methods have to be applied to more than one single image. This is still an open problem in image processing. The parameters involved in the selected processing method have to be adjusted to the specific image. It is often hardly possible to select the parameters for a class of images in such a way that the best result can be ensured for all images of the class. Therefore, methods for parameter learning are required that can assist a system developer in building a model [7] for the image processing task.

While the meta-learning task has been extensively studied for classifier selection, it has not been studied so extensively for parameter learning. Soares et al. [11] studied parameter selection for the identification of the kernel width of a support-vector machine, while Perner [6] studied parameter selection for image segmentation.

The meta-learning problem for parameter selection can be formalized as follows: For a given signal that is characterized by specific signal properties A and domain properties B find the parameters P of the processing algorithm that ensure the best quality of the resulting output signal/information:

$$f : A \cup B \rightarrow P \tag{10.1}$$

Meta-data for images may consist of image-related meta-data (gray-level statistics) and non-image related meta-data (sensor, object data) [8]. In general, the processing of meta-data from signals and images should not require

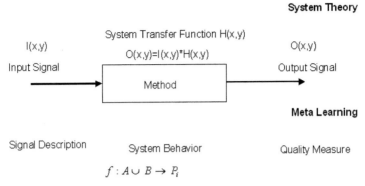

Figure 10.1 Problem description in modeling

too heavy processing and should allow characterizing the properties of the signal that influence the signal processing algorithm.

The mapping function f can be realized by any classification algorithm, but the incremental behavior of case-based reasoning (CBR) fits best to many data/signal processing problems, where the signal-class cannot be characterized ad-hoc, since the data appear incrementally. The right similarity metric that allows mapping data to parameter-groups and, as a consequence, allows obtaining good output results, should be studied more extensively. Performance measures that allow to judge the achieved output and to automatically criticize the systems performance are another important problem [9].

Abstraction of cases to learn domain theory would allow better understanding the behavior of many signal processing algorithms that cannot anymore be described by means of the standard system theory [16].

The aim of our research is to develop methods that allow us to learn a model for the desired task from cases without heavy human interaction (see Figure 10.1). The specific emphasis of this work is to develop a methodology for finding the right image description for the case that group similar images in terms of parameters within the same group and map the case to the right parameters in question.

In Section 10.2, we review the theoretical and behavioral aspects of the Watershed Transformation and how to use case-based reasoning to form the image segmentation model. The derived image descriptions from the theoretical and behavioral study in Section 10.2 are given in Section 10.3. Finally, we present conclusions in Section 10.4.

10.2 The Watershed Algorithm: What Influences the Result of the Segmentation?

An easy way to explain the idea of the Watershed Transform is the interpretation of the image as 3D landscape. Therefore we allocate for every pixel its grey value as z-coordinate. Now we flood the landscape from the regional minima, after having bored the local minima and sunk the landscape into water. Lakes are created (basins, catchment basins) in correspondence with the regional minima. We build dams (watershed lines or simply watersheds), where two lakes meet. Alternatively, instead of sinking the landscape, the latter can be flooded by rainfall and the watershed lines will be the lines of attracting the rain that will fall on the landscape. Whichever of the paradigmas is used - the immersion or rainfall paradigma - to obtain its simulation two approaches are possible: either the basins are identified, then the watershed lines are obtained by taking a set complement, or the complete image partition is computed, then successively the watersheds are found by detecting boundaries between basins.

Many algorithms were developed for computing the Watershed Transform (for a survey, see e.g. [10]). In this work we deal mainly with the Watershed Transformation scheme suggested by Vincent and Soille in [13], and use this scheme also for the new implementation that we have done for the segmentation algorithm from Frucci et al. [4].

The first definition of the so called Watershed Transform by immersion was given by Vincent and Soille [9]. Let D be a digital grey value image with $h_{\min}$ and $h_{\max}$ as the minimal and maximal gray value. MIN_h is the union of all regional minima with grey value h with $h \in [h_{\min}, h_{\max}]$. Furthermore, let $B \subseteq D$ and suppose that B is partitioned in k connected sets B_i ($i \in [1, \ldots, k]$). Then, the geodesic influence zone of the set B_i within D $iz_D(B_i)$ is computed as:

$$iz_D(B_i) = \{p \in D \mid \forall j \in [1 \ldots k]/\{i\} : d_D(p, B_i) < d_D(p, B_j)\}, \quad (10.2)$$

where $d_D(p, B_i) = \min_{q \in B_i} d_D(p, q)$ with $d_D(p, q)$ as the minimum path among all paths within D between p and q. The union of the geodesic influence zones of the connected components B_i, $IZ_D(B)$, is computed as follows:

$$IZ_D(B) = \sum_{i=1}^{k} iz_D(B_i) \qquad (10.3)$$

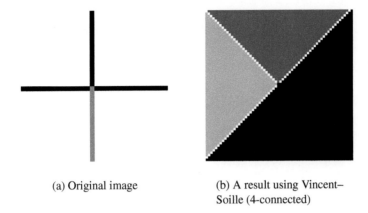

(a) Original image   (b) A result using Vincent–
                            Soille (4-connected)

Figure 10.2 An image and a segmentation result using the Vincent–Soille algorithm (4-connected)

The Watershed Transform by immersion is defined by the following recursion:

$$X_{h_{min}} = \{p \in D | f(p) = h_{MIN}\} = T_{h_{MIN}}$$
$$X_{h+1} = MIN_{h+1} \cup IZ_{T_{h+1}}(X_h), \quad h \in [h_{Min}, h_{Max})$$

(10.4)

where T_h defined as $T_h = \{p \in D \mid f(p) \leq h\}$, is the set of pixels with grey-level smaller than or equal to h, and MIN_h is the union of all the regional minima at level h. The watershed line of D is then the complement of $X_{h_{max}}$ in D.

The Vincent–Soille algorithm, which does not strictly implement equation (10.4) (for details see below and [10]), consists of two steps:

1. The pixels of the input image are sorted in increasing order of grey values.
2. A Flooding step is done, level after level, starting from level h_{min} and terminating at level h_{max}. For every gray value, breadth-first-search is done to determine the label (label of existing basin, new label or watershed label) to be ascribed to the pixel.

The location of regional minima is important for segmentation by Watershed Transform. To extract the objects of interest in the image, the minima should not lie along the border line of the objects (see Figures 10.2(b) and 10.3(b)). On the contrary, regional maxima should be located on the border lines. To correctly identify the minima far from the border lines, in practice

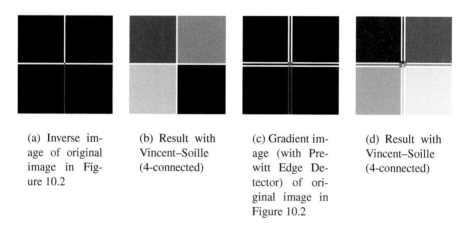

(a) Inverse image of original image in Figure 10.2

(b) Result with Vincent–Soille (4-connected)

(c) Gradient image (with Prewitt Edge Detector) of original image in Figure 10.2

(d) Result with Vincent–Soille (4-connected)

Figure 10.3 An image and the segmentation result using the Vincent–Soille algorithm (4-connected)

3	3	2	2
3	3	1	2
0	0	1	0
0	0	1	0

(a) Original image

W	B	B	B
A	W	B	B
A	A	W	B
A	A	W	B

(b) Result by using theoretical recursion (4-connected)

A	W	W	B
A	A	W	B
A	A	W	B
A	A	W	B

(c) Result by using Vincent–Soille (4-connected)

Figure 10.4 Comparing the results of the theoretical recursion with Vincent–Soille algorithm under condition 1

the gradient image of the original image is used as an input image for the Watershed Transformation (compare Figures 10.2 and 10.3).

The Watershed Transform is also dependent on the connectivity. For example, the gradient image in Figure 10.2 has three regional minima if 4-connectedness is used, but only one minimum if 8-connectedness is used. In the following, we always use 4-connectedness.

An advantage of the Vincent–Soille algorithm is that it runs in linear running time respectively to the number of pixels of the input image (if the used sorting algorithm is linear). To obtain linear computation time, Vincent and Soille modified the theoretical recursion of expression (10.4), so as to access each pixel only once. To this aim, at iteration t, only pixels labeled

2	0	2
0	1	0
2	1	2

W	B	W
A	W	C
A	W	C

W	B	W
A	C	C
W	C	C

(a) Original image

(b) Result by using theoretical recursion (4-connected)

(c) Result by using Vincent–Soille (4-connected)

Figure 10.5 Comparing the results of the theoretical recursion with Vincent–Soille algorithm under condition 2

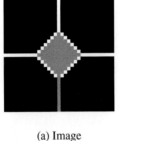 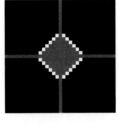

(a) Image

(b) Result (grey: Watershed lines)

Figure 10.6 A segmentation result using the Vincent–Soille algorithm (4-connected)

h are checked, while expression (10.4) requires that pixels labeled less than or equal to h are checked. This modification to the recursion schemes introduces some drawbacks. Specifically, watershed lines thicker than just one pixel can be created and differently labeled basins (see Figure 10.4). A pixel that becomes during an iteration t the watershed label and is neighbor of 2, 4, or more odd-numbered pixels with different labels, can get the wrong label during the same iteration after having visited all adjacent basins (see Figure 10.5).

Another example of thick watershed lines created by the above modification is shown in Figure 10.6.

Theoretically, the results of the Watershed Transformation do not depend on the order in which the neighbors of a pixel are visited. Because of the constraints introduced by Vincent–Soille into the implementation of the algorithm, the order of visitation heavily influences the result. For example, if for a pixel (x, y) its neighbors are visited, in the order

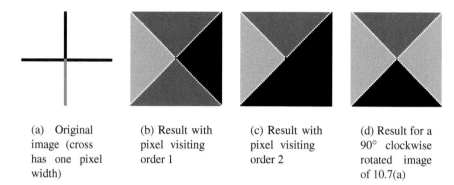

(a) Original image (cross has one pixel width)

(b) Result with pixel visiting order 1

(c) Result with pixel visiting order 2

(d) Result for a 90° clockwise rotated image of 10.7(a)

Figure 10.7 Cross image and segmentation results using the Vincent–Soille algorithm (4-connected) when different pixel visiting orders are used

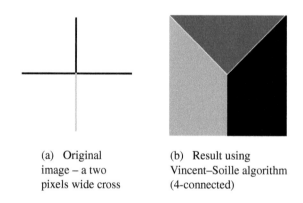

(a) Original image – a two pixels wide cross

(b) Result using Vincent–Soille algorithm (4-connected)

Figure 10.8 A scaled image and the segmentation result using the Vincent–Soille algorithm (4-connected)

$\{(x - 1, y), (x, y + 1), (x + 1, y), (x, y - 1)\}$ (see Figure 10.7(b)), or in the order $\{(x, y + 1), (x - 1, y), (x, y - 1), (x + 1, y)\}$ (see Figure 10.7(c)), different results will be obtained.

This property also explains why the algorithm is not invariant with respect to rotation (see Figure 10.7) and scaling (see Figure 10.8).

The cross in image Figure 10.8 has double the width in comparison with the cross in Figure 10.7. The light-colored pixel at the intersection of the light-colored and dark-colored branches of the cross has now only two darker

A	1	2	3	4	B
A	1	2	3	4	B
A	1	2	3	4	B
A	1	2	3	4	B

A	4	3	2	1	B
A	4	3	2	1	B
A	4	3	2	1	B
A	4	3	2	1	B

(a) Image   (b) Portion of 10.9(a) with the geodesic distance from pixel of the stripe to basin A   (c) Portion of 10.9(a) with the geodesic distance from pixel of the stripe to basin B   (d) Result

Figure 10.9 An image, the geodesic distances from pixel of the stripe to the different basins and the segmentation result using the Vincent–Soille algorithm

neighbors. In such a condition we obtain the correct result for the Watershed Transformation.

Scaling algorithms generally use interpolations, which can produce more regional minima in the rescaled image than in the original image. As a consequence, a larger number of components can be obtained for the scaled image, which will result in a different segmentation from that of the original image.

The Vincent–Soille algorithm does not generate always connected watershed lines. A reason for that is the constraint demonstrated in Figure 10.5. No watershed line separates the two basins A and B, if all the highest grey pixels between the two basins belong to the geodesic influence zone of either A or B. For example, consider Figure 10.9(b), showing an image completely black except for a grey stripe that is four pixels wide. For this image the Vincent–Soille algorithm creates no watershed line between the two basins, because no pixel of the stripe has the same distance to the two basins (compare Figures 10.9(b) and 10.9(c)).

To solve the two above drawbacks, Roerdink and Meijster [10] proposed a slightly modified algorithm, which is, however, remarkably more expensive with regard to computation, since its cost is $O(N^2)$, where N is the number of pixels in the image.

As we already pointed out, watershed segmentation performs better when using the gradient image of the input image. During our study we tested different edge detectors. The Prewitt (with Chessboard Distance) and the Sobel (with Euclidian Distance) edge detectors produced the best results coupled with the use of the new implementation of algorithms [4] with the Vincent–

Soille Watershed Transformation. But neither of them was clearly better than the other. All of the following tests were performed with the Prewitt detector, because we obtained the best results with it when working with our test images.

Furthermore, independently of the selected approach, Watershed Transformation tends to highly oversegment due to many regional minima that can possibly be interpreted as noisy minima. To solve this problem different approaches exist, like intensive preprocessing (e.g. [15]), marker controlled watershed (e.g. [12]) or region-merging (e.g. [2, 3]). For preprocessing in order to reduce the number of local minima, smoothing algorithms or extended edge detectors eliminating unnecessary edges, sharpen edges or produce edges with less gaps are adopted. Often a combination of different preprocessing methods is used. A problem of the majority of the preprocessings methods is the dependence on the result of the particular kind of images. By the marker-controlled watershed a set of regions called markers are used in place of the set of minima. These regions are often manually determined by the user. Therefore this Watershed Transform approach is often qualified for interactive use and less for automation.

To solve the problem of oversegmentation, Frucci [3] combines an iterative computation of the Watershed Transform with processes called digging and flooding. Flooding merges adjacent basins. This is achieved by letting water increase the level, so that it can overflow from one basin into an adjacent one if the level of water is higher than the lowest height along the watershed line separating the two basins. Also digging merges adjacent regions. In this case, to merge a basin A, regarded as non- significant, with a basin B, a canal is dug in the watershed line separating A and B to allow water to flow from B into A. The effect of merging is that the number of local minima found at each iteration diminishes. Flooding and digging are iterated until only significant basins are left.

The algorithm involves expensive computation, but it has the advantage of producing acceptable segmentation results independently of the kind of image. In fact, the approach followed in [3] rather than identifying the watershed lines during the Watershed Transformation, first builds the complete partition of the image into basins and only after that detects the watershed lines by boundary detection. In this way, both drawbacks affecting the Vincent–Soille approach are overcome at the expense of higher computation costs. Therefore, neither the Frucci algorithm presented in [3], nor the extension of the Frucci algorithm introduced in [4], are suitable for real-time processing.

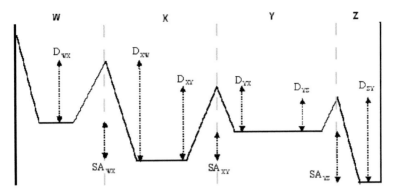

Figure 10.10 Similarity parameter and relative depth

We have used the new implementation of the algorithm [4] for our case-based reasoning studies, since the computation costs are much lower then, due to the use of the Watershed Transform computed by the Vincent and Soille scheme.

Let X and Y be two adjacent basins in the watershed partition of a grey-level image D and W and Z some other basins. As in [3], we denote LO_{XY}, *the local overflow of X with respect to Y*, as the pixel with the minimal grey value along the border line between X and Y. The value of the pixel with the lowest value is the *overflow of X, O_X*. Furthermore, R_X is the grey value of the regional minimum of X, $SA_{XY} = |R_X - R_Y|$ denotes the similarity parameter and $D_{XY} = LO_{XY} - R_X$ defines the relative depth of X at LO_{XY} (see Figure 10.10).

In order to determine if a basin X has to be merged with a basin Y, Frucci et al. [4] introduce the notion of *relative significance of X with respect to Y* and perform the following check:

$$\frac{1}{2}\left(a\frac{SA_{XY}}{At} + b\frac{D_{XY}}{Dt}\right) \geq T, \text{ with } a, b, T \geq 0 \qquad (10.5)$$

where At and Dt are the threshold value (for the automatic computation see [3]) a, b and T constants, and the left-hand side of (10.5) is the relative significance of X with respect to Y.

If condition (10.5) is satisfied, the basin X is significant with respect to Y. The relative significance X is evaluated with respect to all its adjacent basins. Three cases, as shown in Table 10.1, are possible. When merging is possible, Table 10.1 also specifies if digging or flooding has to be performed. After flooding or digging have been performed for all basins that are not

Table 10.1 Cases for the significance of X by taking into account the relative significance with respect to all its adjacent basins

X is denoted	if	type of merging
strongly significant	X is significant with respect to each adjacent region Y.	no merging
non-significant	X is non-significant with respect to each adjacent region Y.	flooding
partially significant	all other cases	digging

strongly significant, the Watershed Transformation is executed again with an obviously smaller number of local minima.

Since we have used the Vincent–Soille Watershed Transformation, which is not invariant for image rotation and scaling, in our implementation of algorithms [4], the best value of a, b and T can be different for two images after scaling or rotation and the chosen image description should reflect this situation.

10.3 Image Description by Central Moments

We studied central moments for the image description. Other image descriptions such as statistical and texture features, image description as marginal distribution for column and lines, image description by similarity between the regional minima of two images have been studied in [1].

An image can be interpreted as a two-dimensional density function. So we can compute the geometric moments

$$M_{pq} = \int \int x^p y^q g(x, y), \quad p, q = 0, 1, 2, \ldots \tag{10.6}$$

with continuous image function $g(x, y)$.

In the case of digital images, we can replace the integrals by sums:

$$m_{pq} = \sum_x \sum_y x^p y^q f(x, y), \quad p, q = 0, 1, 2, \ldots \tag{10.7}$$

where $f(x, y)$ is the discrete function of the gray levels.

The seven-moment invariants from Hu [5] have the property of being invariant to translation, rotation and scale. Since our implementation of an algorithm [4] is not invariant with respect to rotation and scale, the invariant moments are unsuitable for our image description.

We can consider the central moments, which are the only ones translation invariant:

$$m_{pq} = \sum_x \sum_y (x - x_c)^p (y - y_c)^q f(x, y), \quad p, q = 0, 1, 2, \ldots \quad (10.8)$$

where

$$x_c = \frac{\sum_x \sum_y x f(x, y)}{\sum_x \sum_y f(x, y)} \quad \text{and} \quad y_c = \frac{\sum_x \sum_y y f(x, y)}{\sum_x \sum_y f(x, y)} \quad (10.9)$$

For our study we use central moments m_{pq} with p and q between 0 and 3 and as input image the binarized gradient image obtained with the thresholding algorithm of Otsu. To determine the similarity between two images we use the normalized city-block distance.

Figure 10.11 shows the dendrogram of this test. If we virtually cut the dendrogram by the cophenetic similarity measure of 0.0129, we obtain the following groups G1=Monroe, parrot, G2=neu3, neu4, G3=gan128, G4=neu4_r180, G5=neu2, G6=neu1, and G7=cell.

Compared to the statistical and texture features for image description we obtain one more group, but neu4_r180 gets separated from neu4. The resulting groups seem to represent better the relationship between the image characteristic and the image segmentation parameters.

Problematic is only neu3, which appears in the same group as neu4. Figure 10.12(b) shows the best possible segmentation result and in Figure 10.12(c) the segmentation result with the parameter (0.75, 2, 1). Figure 10.12(c) is more oversegmented than Figure 10.12(b). To get a better result, we can use a threshold for small edges.

CMDescript is not rotation and scale-invariant. The obtained groups for this description are:

CMDescript
G1={Monroe, parrot}, G2={neu3, neu4}, G3={gan128}, G4={neu4_r_180}, G5={neu2}, G6={neu1}, and G7={cell}.

In conclusion, we can say that the central moments can be a good image description for our problem if we accept that rotated and rescaled image can make up a new case.

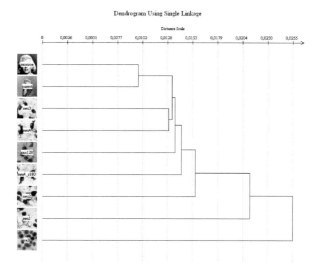

Figure 10.11 Dendrogram for CBR based on central moments (on binary gradient image)

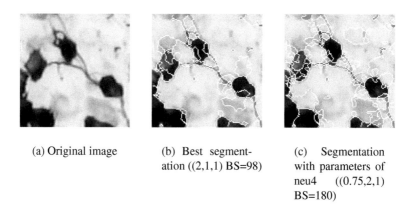

(a) Original image

(b) Best segment-
ation ((2,1,1) BS=98)

(c)   Segmentation
with parameters of
neu4   ((0.75,2,1)
BS=180)

Figure 10.12 An image and a segmentation result using the Vincent–Soille algorithm (4-connected)

10.4 Conclusion

We have proposed a framework for model building in image processing by meta-learning based on case-based reasoning. For such a model building process we need a good image description that allows to describe the signal/image characteristics in relationship to the signal-processing parameters.

To find such a suitable image description requires first to study the theoretical, structural, and syntactical behavior [14] of the chosen image processing algorithm. Based on this analysis we can propose several signal/image descriptions. The selected image description should summarize the cases into groups of similar cases and map these similar cases to the same processing parameters. Having found groups of similar cases, these should be summarized by prototypes that allow fast retrieval over several groups of cases. This generalization process should allow building up the model over time based on the incrementally obtained data stream.

Our theoretical study of different algorithms of the Watershed Transformation showed that the WT produces different results if the image is rotated or rescaled. The particular implementation of the algorithm puts constraints on the behavior of the algorithm. As a result of our study we conclude that we need an image description that describes the distribution of the regional minima and that is not invariant against rotation and scaling.

We studied central moments as image description in this work. As result we can conclude that the central moments can be a good image description for the Watershed Transformation. These image descriptions seem to well represent the relationship between the image characteristics of the particular image and the segmentation parameters. Cases having the same segmentation parameters could be grouped based on the image descriptions into the same group. This will make possible a generalization on these groups of cases that will lead to a complete image-segmentation model. However, to automatically choose the right weight values for the statistical and texture feature description is up to now an open question.

References

[1] A. Attig and P. Perner. A study on the case image description for learning the model of the watershed segmentation. *Tran. CBR*, 2(1):41–53, 2009.

[2] A. Bleau and L.J. Leon. Watershed-based segmentation and region merging. *Computer Vision and Image Understanding*, 77(3):317–370, 2000.

[3] M. Frucci. Oversegmentation reduction by flooding regions and digging watershed lines. *IJPRAI*, 20(1):15–38, 2006.

[4] M. Frucci, P. Perner, and G. Sanniti di Baja. Case-based-reasoning for image segmentation. *IJPRAI*, 22(5):829–842, 2008.

[5] Ming-Kuei Hu. Visual pattern recognition by moment invariants. *IEEE Trans. Information Theory*, 8:179–187, 1962.

[6] P. Perner. An architecture for a CBR image segmentation system. *Engineering Applications of Artificial Intelligence*, 12(6):749–759, 1999.

[7] P. Perner. Why case-based reasoning is attractive for image interpretation. In D.W. Aha and I. Watson (Eds.), *ICCBR*, volume 2080 of Lecture Notes in Computer Science, Vol. 2080, pages 27–43. Springer, 2001.

[8] P. Perner. Case-based reasoning for image analysis and interpretation. In C. Chen and P.S.P. Wang (Eds.), *Handbook on Pattern Recognition and Computer Vision*, 3rd Edition, pages 95–114. World Scientific Publisher, 2005.

[9] P. Perner. Case-based reasoning and the statistical challenges. *Quality and Reliability Eng. Int.*, 24(6):705–720, 2008.

[10] J.B.T.M. Roerdink and A. Meijster. The watershed transform: Definitions, algorithms and parallelization strategies. *Fundamenta Informaticae*, 41:187–228, 2001.

[11] C. Soares, P. Brazdil, and P. Kuba. A meta-learning method to select the kernel width in support vector regression. *Machine Learning*, 54(3):195–209, 2004.

[12] P. Soille (Ed.). *Morphological Image Analysis, Principles and Applications*. Springer, Berlin, 2004.

[13] L. Vincent and P. Soille. Watersheds in digital spaces: An efficient algorithm based on immersion simulations. *IEEE Trans. Pattern Analysis and Machine Intelligence*, 13(6):583–598, 1991.

[14] R. Vogt. *Die Systemwissenschaften – Grundlagen und wissenschaftstheoretische Einordnung*. Haag-Herchen Verlag, Frankfurt am Main, 1983.

[15] D. Wang. A multiscale gradient algorithm for image segmentation using watershelds. *Pattern Recognition*, 30(12):2043–2052, 1997.

[16] G. Wunsch. *Systemtheorie*. Akademische Verlagsgesellschaft, Leipzig, 1975.

11

Image Segmentation: The View from Inside the Image

Ajay Mishra and Yiannis Aloimonos

Computer Vision Laboratory, Institute of Advanced Computer Studies, University of Maryland, College Park, MD 20742, USA; E-mail: mishraka@umiacs.umd.edu, yiannis@cs.umd.edu

Abstract

Humans observe and understand a scene/image through a series of fixations. Each fixation point lies inside a particular region of arbitrary shape and size in the scene which can either be an object or a part of it. We define the basic segmentation problem as segmenting that region containing the fixation which is equivalent to finding the enclosing contour – a connected set of boundary edge fragments in the edge map of the scene – around the fixation. This enclosing contour should be a depth boundary. In the first part of the chapter, we present a novel algorithm to find this bounding contour and achieves the segmentation of one object, given the fixation. The proposed segmentation framework combines monocular cues (color/intensity/texture) with stereo and/or motion, in a cue independent manner. Our approach is different from current approaches. While existing work attempts to segment the whole scene at once into many areas, we segment only one image region, specifically the one containing the fixation point. Experiments with real imagery collected by our active robot and from the known databases demonstrate the promise of the approach. After this description of a bottom up segmentation, in the second part of the chapter, we investigate the role of prior object knowledge in achieving segmentation. Clearly, segmentation and recognition have the form of chicken-egg problems. It appears as if in order

Patrick Shen-Pei Wang (Ed.), Pattern Recognition and Machine Vision – In Honor and Memory of Professor King-Sun Fu, 165–179.

to recognize one first needs to segment. But segmenting (and simultaneously recognizing) an object we are searching for can be facilitated by knowledge of an object model. We present a novel approach for segmenting a known object in challenging inputs containing clutter, and we discuss the relationship of our approach to the seminal work of Fu on syntactic approaches to image segmentation.

Keywords: middle level vision, segmentation, recognition, fixation, active vision, attention.

11.1 Introduction: What Is Visual Segmentation?

The problem of image segmentation has occupied many disciplines. Vision scientists, philosophers and engineers have worked on it for many years, with very interesting results. The problem still remains open and its solution represents an enabling technology. Progress in segmentation means progress in a large variety of areas.

But what does it really mean to segment an image or part of it? Is segmentation a bottom up process resulting from analysis of the perceptual input without any specialized knowledge about the world? Or is it perhaps a top-down process where we recognize the object and thus we simultaneously segment it? Or could it be a synergistic process where bottom up processes interact with top down attention mechanisms? Questions like these remain a challenge even today.

The most prominent definition of segmentation in the literature amounts to dividing the image into regions with some homogeneous property. This is a very general definition and it clearly includes the case where the system processes only a single image. A large part of today's literature is devoted to segmenting single images. This is not surprising as many applications in todays environment are driven by image databases and that field would certainly benefit from single image segmentation and other visual computing. It is nevertheless astonishing that the capability of segmenting a single image is solely a human (primate) capability – only humans appear to segment and "understand" pictures (because only when we see a photograph we see only one image).

Most of the machine vision systems of our immediate future however, like the cognitive systems being built in Europe under the 7th Framework Program or the Cyber physical Systems sponsored by the National Science Foundation in the US, do not look at single images, just like biological systems. They are active agents. They do not "see", they look. They explore and they change the geometric parameters of their sensory apparatus. Their eyes converge or diverge. In general they receive binocular video and fixate at different parts of the scene [1,4,5]. Because of that, such systems are able to understand the spatial layout of the scene, they can understand boundaries and occlusions. This gives us the freedom to revise the definition of segmentation, to a more concrete and physical one.

We define segmentation as the division of the image (or view) into regions that correspond to different surfaces. Thus, the boundaries of the segmentation will be depth boundaries. Note that this new definition is sound because every object occupies some volume of space and has boundaries. One may of course find a picture on the wall where the boundaries of the segment (the picture) are not depth boundaries, but for the majority of the objects in our world it is certainly true that segmentation boundaries are also depth boundaries.

11.2 Segmentation with and without Knowledge of Objects

Note that the above definition is the best one can do without any prior knowledge of object models. The prior knowledge in this case is generic – it is knowledge about visual surfaces and their boundaries. If however knowledge about object models is available, then the situation is different. Such knowledge can be incorporated into the segmentation process so that object detection and recognition can happen simultaneously.

One of the success stories in Active Perception and navigation is SLAM Simultaneous Localization and Mapping, a successful example of loop-closing in another domain. We propose to do the same for the recognition problem and treat it together with segmentation. Traditionally, it has been thought that segmentation provides input to recognition (in a bottom up sense). In contrast, we perceive object models (memory, context) as another cue that contributes to segmentation, like the cues of color and texture, stereo and motion.

The organization of the chapter reflects this treatment of bottom up active segmentation and top down segmentation using object models.

11.3 Prior Art

There is a huge literature on segmenting regions in images and videos. All segmentation algorithms depend upon some form of user inputs, without which the definition of the optimal segmentation of an image is ambiguous. Also, these algorithms always segment the entire image at once. There are two broad categories: first, the segmentation algorithms [11, 19, 20] that need the user-specified global parameters such as the number of regions and thresholds to stop the clustering; second, the interactive segmentation algorithms [3, 7, 18] that always segment the entire image into only two regions: foreground and background. Boykov and Jolly [7] posed the problem of foreground/background segmentation as a binary labeling problem which is solved exactly using the maxflow algorithm [8]. It, however, requires users to label some pixels as foreground or background to build their color models. Blake et al. [6] improved upon [7] by using a Gaussian mixture Markov random field to better learn the foreground and background models. Rother et al. [18] required users to specify a bounding box containing the foreground object. Arbelaez and Cohen [2] required a seed point for every region in the image. For foreground/background segmentation, at least two seed points are needed. Although these approaches give impressive results, they cannot be used as an automatic segmentation algorithm as they critically depend upon the user inputs. Yu and Shi [22] tried to automatically select the seed points by using spatial attention based methods and then used these seed points to introduce extra constraints into their normalized cut based formulation.

Unlike the interactive segmentation methods mentioned above, in [3, 21] only a single seed point is needed from the user. Veksler [21] imposed a constraint on the shape of the object to be a star, meaning the algorithm prefers to segment the convex objects. Also, the user input for this algorithm is critical as it requires the user to specify the center of the star shape exactly in the image. In [3] only one seed point is needed to be specified on the region of interest and segment the foreground region using a compositional framework. But the algorithm is computationally intensive. It runs multiple iterations to arrive at the final segmentation.

11.4 Active Segmentation without Knowledge of Object Models

Consider an active visual agent, like the one of Figure 11.1, consisting of a platform and an active head/eye system consisting of four cameras. The

Figure 11.1 Our robot with a Quad camera vision system mounted on its top. The arrow indicates the line of sight as it fixates on an object in the scene

system is capable of moving in its environment and exploring the scene in view. Is our definition of segmenting the scene into regions separated by depth boundaries a good one? We know that when humans look at a scene, they do not segment the whole scene at once. Rather, they only segment the region on which they fixate. In fact, the structure of the human retina is such that only the small neighborhood around the fixation points is captured in high resolution by the fovea, while the rest of the scene is captured in lower resolution by the sensors on the periphery of the retina. This is the well known figure vs ground problem [10,16]. Why then would we want our segmentation algorithms to segment the whole scene? Is this not too ambitious? The human visual system observes and understands a scene/image by making a series of fixations. Every "fixation point" lies inside a particular region of interest in the scene which can either be an object or just a part of it. Our robot can make fixations as well. In this chapter, we define segmenting the region containing the fixation point as a basic segmentation problem. Since the early attempts on Active Vision, there has been a lot of work on problems surrounding fixation, both from a computational and psychological perspective [1,4,5,9,17]. Despite all these developments however, the operation of fixation never really made it into the foundations of computational vision. Specifically, the fixation point has not become a parameter in the multitude of low and middle level operations that constitute a big part of the visual perception process. This is the avenue we pursue in this paper. It is only natural to make fixation part and parcel of any visual processing. First fixate, then segment the surface containing the fixation point.

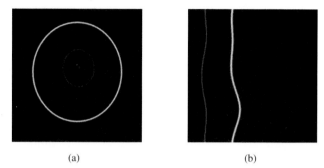

(a)   (b)

Figure 11.2 Left: two circles of different sizes and different pixel intensities around the fixation point F. Right: the transformed polar map of the left image with F (the red dot at the center) as the pole

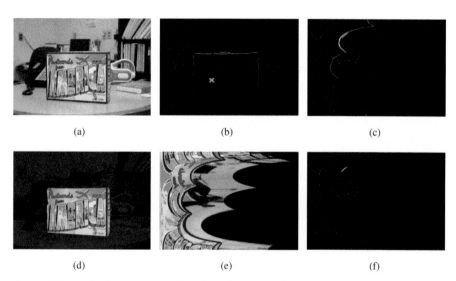

(a)   (b)   (c)

(d)   (e)   (f)

Figure 11.3 (a) An image captured by the robot shown in Figure 11.1. The fixation point is shown by the symbol "X". (b) The final probabilistic boundary edge map as obtained in Section 11.4.2. (c) The polar image of the boundary edge map for the fixation. (d) The polar edge map with the optimal path shown by the blue curve. (e) The color image after the polar transformation with the optimal path (the blue curve) superimposed on it. (f) The region segmented by our algorithm containing the fixation point

11.4.1 Our Approach: Polar Space Is the Key!

Figure 11.3a shows an image that our semantic robot sees in the room when it is summoned to find objects (fixation is shown by the symbol "X"). Figure 11.3b shows the boundary edge map of the image provided by the semantic edge detector [14] which has learned from an annotated database how the edge at depth boundary looks like. Thus, in Figure 11.3b, edges along the depth boundaries are bright, whereas the texture (internal) edges are dim. Brightness of the pixel indicates its probability to be at a depth boundary. We must emphasize that this is not all we have: we also have a disparity stereo map and an optical flow field at each time stamp to further improve this boundary edge map as discussed in Section 11.4.2. However, for now let us concentrate on a single image.

The goal is to find the region containing the fixation which is equivalent to finding the enclosing closed contour around the fixation. But as we can see in Figure 11.3b, there are many ways to enclose the fixation point due to the presence of internal contours. Thus, we define the "optimal" contour as the one with minimum cost where the cost of tracing an enclosing contour is an accumulated cost of adding all of its edge pixels. The cost depends upon on the number of edge pixels in the contour or the length of the contour in the Cartesian space. So, tracing contour with this cost in the Cartesian space will inherently prefer small contours over long ones. It is important to normalize the lengths of all possible contours before comparing their cost to find the "optimal" contour.

So, let us consider two enclosing contours around the fixation F, as shown in Figure 11.2a, of length 40 and 100 pixels and of constant brightness 150 and 200 respectively, where 255 is the maximum brightness. The accumulated cost of tracing the small and the long contour is $4200 (= (255 - 150) \times 40)$ and $5500 (= (255 - 200) \times 100)$ respectively. The small contour costs less and hence will be declared optimal. However, the long contour is brighter than the short contour and should actually be the optimal boundary around the fixation point F. But, for that to happen, lengths of the possible closed paths around the fixation should be normalized such that their accumulated cost does not depend on their length.

To achieve this, the boundary edge map, given in Figure 11.2a, is transformed from the Cartesian co-ordinate system to the polar co-ordinate system with fixation as the pole. (Our convention is that the angle ($\theta \in [0°, 360°]$) is represented along the vertical axis and increases from top to bottom and the radial distance (r) is represented along horizontal axis and increases from left to right.) Now, in the polar space (Figure 11.2b), both enclosing contours become the open curves of normalized length (360 pixels). The brightness of the edge pixels in the polar space remain unchanged. The cost of tracing the dimmer contour and the brighter contour is $3600 (= 360 \times (255 - 150))$ and $1800 (= 360 \times (255 - 200))$ respectively. The brighter contour now costs less and becomes the optimal closing contour around the fixation point when transformed back to the Cartesian co-ordinate system.

In this chapter, we propose a two step process to segment the region for a given fixation: first, the boundary edge map of the image/scene is generated by using all available low level cues; second, the edge map is transformed to the polar space with the fixation as the pole. The optimal path passing through the polar edge map is found using two different techniques, one based on graph cut [15] and the other based on a shortest path algorithm.

11.4.2 Calculating the Boundary Edge Map by Combining Cues

To be able to correctly trace the depth boundary around the fixation point, all the edge pixels along the depth boundary in the edge map should be bright and the rest of the edge pixels should be all dim. But, the boundary edge maps given by Martin et al. [14] have some strong (or brighter) internal edges and some weak (or lighter) boundary edges. We can use stereo and/or motion cues to modify the boundary edge map such that the boundary edges are almost always brighter than the internal edges. It is well known that there is a greater step change in the disparity or flow at the boundary edges than at the internal edges. This can be used to differentiate the boundary edges from the internal edges.

We break our boundary edge map into straight line segments and select two rectangular regions parallel to the segment on its both sides. The average disparity and/or average flow is calculated for these two regions. The difference between average disparity and the magnitude of average flow in the two regions are calculated and represented by ΔD and ΔF. As the flow or disparity values are corrupted at the depth boundary in the image, it is important that the rectangular regions selected on both sides should not be immediately close to the edge. Hence, it should be at some distance away from the boundary to avoid sampling corrupted values. The brightness of the edge segment should be changed according to the amount of change in the disparity or flow magnitude. Clearly, if the difference in the values on its both sides is large, the edge segment is at the boundary. A new boundary edge map is created by changing the intensity of the edge pixels to ΔD or ΔF.

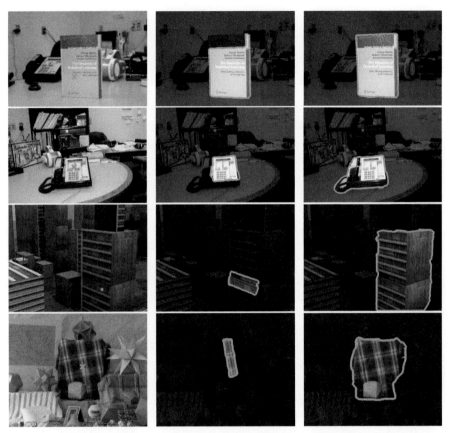

Figure 11.4 Column 1: the original images with fixations (the symbol "X"). Column 2: Our segmentation results for the fixation using monocular cues only. Column 3: Our segmentation results for the same fixation after combining motion or stereo cues with monocular cues

After doing so for all the edge segments, the boundary edge map is scaled to have intensity values between 0 and 1. This new boundary edge map resulted from using stereo or motion cues is combined linearly with the original color and texture based boundary edge map to give an improved boundary edge map (Figure 11.3f) wherein the internal edge are dim and the boundary edges are bright. With the improved boundary edge map, our algorithm traces the real boundary of the region containing the fixation (Figure 11.3d). In Figure 11.4, we show several other examples where optical flow or stereo is used to improve the edge map which results in the desired segmentation output.

The source code of the implementation of this algorithm can be found at http://www.umiacs.umd.edu/~mishraka/activeSeg.html.

11.5 Segmentation Using Object Models: Where Is Waldo?

For simple objects like apple, a single region seems sufficient to represent it completely. On the other hand, we need multiple segments to represent complex objects. For example, an image of the side view of a car needs two regions for the windows, two regions for the wheels, and one region for entire car to sufficiently describe it. Such a description can be developed for individual objects which is then used to detect and segment it if present in a new scene or image.

This description has two components. First, the object part, which is a closed region, is described by the log-polar representation of the its boundary contour with respect to the fixation (the centroid of the region); Second, the relative spatial information among all the parts of the object needs to be learned. It constitutes learned the distribution of relative displacement vectors between the centroids of all pairs of regions (parts) and their relative scale ratio. We use Gaussian functions to learn these distributions.

Now, the question becomes how do we use this information to segment the object we know? Again, for simple objects, the solitary region representing it will be segmented based on the low-level information only. But, for the complex objects, the segmentation process will be carried out in a sequence of steps involving feedback between the segmentation and recognition modules.

We start with a number of fixations at different locations in the image. Then, our segmentation algorithm [15] outputs a region for each of the chosen fixation point. The log-polar representation of the contour of that region is matched with that of all the regions representing different object parts learned during the training phase. If a match is found, an hypothesis is made that the entire object is present at that location. Now, to prove the hypothesis correct, the other parts of that object should also be present at the locations and scales which ensure structural consistency with the detect object. If any of the other part is not detected, the object is either declared "not detected" or partially detected.

After the detection of the first part of the object, detecting other parts of that object is a constrained problem. The scale and the centroid of the first part already detected are used to estimate the centroid and the scale of the other parts. Now, to detect the other parts, the distribution of their shapes learned during training is used to bias the edge map of the scene. This

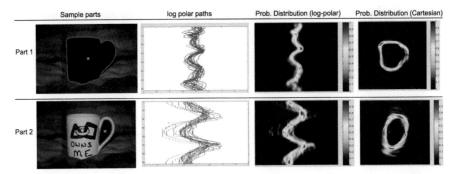

Figure 11.5 Row 1 and Row 2 correspond to the log polar representations and related distribution for the two parts of the object (Mug) from the ETHZ database [12]. Column 1 has samples of the parts with their centroids marked by the symbol "X". Column 2 has collection of all log-polar paths from the training set. Column 3 shows the probabilistic distribution of the scale normalized log-paths built using a non-parametric method, parzen window technique. Column 4 shows the probabilistic distribution in the Cartesian space. The same scale is chosen for both parts to show the distribution in the Cartesian space

means the edge fragments lying around the expected contour of that object are weighed up and the rest weighed down. The biased probabilistic boundary edge map is then used to segment the part with its centroid as the fixation. See Figure 11.7 for an example of how the two parts of a mug are detected using this sequential process. Finally, if the resulting segmentation has the expected shape, the part is declared and thus validates further the hypothesis about the presence of the object.

In this work, we analyze three of the five objects from the ETHZ shape database namely Mugs, Bottles, and AppleLogos. We manually segment the images of these objects into two parts. See Figure 11.7 for the samples of the two parts of these objects. We collect all the regions corresponding to the object part and build a probabilistic distribution of the log-polar representation of their closed contours. The log-polar representation is also normalized with the scale $\sqrt{area/\pi}$ of the corresponding regions to obtain scale-invariant representation (see Figure 11.5). A kernel based non-parametric technique is used to build the distribution of the log-polar paths for a part.

The relative arrangement between object parts is captured using the displacement vectors between the centroids of a pairs of object parts, and their relative scales. The structural information is stored in the memory which is accessed to estimate the scale and the centroid of the other parts of an object.

(a)   (b)   (c)   (d)

Figure 11.6 (a) Image with consecutive fixations. (b) Segmentation output for the first fixation on the low-level cue alone. (c) Segmentation output for the second fixation using the knowledge of how the mug handle looks like. (d). The final segmentation of the mug as it finds the two closed regions that define it at the expected locations and of expected scales in relation to each other

Figure 11.7 Columns 1, 2, and 3 contain the detection/segmentation results of Mugs, Bottles and Applelogos from the ETHZ database respectively. The two closed contours show the two object parts

11.6 The Compositionality of Vision: Brain Mind Cycles

To make major advances, we need to take the presence of feedback seriously both for biological and machine vision. Feed forward processing is the exception in the nervous system rather than the rule. We now know that the idea that a neuron works in a unidirectional manner – performing a weighted sum of its inputs – is an illusion. The spikes of a cortical pyramidal cell have been seen to extend into its dendritic tree. Thus, through the action of voltage-dependent channels, they affect the integration of signals from synapses. But feedback has proved difficult to tackle experimentally.

How do computational studies incorporate feedback? To begin with, in biological vision there is an all-encompassing cycle that links perception with action. Then there are large re-entrant connections relating different levels in the visual hierarchy, such as between object recognition in inferotemporal cortex and early vision in V4. And finally there are smaller re-entrant circuits, like the ones between and among V1, V2, V4, and MT. If any brain area we study is potentially part of a large cycle involving cognition and recognition, then it becomes hard to take feedback into account, because we would need a model of virtually the entire visual system. This has led many of today's computational and experimental vision scientists to shift their focus to the problem of learning and attention. It makes sense to seek general mechanisms and ignore the architecture of the brain if it appears hopeless to model the whole process.

In this chapter, however, we took a different approach, by combining the bottom up processes with the top down attentional knowledge of object models. Rather than ignoring re-entrant brain and computer vision architectures, we focused first on biological and computational processes that build a 3D model of the visual world. This is the subsystem that estimates depth and motion boundaries and creates what appears to be a segmentation of the scene into different surfaces from the cues of stereo, motion, intensity/color and texture. This subsystem is primordial and precognitive - it does not require substantial guidance from higher order representations, allowing us to model it as a physical and mathematical problem. This is our bottom up approach.

This does not mean of course that the object recognition system is not interconnected with the surface segmentation substrate. On the contrary: recognition processes bring top down knowledge into the segmentation process, as we demonstrated in Section 11.4. This knowledge is exhibited by a series of fixations whose goal is to extract closed contours denoting an object. Those fixations, together with relative scale and spatial location information, constitute an active object model.

11.7 Conclusion

We proposed here a novel formulation of bottom up segmentation in conjunction with fixation, as well as a top down approach where one learns models of objects and uses them to achieve segmentation (looking for an object in a scene). Our bottom up framework combines monocular cues with motion and/or stereo to disambiguate the internal edges from boundary edges. The approach is motivated by biological vision and it may have connections to neural models developed for the problem of border ownership in segmentation. Although the framework was developed for an active observer, it applies to image databases as well, where the notion of fixation amounts to selecting an image point which becomes the center of the polar transformation. Our contribution here was to formulate an old problem – segmentation – in a different way and show that existing computational mechanisms in the state of the art computer vision are sufficient to lead us to promising automatic solutions. In addition, attentional mechanisms that guide fixation selection become a component of the problem. Our top down framework concentrated on using learned object models to facilitate the segmentation of those objects in general scenes. Our approach amounts to some form of parsing the contour in the image, to make sure that the detected object is from the learned class. Although our techniques have a statistical flavor, we are one step away from creating a grammar describing the object. Creating such grammatical frameworks has been the quest and vision of the late Professor K.S. Fu [13] and constitutes one of our future research goals. The advances in low and middle level visual processing between the 1980s and now, make this vision possible.

References

[1] J. Aloimonos, I. Weiss, and A. Bandyopadhyay. Active vision. *International Journal of Computer Vision*, 1(4):333–356, January 1988.

[2] P. Arbelaez and L. Cohen. Constrained image segmentation from hierarchical boundaries. In *CVPR*, pages 454–467, 2008.

[3] S. Bagon, O. Boiman, and M. Irani. What is a good image segment? A unified approach to segment extraction. In *ECCV*, Vol. 5305, pages 30–44, 2008.

[4] R. Bajcsy. Active perception. *Proc. IEEE, Special Issue on Computer Vision*, 76(8):966–1005, August 1988.

[5] D.H. Ballard. Animate vision. *Artificial Intelligence Journal*, 48(8):57–86, August 1991.

[6] A. Blake, C. Rother, M. Brown, P. Perez, and P. Torr. Interactive image segmentation using an adaptive GMMRF model. In *ECCV*, pages 428–441, 2004.

[7] Y.Y. Boykov and M.P. Jolly. Interactive graph cuts for optimal boundary and region segmentation of objects in N-D images. In *ICCV*, Vol. I, 105–112, 2001.

[8] Y. Boykov and V. Kolmogorov. An experimental comparison of min-cut/max-flow algorithms for energy minimization in vision. *T-PAMI*, 26:359–374, 2004.

[9] K. Daniilidis. Fixation simplifies 3D motion estimation. *Computer Vision and Image Understanding*, 68(2):158–169, 1997.

[10] J.-O. Eklundh. Vision in robotics: How a robot can segment figure from ground. In *ECAI*, pages 689–693, 1998.

[11] P.F. Felzenszwalb and D.P. Huttenlocher. Efficient graph-based image segmentation. *International Journal of Computer Vision*, 59(2):167–181, 2004.

[12] V. Ferrari, T. Tuytelaars, and L. Van Gool. Object detection by contour segment networks. In *Proceeding of the European Conference on Computer Vision*, Lecture Notes in Computer Science, Vol. 3953, pages 14–28. Springer, June 2006.

[13] K.S. Fu. Syntactic models for image analysis. In *DAGM-Symposium*, pages 271–295, 1981.

[14] D. Martin, C. Fowlkes, and J. Malik. Learning to detect natural image boundaries using local brightness, color and texture cues. *T-PAMI*, 26(5):530–549, May 2004.

[15] A. Mishra and Y. Aloimonos. Active segmentation with fixation. In *ICCV*, 2009.

[16] P. Nordlund and J.-O. Eklundh. Real-time maintenance of figure-ground segmentation. In *ICVS '99: Proceedings of the First International Conference on Computer Vision Systems*, London, UK, pages 115–134, 1999. Springer-Verlag.

[17] K. Pahlavan, T. Uhlin, and J.-O. Eklundh. Dynamic fixation and active perception. *International Journal of Computer Vision*, 17(2):113–135, 1996.

[18] C. Rother, V. Kolmogorov, and A. Blake. "Grabcut": Interactive foreground extraction using iterated graph cuts. *ACM Trans. Graph.*, 23(3):309–314, 2004.

[19] J. Shi and J. Malik. Normalized cuts and image segmentation. *T-PAMI*, 22(8):888–905, 2000.

[20] Z.W. Tu and S.C. Zhu. Mean shift: A robust approach toward feature space analysis. *T-PAMI*, 24(5):603–619, May 2002.

[21] O. Veksler. Star shape prior for graph-cut image segmentation. In *ECCV (3)*, pages 454–467, 2008.

[22] S.X. Yu and J. Shi. Grouping with bias. In *NIPS*, 2001.

12

3D Average-Shape Atlas of Drosophila Brain and Its Applications

Guan-Yu Chen and Yung-Chang Chen

Department of Electrical Engineering, National Tsing Hua University, 101, Sec. 2, Kuang Fu Rd., Hsinchu 30013, Taiwan, ROC; E-mail: cgu@benz.ee.nthu.edu.tw, ycchen@ee.nthu.edu.tw

Abstract

If all connections among neurons could be mapped, the resulting spatial and temporal circuit diagrams would help us find mechanisms underlying the brain's operations. A pattern recognition system for this purpose, which usually involves data acquisition and data representation, is under construction. In this chapter, we introduce a new idea of creating a template to be an atlas which has average anatomic shapes and their positions, instead of probability maps only.

Keywords: neuron, template, atlas, shape.

12.1 Introduction

One of the most important and formidable challenges of neuroscience is finding mechanisms underlying the brain's operations. If all connections among neurons could be mapped, together with genes expressed in these neurons at different times, the resulting spatial and temporal circuit diagrams should reveal some information.

Advances in imaging and molecular tagging are opening up exciting new ways to visualize gene expression patterns and follow the process of protein

Patrick Shen-Pei Wang (Ed.), Pattern Recognition and Machine Vision – In Honor and Memory of Professor King-Sun Fu, 181–195.

interactions in time and space. Due to the complexity and variety of inform-ation, it is necessary to integrate this multidimensional information in a way that will help biologists find meaningful relationships and formulate creative hypotheses that can be tested by further observations and experiments.

In the past, neuron segmentation and classification were done by experts who spent a lot of time and manpower. With the remarkable advances in com-puter, it is possible to teach machines how to distinguish neurons of interest and how to analyze from large amount of data.

12.2 Standard Coordinate

A pattern recognition system usually involves data acquisition and data rep-resentation. In order to help biologists gather data from different experiments, which only get several neurons in each experiment, it is necessary to build up a universal representation system to describe these data. With data described by the same representation system, analysis, classification, and identification can be done by biologists or computers.

A universal coordinate is one kind of data representation system. By pla-cing each circuit from different experiments into the universal coordinate, we can acquire the spatial and temporal relationships among neurons. After collecting all neurons, the final spatial and temporal circuit diagrams should reveal mechanisms underlying the brain's operations. For this purpose, a reference template or atlas is needed to act as a universal coordinate which experiment data can be warped or projected into.

A probabilistic atlas has been proposed as a representation system for the honeybee brain [1]. It provides a boundary for statistical confidence instead of an absolute anatomic shape and its position. The probabilistic atlas is versatile and suitable for distinguishing normal brains from abnormal ones. But it is not suitable for serving as a common coordinate system. Heisenberg et al. [5] proposed a famous standard Drosophila brain model which is also a voxel-based probabilistic atlas. It is constructed from the superposition of rigid-registered neuropils but no coordinates.

For systematic collections of individual neuronal images warped in a common template, a deterministic reference template which has absolute ana-tomic shapes and their positions is needed instead of the probabilistic atlas. Since the result of warping is highly dependent on the disparity between the template and the individual, anatomic shapes and their positions inside a good reference template should have as small average disparity to all the individu-als as possible. Based on these criteria, a shape-based averaging algorithm for

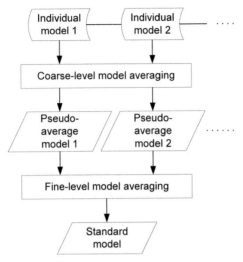

Figure 12.1 The two-level model averaging algorithm

creating the average template is proposed [6]. This new algorithm generates the standard Drosophila brain model for serving as a common coordinate system for the fly brain neurology. The standard Drosophila brain comprises the standard cortex and the standard neuropils which are located within the standard cortex with the average positions and orientations.

12.2.1 Two-Level Model Averaging Techniques

Two-level model averaging algorithm is a hierarchical method shown in Figure 12.1. The coarse-level stage averages the global characteristics of models, such as orientations, positions, sizes, and angles between structures. And the fine-level stage obtains the standard model by determining the average shape of the globally-registered individual models. In order to generate a good average model with precise boundaries, each input neuropil (e.g. cortex, mushroom bodies, and antennal lobe) is labelled by experienced experts. Precise surface model of each neuropil can be constructed from labelled data with some existing algorithms or commercial softwares.

12.2.2 Coarse-Level Model Averaging

Instead of averaging the global characteristics, we try to unify these characteristics at the coarse-level stage to avoid finding them in different samples. The model which has the volume size closest to the average volume size of input models is chosen to be the reference model and the global characteristics of this model is also chosen to be the reference ones. By warping input models to fit the reference one, we can assume that all the global characteristics of input models are unified.

Among existing warping techniques, such as rigid transform, affine transform, and freeform transform, we choose traditional rigid transform which only deals with rotation and translation to unify global characteristics of input models. And the rigid transform is defined as

$$x' = R(x - C) + C + T, \tag{12.1}$$

where $C = [C_x, C_y, C_z]^T$ is the center of rotation and $T = [T_x, T_y, T_z]^T$ is the translation matrix. x is the original position of each point on the surface model and x' is the warped position. R is the rotation matrix which can be represented by the product of three elementary rotation matrixes $R = R_x R_y R_z$, where R_x, R_y, and R_z are defined as

$$R_x = \begin{bmatrix} 1 & 0 & 0 \\ 0 & \cos\theta_x & \sin\theta_x \\ 0 & -\sin\theta_x & \cos\theta_x \end{bmatrix}, \quad R_y = \begin{bmatrix} \cos\theta_y & 0 & \sin\theta_y \\ 0 & 1 & 0 \\ -\sin\theta_y & 0 & \cos\theta_y \end{bmatrix},$$

$$R_z = \begin{bmatrix} \cos\theta_z & \sin\theta_z & 0 \\ -\sin\theta_z & \cos\theta_z & 0 \\ 0 & 0 & 1 \end{bmatrix}. \tag{12.2}$$

The warping algorithm according to rigid transform is trying to minimize the difference between the warped target model and the reference one. The average distance between the warped target model and the reference one is used to measure the difference and is defined as

$$D = \frac{1}{N} \sum_{i=1}^{N} d^2(x_i'), \tag{12.3}$$

where $d(x_i')$ is the shortest distance between x_i', which is a point on the target surface, and the reference model. N is the total number of points on the target surface.

A 3D distance field [4] is generated from the reference model to facilitate the computation of the distance between two models. This field records the

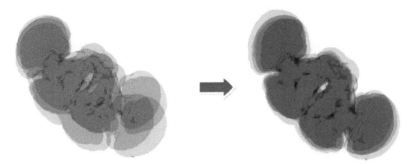

Figure 12.2 The result of rigid transformation. Models fit others in position and orientation

distance from each voxel inside the whole space to its nearest point on the reference surface. When placing the target model inside the 3D distance field of the reference model, distance between the reference and target models can be represented by these values recorded in this field. And the warping algorithm is now modified to minimize the average value from the 3D distance field which is created once.

The warping algorithm which minimizes Equation (12.3) can be considered as a function of variables $\text{var} = [\theta_x, \theta_y, \theta_z, T_x, T_y, T_z]$. The solution of the warping algorithm can be computed with gradient descent method,

$$\text{var}' = \text{var} - \delta \nabla D(\text{var}), \tag{12.4}$$

where δ is the step size, and

$$\nabla D(\text{var}) = \left[\frac{\partial D(\text{var})}{\partial \theta_x}, \frac{\partial D(\text{var})}{\partial \theta_y}, \frac{\partial D(\text{var})}{\partial \theta_z}, \frac{\partial D(\text{var})}{\partial T_x}, \frac{\partial D(\text{var})}{\partial T_y}, \frac{\partial D(\text{var})}{\partial T_z} \right].$$

$$\tag{12.5}$$

Because the rotation matrix is a product of R_x, R_y, and R_z, the variables $\theta_x, \theta_y, \theta_z$ are not independent. We compute each variable individually and iteratively to avoid complex computation and the result is shown in Figure 12.2.

12.2.3 Fine-Level Model Averaging

Fine-level averaging, which is also called shape-based averaging, tries to figure out the average surface from the ones which come from coarse-level averaging and have same global characteristics. The shape-based averaging decides where the average surface should locate. To explain how this works,

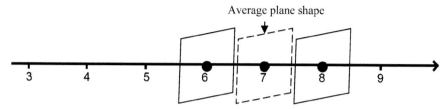

Figure 12.3 The average plane surface of two planar ones

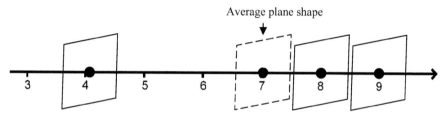

Figure 12.4 The average plane surface of three planar ones

we first take two simple plane surfaces into consideration. The average surface of two simple plane surfaces should be a plane surface also and locate at the middle position between two planes (Figure 12.3). If we add one more plane surface, the location of the average plane should move to the average position of these three surfaces (Figure 12.4).

But what if the input surfaces are not planar ones? We now take three round shapes for example (Figure 12.5). The average one should have average radius. Then, if taking some more complex surfaces into consideration, how do we compute the average position? We now rewrite the solutions of averaging plane and round surfaces into new forms (Equation (12.6)). We can observe that the average surface locates at the position where the sum of shortest distances to each input surface is zero.

$$7 = \frac{6+8}{2} \Rightarrow (6-7) + (8-7) = 0,$$

$$7 = \frac{4+8+9}{3} \Rightarrow (4-7) + (8-7) + (9-7) = 0,$$

$$3 = \frac{1+2+6}{3} \Rightarrow (1-3) + (2-3) + (6-3) = 0. \qquad (12.6)$$

The shape-based averaging is constructed according to this assumption. Fortunately, the 3D distance map, which is mentioned before that can record

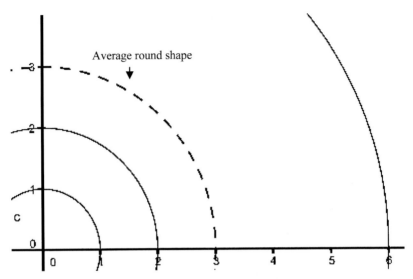

Figure 12.5 The average round surface of three round ones

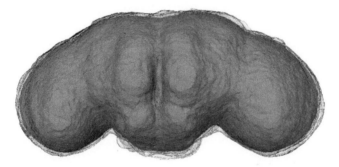

Figure 12.6 The average cortex (opaque) and the superposition of nine pseudo-average cortices (transparent)

the distance from each voxel to its nearest surface point, can be used in Equation (12.6). For convenience, a signed distance map, which records a positive distance for a voxel outside the surface and a negative distance for a voxel inside the surface, is used to help summing the distances. The algorithm of shape-based averaging takes the following steps: (1) Create signed distance map of each input surface model. (2) Cumulatively sum up the signed distance maps and find the zero areas as the average shape (Figure 12.7). And an average surface from nine input cortex surfaces of Drosophila is shown in Figure 12.6.

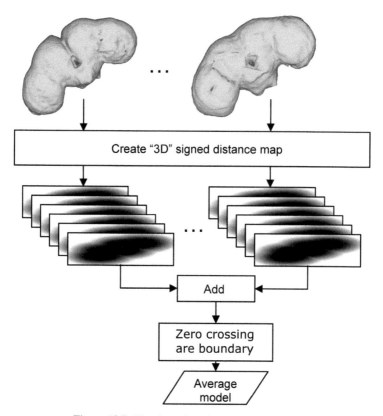

Figure 12.7 The shape-based averaging algorithm

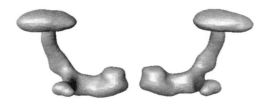

Figure 12.8 The average model of the mushroom bodies which is averaged by the original method

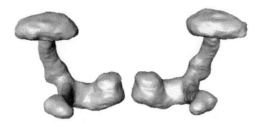

Figure 12.9 The average model of the mushroom bodies which is averaged by the special method

12.2.4 Special Coarse-Level Model Averaging

The shape-based averaging algorithm deals with models which have same global characteristics. In Section 12.2.2, the traditional rigid transform is used to unify the global characteristics, such as rotation and translation. But not all the anatomic structures can be handled very well in this way. Some structures which consist of several small geometric structures, like mushroom bodies, may mismatch with others due to the variations that come from these small structures. Figures 12.8 and 12.9 show two different average surfaces by using the original method and the special method that will be discussed later. Obviously, the average one from the original method is thinner than the other one.

The reason is that the cross area from two mismatched structures is inversely proportional to the variation and a smaller cross area leads to a thinner average shape.

To solve this problem, the model is first separated into several small structures. The same part from each input model is then warped individually. For preserving the original geometry, a more complex transformation [3] is needed to replace original rigid transform.

For mushroom bodies, six lobes are labeled manually. With the concept of unifying the global characteristics discussed in Section 12.2.2, each lobe is analyzed by PCA to get its characteristics which include the principle axis and the length of this axis. Then, these characteristics are averaged and the input models are all warped to fit these average characteristics. Figure 12.10 shows the procedures of creating a pseudo-average model of mushroom bodies.

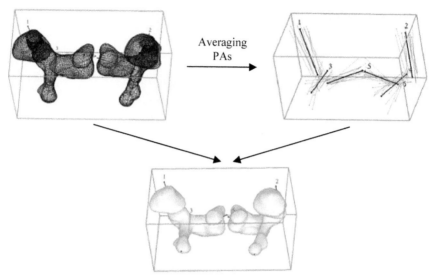

Figure 12.10 Pseudo-average model of the mushroom bodies that has the average characteristics which are calculated by averaging PAs

12.2.5 Position and Rotation Averaging

Though we have constructed the average models of neuropils, it is still not enough to be an atlas. A deterministic reference template should have absolute anatomic shapes and their positions. So, an average brain model should comprise the average cortex and the average neuropils inside the cortex with average locations.

To compute the average locations, each individual dataset should consist of a cortex model, which is used to be a landmark, and neuropil models inside this cortex. The individual cortex is first registered to the average cortex with affine transform and the individual neuropils are then transformed with same transformation. After this procedure, the same neuropils from different datasets are all collected into the same coordinates. In other words, the coordinates of the average cortex are warped into individual dataset.

The average neuropil is then registered to the transformed neuropil with rigid transform to find the position and orientation of this neuropil inside each dataset. Each result can be regarded as a suggestion about the position and orientation of the average neuropil within the average cortex. The flowchart of computing the average position and orientation of one average neuropil is depicted in Figure 12.11.

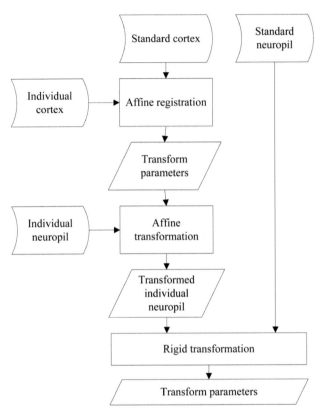

Figure 12.11 The flowchart of finding the suggestion about the position and orientation of an average neuropil

For each average neuropil, the position and orientation in the average cortex can be calculated by averaging the suggestions from input datasets. The average translation can be obtained by

$$\bar{T} = \frac{1}{N} \sum_{i=1}^{N} T_i, \tag{12.7}$$

where N is the number of input datasets. T_i is the translation suggested by the ith individual dataset.

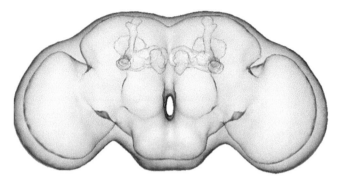

Figure 12.12 The result of combining the average cortex and the average neuropils

To find the average rotation, we solve an optimal $\bar{R}$ that minimizes the following equation:

$$\frac{1}{N}\sum_{j=1}^{N}\left(\frac{1}{m}\sum_{i=1}^{m}\|y_{ji} - \bar{R}x_{ji}\|^2\right),\tag{12.8}$$

where N is the total number of input datasets. y_{ji} is the corresponding point of x_{ji} in the jth individual dataset. And m is the number of points we take into consideration. For the jth individual dataset, the three corresponding points are defined as

$$\begin{cases} x_{j1} = (1, 0, 0)' \leftrightarrow y_{j1} = R_jx_{j1}, \\ x_{j2} = (0, 1, 0)' \leftrightarrow y_{j2} = R_jx_{j2}, \\ x_{j3} = (0, 0, 1)' \leftrightarrow y_{j3} = R_jx_{j3}, \end{cases}\tag{12.9}$$

where R_j is the rotation suggested by the jth individual dataset. Figure 12.12 also shows the result of combining the average cortex and the average neuropils.

12.3 Neuropil Segmentation

Template matching is one of the earlier approaches to pattern recognition. Matching is an operation which is used to measure the similarity between two objects. Therefore, template matching is always used to indentify the same object like the template from input data. The more representative a template is, the better the matching result will be. Based on this idea, the average surface computed from the data can be seen as a good template and can be used in template matching to recognize the same object.

Figure 12.13 A matching result of mushroom bodies

By taking the average Drosophila mushroom bodies from the two-level averaging technique to be a template, an automatic segmentation has been proposed [2]. To measure the similarity, it is better to use edges from image data instead of volume data, because a surface model is constructed from contours of an object. These edges can be extracted by several traditional methods and can be translated into a distance map which records the shortest distance from each voxel inside the volume to the nearest edge. With a traditional rigid transform discussed in Section 12.2.2 which can translate and rotate the surface model to minimize the sum of distance of each node on the whole surface, the template is considered to match an object which looks alike.

Then, with a special design to push six lobes of the average mushroom bodies to fit edges individually, which is the inverse procedure of creating a pseudo-average model, the matching result will be improved because of the correction of individual disparity. Figure 12.13 shows a matching result of mushroom bodies.

Combining with the matching result, traditional automatic segmentation which is always bothered by unexpected noises can be improved because of much more prior information. Figure 12.14 shows a result of an automatic segmentation of the mushroom bodies with the template matching procedure. It can be observed that some larger noises near mushroom bodies which may come from other neuropils are removed.

12.4 Statistical Classification

With a standard coordinate, we can now integrate different neurons from different experiments and analyze or classify these data. Due to individual

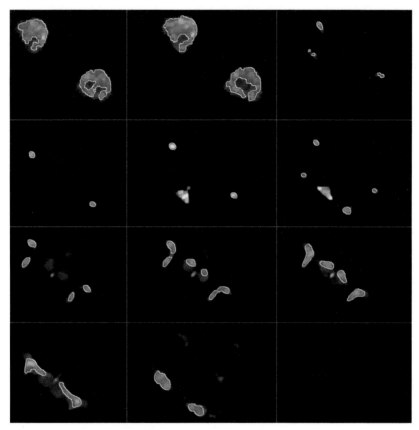

Figure 12.14 Segmentation result with template matching

disparity between samples, the same neurons from different samples may have difference in shape, but cross or connect with same neuropils. Therefore, a new type of analysis based on connections, passing regions, or passing neuropils may be considered to replace traditional methods.

12.5 Conclusions

A new idea of a 3D shape-based atlas which includes absolute anatomic shapes and their positions is constructed. It is possible now to gather neurons that come from different experiments and connect them to find mechanisms underlying the brain's operations. Furthermore, with data described by the same representation system, analysis, classification, and identification can be done by biologists or computers.

The advantages of two-level averaging technique are correcting global characteristics and averaging local differences which come from individual disparity. Though there are still a lot of neuropils which are not averaged yet, we believe that, with two-level averaging technique, most of neuropils could be averaged if we can effectively handle the global characteristics of neuropils which have much more complex structures.

References

[1] R. Brandt, T. Rohlfing, J. Rybak, S. Krofczik, A. Maye, M. Westerhoff, H.-C. Hege, and R. Menzel. Three-dimensional average-shape atlas of the honeybee brain and its applications. *Journal of Comparative Neurology*, 492(1):1–19, 2005.

[2] G.Y. Chen, Y.C. Chen, C.F. Lin, A.C. Hu, C.C. Wu, and Y.C. Chen. Template-based automatic segmentation of Drosophila mushroom bodies. *Journal of Information Science and Engineering*, 24(1):99–113, 2008.

[3] Y.C. Chen, Y.C. Chen, and A.S. Chiang. Two-level model averaging technique in Drosophila brain imaging. In *Proceedings of 2002 IEEE International Conference on Image Processing*, Vol. 2, pages 941–944, 2002.

[4] C.R. Maurer Jr., R. Qi, and V. Raghavan. A linear time algorithm for computing exact Euclidean distance transforms of binary images in arbitrary dimensions. *IEEE Trans. on Pattern Analysis and Mechine Intelligence*, 25:265–270, 2003.

[5] K. Rein, M. Zockler, M.T. Mader, C. Grubel, and M. Heisenberg. The Drosophila standard brain. *Current Biology*, 12:227–231, 2002.

[6] C.C. Wu, G.Y. Chen, Y.C. Chen, H.C. Shao, H.M. Chang, and Y.C. Chen. Creation of the standard Drosophila brain model and its coordinate system. In *Proceedings of VIE2008*, pages 478–483, 2008.

PART III

MACHINE LEARNING
AND IMAGE RECOGNITION

13

Representation Models and Machine Learning Techniques for Scene Classification

Giovanni Maria Farinella and Sebastiano Battiato

Image Processing Lab, Dipartimento di Matematica e Informatica,
Università degli Studi di Catania, Viale A. Doria 6, 95125 Catania, Italy;
E-mail: {gfarinella, battiato}@dmi.unict.it

Abstract

Scene classification is a fundamental process of human vision that allows us to efficiently and rapidly analyze our surroundings. Humans are able to recognize complex visual scenes at a single glance, despite the number of objects with different poses, colors, shadows and textures that may be contained in the scenes. Understanding the robustness and rapidness of this human ability has been a focus of investigation in the cognitive sciences over many years. These studies have stimulated researches in computer vision in building artificial scene recognition systems. Motivations beyond that of pure scientific curiosity are provided by several important computer vision applications in which scene classification can be exploited (e.g., robot navigation systems). Different methods have been proposed to model and to describe the content of a scene. Different machine learning procedures have been employed to automatically learn commonalities and differences between different classes. In this chapter we survey some of the state of the art approaches for scene classification. For each approach we report a description and a discussion of the most relevant peculiarities.

Keywords: scene representation, classification.

Patrick Shen-Pei Wang (Ed.), Pattern Recognition and Machine Vision – In Honor and Memory of Professor King-Sun Fu, 199–214.

13.1 Introduction and Motivation

Vision is perhaps our most important sense. The recognition of the scenes' category is a fundamental process of the human visual system that consist in processing images of a scene to make explicit what needs to be known about its content. Seminal studies in computational vision [27] have portrayed scene recognition as a progressive reconstruction of the input from local measurements (e.g., edges, surfaces). In contrast, some experimental studies have suggested that the recognition of real world scenes may be initiated from the encoding of the global configuration of the scene, ignoring most of the details about concepts and object information [6]. This ability is achieved mainly by exploiting the holistic cues of scenes that can be processed as single entity over the entire human visual field without requiring attention to local features [30, 32]. Recent studies suggested that humans rely on local information as much as on global information to recognize the scene category. Specifically, the human visual system seems to integrate both type of information during the categorization of scenes [44]. The human visual system has inspired computer vision researchers to develop scene classification engines that can be useful task for many relevant applications, such as robot navigation systems [38], content-based image retrieval (CBIR) [43], semantic organization of databases of digital pictures [19], scene depths estimation [1, 40], multimedia marketing services [2, 3]. The context of a scene have been employed as rich source of information about an objects identity, location and scale [35]. Indeed, the structure of many real-world scenes is governed by strong configurational rules akin to those that apply to a single object. In the real world, there exists a strong relationship between the environment and the objects that can be found within it. So the knowledge about the category of a scene give information that can be used, for instance, to change the prior probability to find a specific object in the context of the observed scene [35].

Despite humans can recognize the class of a scene effortlessly, this task is a daunting challenge for computers-based scene classification systems. Many source of variability exists (Figure 13.1): changes in viewing angle, lighting, scale, object inside, and environmental factors. Furthermore, for scene classification, commonalities must be found to generalize across the variability within the class, while determining differences to discriminate between classes (Figure 13.2).

Figure 13.1 Some examples of images that can be classified as "industrial scene". Computer-based scene classification systems must deal with visual variability arising from the viewing angle, the lighting of the scene, widely varying scales, environmental conditions, variability of object inside the scene

In building scene recognition systems some consideration about the level of description of the scene should be taken into account. According with the terminology used in [29], the description of an environmental scene can be done at three different levels (Figures 13.3, 13.4 and 13.5):

- *Subordinate*: This level of description requires the analysis of local structures, the recognition of objects or the labeling of regions in the image (e.g., *grass, sky, building, people, car*);
- *Basic*: The basic level categorization corresponds to the most common categorical representation (e.g., *forest, mountain, street, city*). Members of a basic level category usually have a similar components and share the same function.
- *Superordinate*: This level of description corresponds to the highest level of abstraction, and therefore, it has the lowest visual category resemblance (e.g., *Open* vs. *Close, Natural landscape* vs. *Artificial landscape*, etc.).

These three descriptions provide different levels of abstraction and thus, different semantic information. Levels of description that use precise semantic names to categorize an environment (e.g., *beach, street, forest*) do not explicit refer to the scene structure. Hence the spatial envelop of a scene should

Figure 13.2 Within-class and between-class variability. Above: example images of *tall buildings* and *open countries* illustrate the wide within-class variability in appearance of scenes from particular categories. Below: depending on the application, *highway* and *street* inside a city could be considered (i) different classes, in which case their visual differences are important, or (ii) the same class (*road*), in which case their visual similarities are important

Figure 13.3 Examples of images of scenes belonging to different classes. The scenes are grouped taking into account different levels of description. The classes at the same level of description are shown within the group. For instance, the *Naturalness* of a scene refers to the origin of the components used to build the scene (*Manmade* or *Natural*). The *Naturalness* is a general property that applies to any picture

Figure 13.4 The three different levels of description of the scene

Figure 13.5 Examples images in which the scene depicted can be described abstractly as *Close* or *Open*. Scenes with small perceived depth are referred as *Close* whereas scenes in which the depth is perceived are referred as *Open*. The *Openness* of a scene refers to the sense of enclosure of the space. It opposes *indoor* scenes (enclosed spaces) to *open landscapes*. The *Openness* characterizes places and it is not relevant for describing single objects or close-up views. This type of description refers to the volume of the scene [29]

be taken into account and it should be encoded in the scene representation independently from the required level of scene description. Moreover, the scene representation model and the related computational approach typically depend on the task to be solved and the level of description required. For instance, a software engine able to automatically infer information about the category of a scene could be helpful to drive different tasks performed by single sensor imaging devices during acquisition time (e.g., Autofocus, AutoExposure, White balance) or during post acquisition time (e.g., image enhancement, image coding). Of course, different problems should be considered in transferring the ability of scene recognition to imaging devices domain: memory limitation, computational power limitation and the data format to be used in scene recognition task. In single sensor imaging domain, for some tasks (e.g. White balancing) a Superordinate level of description of the scene can be enough to drive further tasks of the imaging pipeline [5].

Different methods have been proposed in literature to model the scene to obtain a powerful description of its content. Existing methods mainly work extracting local concepts directly on spatial domain [22, 32, 43] or in frequency domain [15, 29]. A global representation of the scene may be obtained grouping together these information in different ways. Recently, the spatial layout of the local information [4, 20] as well as metadata information collected during acquisition time [11] have been used to improve the classification quality. The final descriptor of the scene is eventually exploited by machine learning algorithm to infer the scene category, skipping the recognition of the objects that are present in the scene [29]. The machine learning procedures are employed to automatically learn commonalities and differences between different classes.

In the rest of this chapter we will illustrate in more detail some of the state of the art approaches for scene classification.

13.2 Scene Classification in Spatial Domain

Several studies in Computer Vision have considered the problem of discriminating between different classes of scenes by representing images with features extracted on spatial domain. A wide class of scene recognition algorithms working at superordinate level of description, use color, texture or edge feature. Gorkani et al. used statistics of orientation in the images to discriminate scene into two categories (*cities* vs. *natural landscapes*) [16]. *Indoor* vs. *Outdoor* classification based on color and texture was addressed by Szummer et al. [34]. Many authors proposed to organize and classify images by using the visual content encoded on spatial domain. The methods we review comprise the state of the art methods of the last five years with respect to the basic level of description of the scene. Other related approaches are reviewed in [8].

The fist work we consider here employs an holistic representation of the scene to recognize its category [32]. The rationale beyond the method proposed by Renninger et al. was that human visual system is able to process holistic cues in parallel over the entire human visual field without requiring attention to analyze local features allowing the recognition of the category of the scene. Taking into account that humans can process texture quickly and in parallel over the visual field, a global representation based on Textons [18, 21] was built. The main process used to encode textures started by building a vocabulary of distinctive patterns, able to identify properties and structures of different textures present in the scenes. To this aim a bank of Gabor filters and K-means clustering have been employed. Using the built vocabulary each image is represented as a frequency histogram of Textons. Images of scenes used in the experiments were within ten basic-level categories: *beach, mountain, forest, city, farm, street, bathroom, bedroom, kitchen* and *livingroom*. A χ^2 similarity measure was coupled with a K-nearest neighbors algorithm to perform classification. The performances of the proposed model stayed nearly at 76% correct.

Li and Perona [22] suggested an approach to learn and recognize natural scene categories with the interesting peculiarity that it does not require any experts to annotate the training set. The dataset involved in their experiments contained thirteen basic level categories of scenes:[1] *highway*, inside of cities*, tall buildings*, streets*, forest*, coast*, mountain*, open country*, suburb residence, bedroom, kitchen, living room* and *office*. The images of scenes were modeled as a collection of local patches automatically detected on scale invariant points and described by a features vector invariant to rotation, illumination and 3D viewpoint [23]. Each patch was represented by a codeword from a large vocabulary previously learned through K-means clustering on a set of training patch. In the learning phase a model that best represented the distribution of the involved codewords in each category of scenes was built by using a learning algorithm based on Latent Dirichlet Allocation [7]. In recognition phase, first the identification of all the codewords in the unknown image was done. Then the category model that best fitted the distribution of the codewords of a test image was inferred comparing the likelihood of an image given each category. The performances obtained by authors reaches 65.2% of accuracy.

The goal addressed by Bosch et al. was to discover latent topics (i.e., object categories) in each image in an unsupervised manner, and to use the

---

[1] The starred categories originate from [29].

distribution of the learned topics to perform scene classification[9]. To this aim, probabilistic Latent Semantic Analysis (pLSA) [17] was applied to a bag of visual words representation [14] for each image. A new visual vocabulary for the bag of visual word model exploiting the SIFT descriptor [23] on HSV color domain has been proposed. Again, the K-means was employed to build the vocabulary. The scene classification on the object distribution was carried out by a K-nearest neighbors classifier. The combination of (unsupervised) pLSA followed by (supervised) nearest neighbour classification proposed by Bosch et al. outperformed previous methods. For instance, the accuracy obtained was 8.2% better with respect to the method proposed by Li and Perona [22] when compared on the same dataset.

One of the most complete scene category dataset at basic level of description was exploited by Lazebnik et al. [20]. The dataset is an augmented version of the dataset used by Li and Perona [22] and Bosch et al. [9] in which two basic level categories have been added: *industrial* and *store*. The proposed method exploits a spatial pyramid image representation. For a specific visual word, the first process is to identify where spatially the visual word appears in the image. The authors employed a Support Vector Machine (SVM) using the one-versus-all rule to perform the recognition of the scene category. This method obtained 81.4% when SIFT descriptors of 16×16 pixels patches computed over a grid with 8 pixels spacing were employed in building the visual vocabulary through K-means clustering.

Battiato et al. proposed a method to recognize scene categories using bags of visual words obtained hierarchically partitioning into sub-region the input images [4]. Specifically, for each subregion the Textons distribution and the extension of the corresponding subregion are taken into account. The bags of visual words computed on the subregions are weighted and used to represent the whole scene. The classification of scenes is carried out by a Support Vector Machine. A K-nearest neighbors algorithm and a similarity measure based on Bhattacharyya coefficient are used to retrieve from the scene database those that contain similar visual content to a given a scene used as query. The performance obtained was 84% on the 13 different scene categories used in [22].

Vogel et al. [43] explored the problem of identifying the category of six different basic level category that can be considered as *Natural* scenes at the major level of description. The basic category involved in the experiments was related to *costs, rivers/lakes, forests, plains, mountains, sky/clouds*. A novel image representation was introduced. The scene representation model take into account nine local concepts that can be present in *Natural* scenes (*sky, water, grass, trunks, foliage, field, rocks, flowers, sand*) and combine them to a global representation used to address the category of the scenes. The descriptor for each image scene is built in two stages. In the first stage local image regions are classified by a concept classifier taking into account the nine semantic concept classes. In the second stage, the region-wise information of the concept classifier is combined to a global representation through a normalized vector in which each component represents the frequency of occurrence of a specific concept taking into account the image labelled in the first stage. In order to model information about which concept appears at any specific part of the image (e.g. top, bottom), the vector of frequency concepts was computed on several overlapping or non-overlapping image areas. In this manner a semi-local, spatial image representation by computing and concatenating the different frequency vectors can be obtained. To perform concept classification each concept patch of an image was represented by using low level features (HIS color histogram, edge directions histogram and gray-level co-occurrence). A multi-class Support Vector Machine using a one-against-one approach was used to infer local concepts as well as the final category of the scene. The best classification accuracy obtained with this approach was 71.7% for the nine concepts and 86.4% for the six classes of scene.

Recently, Bosch et al. [10] inspired by previous works [9, 20], presented a method in which the pLSA model was augmented using spatial pyramid in building the distribution of latent topics. The final scene classification was performed using the discriminative classifier SVM on the learned distribution obtaining 83.7% of accuracy on the same dataset used by Lazebnik et al. [20].

In sum, most of the scene classification approaches discussed above make use of features extracted on spatial domain and share a basic structure: an image is considered as a distribution of visual words and this holistic representation is used to perform classification. Eventually local spatial constraints are added in order to capture the spatial layout of the visual words within images.

13.3 Scene Classification in Frequency Domain

In the last decade the frequency domain have been exploited as an useful and effective source of information to holistically encode images for scene understanding. The statistics of natural images in frequency domain reveal that there are different spectral signatures for different image categories [41]. As example, let us consider the problem of discriminating *Natural* vs. *Artificial* scene classification. Straight horizontal and vertical lines dominate man-made structures whereas most natural landscapes have textured zones and undulating contours. It is possible to experimentally observe [15, 29, 41] that the spectrum of *Natural* scenes is quite isotropic with no preferred direction, whereas the spectrum of *Artificial* scenes have strong vertical and horizontal axis. Scenes having a distribution of edges (in spatial domain) commonly found in natural landscapes would have a high degree of naturalness whereas scenes with edges (in spatial domain) biased toward vertical and horizontal orientations would have a low degree of naturalness. The spatial domain properties regarding the naturalness of the scenes are well reflected in frequency domain where holistic representations may hence be used to discriminate between *Natural* vs. *Artificial* scenes (i.e., at superordinate level of description). Considering the shape of the spectrum of an image it is possible to address scene category [29, 39, 41], scene depth [40], and object priming [42]. Although these studies have demonstrated impressive levels of performances, just a few researchers have exploited the frequency domain for scene classification purpose. In this section we review some state of the art approaches whose scene representation is built on frequency domain.

As suggested by different studies in computational vision, scene recognition may be initiated from the encoding of the global configuration of the scene, disregarding details and object information. Inspired by this knowledge, Torralba and Oliva have introduced computational procedures to extract global structural information of complex natural scenes looking at the frequency domain [29, 39, 41]. The computational model that they propose works in the Fourier domain where Discriminant Structural Templates (DSTs) are built using the power spectrum. A DST is a weighting scheme over the power spectrum that assigns positive values to the frequencies that are most representative of one of the classes and negative for the other. Specifically, the sign of the DST values indicates the correlation between the spectral components and the "spatial envelope" properties of two classes. When the task is to discriminate between two kinds of scenes (e.g. *Natural* vs. *Artificial*, *Open* vs. *Closed*) a suitable DST is built and used for the classification. The DST is learned in a supervised way using Linear Discriminant Analysis [45]. The classification of a new image can be hence performed by the sign of the correlation between the power spectrum of the image and the DST. A relevant issue in building a DST is the sampling of the power spectrum both at the learning and classification stages. Torralba and Oliva make use of a bank of Gabor filters with different frequencies and orientation. The final classification is performed on the Principal Component of the sampled frequencies. They performed test on a dataset containing about 8000 pictures of environmental scenes covering a large variety of outdoor places (256 × 256 pixels in size, in 256 gray levels). An accuracy of about 94% have been obtained on tests performed to discriminate between classes at the superordinate level of description (e.g., *Natural* vs. *Artificial*, *Open* vs. *Close*, etc.).

Inspired the above method, Luo and Boutell proposed to use Independent Component Analysis rather than PCA for features extraction [25]. In addition they have combined the camera metadata related to the image capture conditions with the information provided by the power spectra to perform classification. In the experiments performed by the authors the results have been similar to results obtained in previous works in [29].

Recently, Farinella et al. proposed exploit features extracted ordering the Discrete Fourier Power Spectra for *naturalness* classification [15]. The new and main contribution of such method is to demonstrate that the ordering of the frequencies is a useful feature for naturalness classification. Ordering the frequencies indeed enables capture the overall shape of the scene in frequency domain. In particular the frequencies that better capture the differences in the energy shapes related to *Natural* and *Artificial* categories are selected and ordered by their response values in the discrete Fourier power spectrum. In this way a "ranking number", corresponding to the relative position in the ordering, is assigned to each discriminative frequency. The vector of the response values of the discriminative frequencies and the vector of the relative positions in the ordering of the discriminative frequencies are used singularly or in combination to provide a holistic representation of the scene. For the *Natural* vs. *Artificial* scene classification problem, an accuracy of 92.6% have been obtained just using the ordering position of the selected discriminative frequencies.

The Discrete Cosine Transform (DCT) domain was explored by Ladret et al. to perform image classification and retrieval [19]. The DCT domain is particularly interesting because it is compatible with JPEG compressed images. When used in JPEG compression, DCT turns an image into a set of 8×8 blocks of components related to the DCT basis. In [19] each block is then processed to understand the main local edge orientation in the corresponding image patch. A weighted orientation histogram is formed from an image and it is used as a global representation of the scene. Scene classification and retrieval are eventually performed using K-nearest neighbors algorithm. Tests were performed on a database consisting of 470 pictures (256 × 256 pixels, gray levels values) from COREL database. The images were described at superordinate level of description. Specifically, tests have been done to discriminate *Natural* vs. *Artificial* scenes (obtaining about 94% of accuracy) as well as to discriminate between four classes of scenes (obtaining about 80.75% of accuracy): *Outdoor, Indoor, Closed, Open.*

The above reported techniques disregard the spatial layout of discriminative frequencies. Torralba et al. [35, 40, 42] proposed to further look at the spatial frequency layout to address more specific vision tasks like object recognition, location detection, scale understanding and scene depth estimation.

13.4 Scene Classification in Constrained Domain

If the task of scene recognition should be performed timely by using hardware with limited resources in term of space and computational power (e.g., in a single sensor imaging device), these constraints should be taken into account in building a suitable scene representation and a fast not memory based recognition engine.

Despite the progress in the area of scene classification in post acquisition time (all the methods reviewed in previous sections implicitly assume that images are already acquired and stored), a framework describing this task during (or right after) acquisition time in a constrained domains (e.g., Digital Still Camera, Mobile Imaging Devices, HD-TV, etc.) is still missing. Clearly, a software engine able to automatically infer information about the category of a scene could be helpful to drive different tasks performed by single sensor imaging devices during acquisition time (e.g., Autofocus, AutoExposure, White balance, etc.) or during post acquisition time (e.g., image enhancement, image coding). Although the Hardware and Software advances (see [24] for a recent review in the field), the consumer single sensor imaging technology is quite far from the ability of recognize scenes and/or to exploit the visual content during (or after) acquisition time. Recent studies empirically demonstrate that better results are achieved when a scene recognition engine is used to drive the process of color constancy [5]. Moreover, in [25, 11] authors have proposed to use digital cameras metadata as features in joint with other image features extracted in spatial [11] or frequency domain [25].

Different problems should be considered in transferring the ability of recognizing the category of a scene in imaging devices domain. Taking into account the limits of the devices in terms of memory space, most of the approach reviewed in the previous sections cannot be used in their original form. For example, a lot of memory space for the so called visual vocabulary is needed and most of the classification algorithms (e.g., Support Vector Machine) need to take all training features vectors in memory to perform the learning phase. Even when the training phase can be done off line, many patterns (e.g., final support vectors) have to be stored into main memory for the final classification task. Moreover the input data should be cover all different aspects and situations of the scene to be recognized in order to guarantee the robustness of the involved solutions. The Color Filter Array (CFA) domain or the DCT domain are already used for classification in applications different than imaging devices [12, 19]. In these domains features such as edge can be extracted in a simple way [13, 33]. By using similar approaches devoted to extract some low level features, properly solution for the recognition of the scene category on imaging devices may be designed.

13.5 Summary and Conclusions

Different application domains could benefit from a scene classification systems. In this chapter, different state of the art approaches for scene classification have been presented. Specifically, in Section 13.2 methods that exploit features on spatial domain have been discussed. Section 13.3 reviewed techniques working on frequency domain, whereas Section 13.4 address some scene recognition challenges in constrained domain. Although recent advances in the field different challenges are still open in this research area; among others we highlight the following topics:

- Studies on how to produce powerful visual vocabularies to better discriminate between different classes are becoming more appealing [26, 28].
- Models that exploit local and global information to better discriminate complex scenes environments (e.g., indoor scenes) are under consideration [31].
- Very large datasets of images are already available [31, 36] and there is a need to develop advanced techniques to scale with large dataset [37].

Acknowledgement

The authors would like to thank Professor Giovanni Gallo for his invaluable support on this work.

References

[1] S. Battiato, S. Curti, M. La Cascia, M. Tortora, and E. Scordato. Depth map generation by image classification. In *Proceedings of SPIE Electronic Imaging - Three-Dimensional Image Capture and Applications VI*, Vol. 5302, pages 95–104, 2004.

[2] S. Battiato, G.M. Farinella, G. Giuffrida, C. Sismeiro, and G. Tribulato. Exploiting visual and text features for direct marketing learning in time and space constrained domains. *Pattern Analysis and Applications*, DOI 10.1007/s10044-009-0145-2, 2009.

[3] S. Battiato, G.M. Farinella, G. Giuffrida, C. Sismeiro, and G. Tribulato. Using visual and text features for direct marketing on multimedia messaging services domain. *Multimedia Tools and Applications*, 42(1):5–30, 2009.

[4] S. Battiato, G.M. Farinella, G. Gallo, and D. Ravì. Scene categorization using bag of textons on spatial hierarchy. In *15th IEEE International Conference on Image Processing, 2008 (ICIP2008)*, pages 2536–2539, October 2008.

[5] S. Bianco, G. Ciocca, C. Cusano, and R. Schettini. Improving color constancy using indooroutdoor image classification. *IEEE Transactions on Image Processing*, 17(12):2381–2392, December 2008.

[6] I. Biederman. Aspects and extension of a theory of human image understanding. In *Computational Processes in Human Vision: An Interdisciplinary Perspective*, 1988.

[7] D.M. Blei, A.Y. Ng, and M.I. Jordan. Latent dirichlet allocation. *J. Mach. Learn. Res.*, 3:993–1022, 2003.

[8] A. Bosch, X. Muñoz, and R. Martí. Review: Which is the best way to organize/classify images by content? *Image Vision Comput.*, 25(6):778–791, 2007.

[9] A. Bosch, A. Zisserman, and X. Munoz. Scene classification via pLSA. In *Proceedings of the European Conference on Computer Vision*, Vol. IV, pages 517–530, 2006.

[10] A. Bosch, A. Zisserman, and X. Muñoz. Scene classification using a hybrid generative/discriminative approach. *IEEE Trans. Pattern Anal. Mach. Intell.*, 30(4):712–727, 2008.

[11] M.R. Boutell and J. Luo. Bayesian fusion of camera metadata cues in semantic scene classification. In *International Conference on Computer Vision and Pattern Recognition*, pages 623–630, 2004.

[12] H.S. Chang and K. Kang. A compressed domain scheme for classifying block edge patterns. *IEEE Transactions on Image Processing*, 14(2):145–151, 2005.

[13] C.-H. Chen, S.-J. Chen, and P.-Y. Hsiao. Edge detection on the Bayer pattern. In *APCCAS*, pages 1132–1135. IEEE, 2006.

[14] G. Csurka, C. Dance, L. Fan, J. Willamowski, and C. Bray. Visual categorization with bags of keypoints. In *ECCV International Workshop on Statistical Learning in Computer Vision*, 2004.

[15] G.M. Farinella, S. Battiato, G. Gallo, and R. Cipolla. Natural versus artificial scene classification by ordering discrete fourier power spectra. In *SSPR & SPR '08: Proceedings of the 2008 Joint IAPR International Workshop on Structural, Syntactic, and Statistical Pattern Recognition*, pages 137–146. Springer-Verlag, Berlin/Heidelberg, 2008.

[16] M.M. Gorkani and R.W. Picard. Texture orientation for sorting photos "at a glance". *ICPR-A*, 94:459–464, 1994.

[17] T. Hofmann. Unsupervised learning by probabilistic latent semantic analysis. *Machine Learning*, 42(1-2):177–196, 2001.

[18] B. Julesz. Textons, the elements of texture perception, and their interactions. *Nature*, 290:91–97, 1981.

[19] P. Ladret and A. Guérin-Dugué. Categorisation and retrieval of scene photographs from jpeg compressed database. *Pattern Analysis & Application*, 4:185–199, June 2001.

[20] S. Lazebnik, C. Schmid, and J. Ponce. Beyond bags of features: Spatial pyramid matching for recognizing natural scene categories. In *IEEE Conference on Computer Vision and Pattern Recognition*, Vol. II, pages 2169–2178, 2006.

[21] T.K. Leung and J. Malik. Recognizing surfaces using three-dimensional textons. In *IEEE International Conference on Computer Vision*, pages 1010–1017, 1999.

[22] F.-F. Li and Pietro Perona. A Bayesian hierarchical model for learning natural scene categories. In *CVPR '05: Proceedings of the 2005 IEEE Computer Society Conference on Computer Vision and Pattern Recognition (CVPR'05) - Volume 2*, pages 524–531, Washington, DC, USA, 2005. IEEE Computer Society.

[23] D.G. Lowe. Distinctive image features from scale-invariant keypoints. *Int. J. Comput. Vision*, 60(2):91–110, 2004.

[24] R. Lukac. *Single-Sensor Imaging: Methods and Applications for Digital Cameras*. CRC Press, 2008.

[25] J. Luo and M.R. Boutell. Natural scene classification using overcomplete ica. *Pattern Recognition*, 38(10):1507–1519, 2005.

[26] J. Mairal, F. Bach, J. Ponce, G. Sapiro, and A. Zisserman. Discriminative learned dictionaries for local image analysis. In *Computer Vision and Pattern Recognition, 2008. CVPR 2008. IEEE Conference on*, pages 1–8, June 2008.

[27] D. Marr. *Vision: A Computational Investigation into the Human Representation and Processing of Visual Information*. W.H. Freeman, 1982.

[28] F. Moosmann, B. Triggs, and F. Jurie. Fast discriminative visual codebooks using randomized clustering forests. In *Neural Information Processing System (NIPS)*, 2006.

[29] A. Oliva and A. Torralba. Modeling the shape of the scene: A holistic representation of the spatial envelope. *International Journal of Computer Vision*, 42:145–175, 2001.

[30] A. Oliva and A. Torralba. Building the gist of a scene: The role of global image features in recognition. *Visual Perception, Progress in Brain Research*, pages 251–256, 2006.

[31] A. Quattoni and A. Torralba. Recognizing indoor scenes. In *IEEE Conference on Computer Vision and Pattern Recognition*, pages 413–420, 2009.

[32] L.W. Renninger and J. Malik. When is scene recognition just texture recognition? *Vision Research*, 44:2301–2311, 2004.

[33] B. Shen and I.K. Sethi. Direct feature extraction from compressed images. In *Storage and Retrieval for Image and Video Databases (SPIE)*, pages 404–414, 1996.

[34] M. Szummer and R.W. Picard. Indoor-outdoor image classification. In *IEEE International Workshop on Content-based Access of Image and Video Databases, in conjunction with ICCV'98*, pages 42–51, 1998.

[35] A. Torralba. Contextual priming for object detection. *International Journal of Computer Vision*, 53(2):169–191, 2003.

[36] A. Torralba, R. Fergus, and W.T. Freeman. 80 million tiny images: A large data set for nonparametric object and scene recognition. *IEEE Trans. Pattern Anal. Mach. Intell.*, 30(11):1958–1970, 2008.

[37] A. Torralba, R. Fergus, and Y. Weiss. Small codes and large image databases for recognition. *IEEE Computer Society Conference on Computer Vision and Pattern Recognition*, pages 1–8, 2008.

[38] A. Torralba, K.P. Murphy, W.T. Freeman, and M.A. Rubin. Context-based vision system for place and object recognition. In *Proceedings Ninth IEEE International Conference on Computer Vision*, Vol. 1, pages 273–280, October 2003.

[39] A. Torralba and A. Oliva. Semantic organization of scenes using discriminant structural templates. In *Internation Conference on Computer Vision*, pages 1253–1258, 1999.

[40] A. Torralba and A. Oliva. Depth estimation from image structure. *IEEE Trans. Pattern Anal. Mach. Intell.*, 24(9):1226–1238, 2002.

[41] A. Torralba and A. Oliva. Statistics of natural image categories. *Network: Computing in Neural Systems*, 14:391–412, 2003.

[42] A. Torralba and S. Pawan. Statistical context priming for object detection. In *Internation Conference on Computer Vision*, 2001.

[43] J. Vogel and B. Schiele. Semantic modeling of natural scenes for content-based image retrieval. *International Journal of Computer Vision*, 72(2):133–157, April 2007.

[44] J. Vogel, A. Schwaninger, C. Wallraven, and H.H. Bülthoff. Categorization of natural scenes: Local vs. global information. In *Symposium on Applied Perception in Graphics and Visualization*, 2006.

[45] A.R. Webb. *Statistical Pattern Recognition* (2nd edition). John Wiley & Sons, November 2002.

14

Manifold and Subspace Learning for Pattern Recognition*

Yun Fu[1] and Thomas S. Huang[2]

[1]*Department of Computer Science and Engineering, University at Buffalo (SUNY), 201 Bell Hall, Buffalo, NY 14260, USA; E-mail: raymondyunfu@gmail.com*
[2]*Beckman Institute, University of Illinois at Urbana-Champaign, 405 North Mathews Avenue, Urbana, IL 61801, USA; E-mail: huang@ifp.uiuc.edu*

Abstract

In the last few decades, more and more real-world sensory inputs have been collected to form multivariate data sets. Data have become rich yet may embody large dimensionality redundancy and complicated distributions. Feature extraction along with dimensionality reduction is therefore in great demand for the pattern recognition purpose. Manifold and subspace learning are prevalent tools for the high-dimensional data feature extraction. This chapter reviews the most state-of-the-art manifold and subspace learning methods in the pattern recognition applications. Four subspace learning algorithms derived from manifold criteria, namely Locally Embedded Analysis (LEA), Discriminant Simplex Analysis (DSA), Correlation Embedding Analysis (CEA), and Correlation Tensor Analysis (CTA), are introduced for the real-world application of face recognition.

Keywords: manifold learning, subspace learning, pattern recognition, feature extraction.

---

*This chapter is partly based on the previously published papers in [8, 10, 12, 13].

Patrick Shen-Pei Wang (Ed.), Pattern Recognition and Machine Vision – In Honor and Memory of Professor King-Sun Fu, 215–229.

14.1 Introduction

Conventional subspace learning methods, such as Principal Component Analysis (PCA) [27] and Linear Discriminant Analysis (LDA) [21], favored parametric statistics for feature representation by learning linear subspaces. An optimization formulation is usually used to find the linear global projection assuming the Gaussian distribution in the original sample space. The subspace learning is conducted by measuring a set of fixed parameters, such as mean and variance. When the data distribution is statistically sufficient, those methods are effective and efficient. However, for the small-sample and high-dimensionality case, such linear models are fundamentally limited because the richness in local variations and the nonlinearity of the underlying manifold structure can not be sufficiently discovered and described [7].

Derived from the manifold criteria [1,23,26,28], the nonparametric learning strategy has been widely adopted to design subspace learning algorithms. Since the nonparametric model does not rely on any distribution assumptions, the subspace learning is conducted by measuring the pair-wise data relationship in the local neighborhoods based on the manifold concept. The manifold structure is well captured indicating the intrinsic nature of data. In general, a mapping, either nonlinear or linear, from the original space to the embedded space is constructed by solving a generalized eigenvalue decomposition problem. These methods can effectively deal with the small-sample and high-dimensionality case due to their reliance on fewer assumptions and parameters [30].

The nonlinear methods, such as Locally Linear Embedding (LLE) [23, 24], Laplacian Eigenmaps [1], Isomap [26], Hessian LLE (hLLE) [6], and Semidefinite Embedding (SDE) [28], focus on preserving the geometry structure of the low-dimensional manifold for better data visualization. Since the nonlinear projection is only defined on the training data, the out-of-sample extension [2,24] has to be adopted for representing new data. However, there is still a storage problem which is impractical in real-world applications since the system has to store all the training data for the extension usage. Mitigating the practical problem with an explicit projection, many linearization forms of those nonlinear methods, based on the Graph Embedding (GE) theory [30], have been demonstrated to be effective and efficient for real-world applications, such as Multidimensional Scaling (MDS) [30], Locality Preserving Projections (LPP) [16], Neighborhood Preserving Embedding (NPE) [17], Neighborhood Preserving Projections (NPP) [22], and Locally Embedded Analysis (LEA) [8,9]. The main objective of these algorithms is to learn a global projection that preserves data localities and similarities in the embedded space. As for the purpose of discriminant analysis, they are still insufficient in providing enough discriminating power. The Fisher criterion [10] and Tensorization [30] have been introduced to enhance the discriminating power of graph embedding. Some representative algorithms include Marginal Fisher Analysis (MFA) [30], Conformal Embedding Analysis (CEA) [11], Local Discriminant Embedding (LDE) [4], Locality Sensitive Discriminant Analysis (LSDA) [3], tensor PCA [31], tensor LDA [32], Tensor Subspace Analysis (TSA) [18], and tensor LDE [29]. These methods explicitly aim at the classification capacity and discriminating efficiency of the graph embedding.

This chapter presents a general view of manifold and subspace learning. Four particular criteria–manifold learning, Fisher graph, similarity metric, and high-order data structure–are summarized based on a general framework for novel algorithm design. Four particular subspace learning algorithms corresponding to different criteria are introduced, namely Locally Embedded Analysis (LEA) [8], Discriminant Simplex Analysis (DSA) [10], Correlation Embedding Analysis (CEA) [12], and Correlation Tensor Analysis (CTA) [13]. Extensive experiments and comparisons are provided on real-world face recognition applications, which demonstrate the effectiveness of these methods from different aspects.

14.2 A General View of Manifold and Subspace Learning

14.2.1 Generalized Linear Projection

The generalized linear projection can be explained by a graph embedding theory [9, 30]. Define a one-to-one mapping between set $\mathcal{X} = \{\mathbf{x}_i : i = 1 \ldots n, \mathbf{x}_i \in \mathbb{R}^D\}$ and set $\mathcal{Y} = \{\mathbf{y}_i : i = 1 \ldots n, \mathbf{y}_i \in \mathbb{R}^d\}$. Suppose set $\mathcal{G} = \{\mathbf{X}, \mathbf{W}\}$ is a weighted graph with similarity matrix $\mathbf{W} \in \mathbb{R}^{n \times n}$ and $\mathbf{X}$ is formed by $\mathbf{x}_i$ from set $\mathcal{X}$. Define a Laplacian matrix $\mathbf{L} = \mathbf{D} - \mathbf{W}$, where $\mathbf{D}[i, i] = \sum_j W_{ij}$ and $\mathbf{D}[i, j] = 0$ for $\forall i \neq j$. Then we find the graph embedding of graph $\mathcal{G}$ and low-dimensional representation through $\mathbf{y}^* = \arg \min_{\mathbf{y}} \mathbf{y}^T \mathbf{L} \mathbf{y}$, with the constraint $\mathbf{y}^T \mathbf{B} \mathbf{y} = s$, where $\mathbf{B}$ is a constraint matrix with some particular geometric sense and s is a nonnegative constant. The space projection is defined by a linear transformation $P : \mathbb{R}^D \rightarrow \mathbb{R}^d$, where D denotes the high dimension of the original data and d denotes the reduced low dimension. The D-by-d projection matrix is denoted by $\mathbf{P} = [\mathbf{p}_1, \mathbf{p}_2, \ldots, \mathbf{p}_d]$ which satisfies $\mathbf{y}_i = \mathbf{P}^T \mathbf{x}_i$. The D-to-d projection can

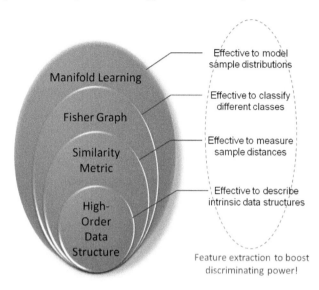

Figure 14.1 General algorithm criteria for feature extraction

be written as a single matrix equation $\mathbf{Y} = \mathbf{P}^T \mathbf{X}$ where $\mathbf{x}_i$ and $\mathbf{y}_i$ are viewed as columns of D-by-n matrix $\mathbf{X}$ and d-by-n matrix $\mathbf{Y}$, respectively. To calculate $\mathbf{P}$, we consider the objective function $\mathbf{p}^* = \arg\min_{\mathbf{p}} \mathbf{p}^T \mathbf{XLX}^T \mathbf{p}$, with the constraint $\mathbf{p}^T \mathbf{XBX}^T \mathbf{p} = s$ or $\mathbf{p}^T \mathbf{p} = s$. The desired projection vector $\mathbf{p}$ is calculated in closed-form by solving the generalized eigenvalue decomposition problem [7].

14.2.2 General Criteria for Algorithm Design

The generalized linear projection can be used to design new subspace learning algorithms with four general criteria – manifold learning, Fisher graph, similarity metric, and high-order data structure. These criteria can be applied either separately or jointly to conduct the algorithm design constrained by the linearized graph embedding. Figure 14.1 shows the basic connections between these criteria, which are described as follows in a pattern recognition view:

- *Consider the entire sample set.* The manifold learning, an effective way to model the nonparametric sample distributions, is the basis for the other three criteria.

- *Consider the sample sets of different classes.* Fisher graph is the fusion of manifold learning criterion and Fisher criterion, which is effective to classify different classes.
- *Consider the sample pairs in the same class.* To extend the Fisher graph, similarity metric criterion may also be considered. By selecting an effective sample distance metric, the discriminating power of the Fisher graph can be significantly boosted.
- *Consider a single sample point.* High-order data structure, especially the multilinear representation (tensor), is effective to describe intrinsic data structures when high-order patterns exist.

Overall, the four criteria are all feasible ways for feature extraction to boost the discriminating power of subspace learning algorithms. We will introduce following four particular algorithms based on each of the four criteria.

14.3 Locally Embedded Analysis (LEA) [8]

Sample distributions can be estimated by the manifold criteria. The Locally Embedded Analysis (LEA), a particular graph-based method, is the linearization variant of Locally Linear Embedding (LLE) [23, 24], which considers the geometric relationship among the high-dimensional data as the measure of linear nearest-neighbor reconstruction.

Find $\mathbf{x}_i$'s k nearest neighbors from the original data set $\mathcal{X} = \{\mathbf{x}_i : i = 1 \ldots n, \mathbf{x}_i \in \mathbb{R}^D\}$, and define $\mathbf{X}^i = [\mathbf{x}_{i_1}, \mathbf{x}_{i_2}, \ldots, \mathbf{x}_{i_k}]$. The objective of locally linear reconstruction is to minimize the following cost function

$$\varepsilon(W) = \sum_{i=1}^{n} \left\| \mathbf{x}_i - \sum_{j=1}^{k} W_{i,i_j} \mathbf{x}_{i_j} \right\|^2, \qquad (14.1)$$

where the weight matrix W is constrained by $\sum_{j=1}^{k} W_{i,i_j} = 1$ and is calculated by solving $W = \arg\min_W \varepsilon(W)$ given $\mathcal{X}$. The weight matrix is an estimation of the data distribution in a nonparametric model.

In the LLE embedding, a low-dimensional data set $\mathcal{Y} = \{\mathbf{y}_i : i = 1 \ldots n, \mathbf{y}_i \in \mathbb{R}^d\}$, where $\mathbf{y}_i$ corresponds to $\mathbf{x}_i$ and $d < D$, is constructed with the same weights. In the LEA embedding, $\mathbf{y}_i = \mathbf{P}^T \mathbf{x}_i$ is applied to find the LEA linear subspace obtained by minimizing the following objective functions

$$\begin{cases} \varepsilon_{\text{LLE}} = \sum_{i=1}^{n} \left\| \mathbf{y}_i - \sum_{j=1}^{k} W_{i,i_j} \mathbf{y}_{i_j} \right\|^2, \\ \varepsilon_{\text{LEA}} = \sum_{i=1}^{n} \left\| \mathbf{P}^T \mathbf{x}_i - \sum_{j=1}^{k} W_{i,i_j} \mathbf{P}^T \mathbf{x}_{i_j} \right\|^2. \end{cases} \qquad (14.2)$$

Without losing generalization, we consider the case when $d = 1$, and then the matrix $\mathbf{P}$ is replaced by the vector $\mathbf{p}$. According to $\mathbf{y} = \mathbf{p}^T\mathbf{X}$, we substitute each y_i with $\mathbf{p}^T\mathbf{x}_i$. Then ε_{LEA} in (14.2) is written as

$$\varepsilon_{\text{LEA}} = \left\|\mathbf{p}^T\mathbf{X} - \mathbf{p}^T\mathbf{X}\mathbf{W}^T\right\|^2 = \mathbf{p}^T\mathbf{X}(\mathbf{I} + \mathbf{W}^T\mathbf{W} - \mathbf{W} - \mathbf{W}^T)\mathbf{X}^T\mathbf{p}. \quad (14.3)$$

14.3.1 LEA Constraint

If we follow LLE, the constraint $\sum_{i=1}^{n} y_i^2 = \sum_{i=1}^{n} \mathbf{p}^T\mathbf{x}_i\mathbf{x}_i^T\mathbf{p} = 1$ can be applied, and NPE [17] and NPP [22] use this same constraint as LLE. A physical interpretation is that the low-dimensional representations of the training samples are scaled to have norm sum of one during the embedding.

We impose a different constraint for LEA as $\mathbf{p}^T\mathbf{X}(\mathbf{I} + \mathbf{W}^T\mathbf{W})\mathbf{X}^T\mathbf{p} = 1$, which is easier to understand after rewritten as

$$\sum_{i=1}^{n}\mathbf{p}^T\mathbf{x}_i\mathbf{x}_i^T\mathbf{p} + \sum_{i=1}^{n}\left(\sum_{j=1}^{k} W_{i,i_j}\mathbf{p}^T\mathbf{x}_{i_j}\right)\left(\sum_{j=1}^{k} W_{i,i_j}\mathbf{p}^T\mathbf{x}_{i_j}\right)^T = 1, \quad (14.4)$$

$$\sum_{i=1}^{n}\mathbf{p}^T\mathbf{x}_i\mathbf{x}_i^T\mathbf{p} + \sum_{i=1}^{n}\mathbf{p}^T\tilde{\mathbf{x}}_i\tilde{\mathbf{x}}_i^T\mathbf{p} = \sum_{i=1}^{n}(y_i^2 + \tilde{y}_i^2) = 1, \quad (14.5)$$

where $\tilde{y}_i = \mathbf{p}^T \sum_{j=1}^{k} W_{i,i_j}\mathbf{x}_{i_j} = \mathbf{p}^T\tilde{\mathbf{x}}_i$ and $\tilde{\mathbf{x}}_i$ is the optimal local reconstruction of sample $\mathbf{x}_i$ from its k nearest neighbors. We can see that the first term in the left-hand side of (14.5) is the norm sum of all the data projections in the LEA space, which is the same as LLE constraint. The second term is the norm sum of projections for all the reconstructed data $\tilde{\mathbf{x}}_i$'s in the LEA subspace. It means that the constraint takes into account both the original data and the reconstructed data from their neighboring samples. A benefit of this consideration is that LEA bridges the two popular algorithm NPE and LPP, and LEA can also be considered as a special case of LPP as justified in the next subsection. Then, the LEA is mathematically formulated as

$$\min_{\mathbf{p}} \mathbf{p}^T\mathbf{X}(\mathbf{I} + \mathbf{W}^T\mathbf{W} - \mathbf{W} - \mathbf{W}^T)\mathbf{X}^T\mathbf{p}, \quad \text{s.t. } \mathbf{p}^T\mathbf{X}(\mathbf{I} + \mathbf{W}^T\mathbf{W})\mathbf{X}^T\mathbf{p} = 1. \quad (14.6)$$

This problem can be solved by using the generalized eigenvalue decomposition method as $\mathbf{X}(\mathbf{I} + \mathbf{W}^T\mathbf{W} - \mathbf{W} - \mathbf{W}^T)\mathbf{X}^T\mathbf{p} = \lambda\mathbf{X}(\mathbf{I} + \mathbf{W}^T\mathbf{W})\mathbf{X}^T\mathbf{p}$, where $\mathbf{p}$ is the eigenvector corresponding to the eigenvalue λ.

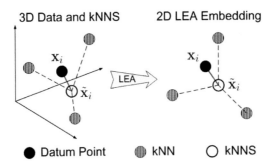

3D Data and kNNS   2D LEA Embedding

● Datum Point   ◉ kNN   ○ kNNS

Figure 14.2 *k*-Nearest-Neighbor Space and *k*NNS approximation

14.3.2 A Novel Perspective on LEA

In the above two subsections, we have introduced that LEA is a linearization of LLE with a specific constraint. Here, we justify from a novel perspective that LEA can also be considered as a specific LPP algorithm, and hence LEA bridges the two popular manifold based subspace learning algorithms.

Before formally introducing this novel perspective, we introduce the concept of k-Nearest-Neighbor Space (kNNS) [5, 20]. A kNNS is defined on k points $\mathbf{x}_1, \mathbf{x}_2, \ldots, \mathbf{x}_k$ as $\mathcal{S}(\mathbf{x}_1, \mathbf{x}_2, \ldots, \mathbf{x}_k) = \left\{ \sum_{i=1}^{k} \ell_i \mathbf{x}_i \mid \sum_{i=1}^{k} \ell_i = 1 \right\}$. Note that kNNS is different from the concept of simplex since the latter requires ℓ_i's to be nonnegative.

The terms $\sum_{j=1}^{k} W_{i,i_j} \mathbf{x}_{i_j}$ and $\sum_{j=1}^{k} W_{i,i_j} \mathbf{y}_{i_j}$ in (14.1) and (14.2) can be considered as the particular instances of $\mathcal{S}(\{\mathbf{x}_{i_j}\}_{j=1}^{k})$ and $\mathcal{S}(\{\mathbf{y}_{i_j}\}_{j=1}^{k})$ respectively. Then, an explanation of (14.1) is that, for each point $\mathbf{x}_i$, we find the nearest point within the kNNS constituted by its k nearest neighbors. Denote the nearest point of $\mathbf{x}_i$ within its kNNS as $\tilde{\mathbf{x}}_i$, then the formulation of LEA in (14.2) becomes

$$\varepsilon_{\text{LEA}} = \sum_{i=1}^{n} \left\| \mathbf{P}^T \mathbf{x}_i - \mathbf{P}^T \tilde{\mathbf{x}}_i \right\|^2. \tag{14.7}$$

Figure 14.2 shows an example of LEA embedding based on the kNNS concept. The geometric structure of the original data distribution is kept via minimizing the distance between datum point and its nearest sample in kNNS, which can be better understood from the following theory:

Theorem. *LEA can be considered as a specific LPP algorithm with the following configurations: (1) the graph vertices consist of the original training data and their nearest points in the kNNS's; and (2) the graph is constructed* *using 1 nearest neighbor based on only the original data, namely, connecting each original datum with its nearest sample in the corresponding kNNS, and the weights are 1's for these pairs.*

Proof. Denote the graph vertices as $\hat{\mathbf{X}} = [\mathbf{x}_1, \mathbf{x}_2, \ldots, \mathbf{x}_n, \tilde{\mathbf{x}}_1, \tilde{\mathbf{x}}_2, \ldots, \tilde{\mathbf{x}}_n]$, and the weight matrix $\hat{\mathbf{W}}_{ij} = 1$ if $i = j + n$ or $j = i + n$, which means the 1-nearest-neighbor graph is constructed based on only original samples. Then, we can have $\mathbf{D}_{ii} = 1$ and $\mathbf{D} = \mathbf{I}_{2n}$. From the LPP definition, the objective function is

$$\frac{1}{2} \sum_{ij} \left\| \mathbf{p}^T \hat{\mathbf{X}}_i - \mathbf{p}^T \hat{\mathbf{X}}_j \right\|^2 \hat{\mathbf{W}}_{ij} = \sum_{i=1}^{n} \left\| \mathbf{P}^T \mathbf{x}_i - \mathbf{P}^T \tilde{\mathbf{x}}_i \right\|^2, \tag{14.8}$$

where $\hat{\mathbf{X}}_i$ is the i-th column vector in matrix $\hat{\mathbf{X}}$. We can see that the objective of LPP in this case is the same as that of LEA as in (14.7).

Also from the definition, the constraint of LPP is

$$\mathbf{p}^T \hat{\mathbf{X}} \mathbf{D} \hat{\mathbf{X}}^T \mathbf{p} = \mathbf{p}^T \hat{\mathbf{X}} \mathbf{I} \hat{\mathbf{X}}^T \mathbf{p} = \sum_{i=1}^{n} (\mathbf{p}^T \mathbf{x}_i \mathbf{x}_i^T \mathbf{p} + \mathbf{p}^T \tilde{\mathbf{x}}_i \tilde{\mathbf{x}}_i^T \mathbf{p}), \tag{14.9}$$

which is the same as the constraint of LEA as in (14.5).

From these two points, we can conclude that LEA is a specific LPP.

14.4 LEA Extensions: Three Algorithms

14.4.1 Discriminant Simplex Analysis (DSA) [10]

In the LEA algorithm, the entire sample set is considered as an unsupervised manner. What if the sample points have labels? The objective of Discriminant Simplex Analysis (DSA) [10] is to keep the class relation reflected by the known labels in a supervised manner. It considers the sample sets of different classes in a Fisher criteria, which keeps neighboring samples close if they have the same label, while preventing samples of other classes from entering the neighborhood. The k-nearest-neighbor simplex [10] is used in the DSA formulations.

There are two types of k-nearest neighbors for each sample in $\mathcal{X}$: *within-locality graph* k_w-NN and *between-locality graph* k_b-NN. Each sample pattern in the training set is associated with a category label l, so we have a label set $\mathcal{L} = \{l_i : l_i \in \mathbb{R}\}_{i=1}^{n_c}$, where n_c denotes the number of classes in the data space. For each sample $\mathbf{x}_i$ with label $l^{(i)}$, we search its within-locality k_w-NNs from $\mathcal{X}$ and obtain a set $\mathcal{X}_w^{(i)} = \{\mathbf{x}_{w(j)}^{(i)}\}_{j=1}^{k_w}$ satisfying

$l_{w(j)}^{(i)} = l^{(i)}$. On the other hand, we find the set of $\mathbf{x}_i$'s between-locality k_b-NNs as $\mathcal{X}_b^{(i)} = \{\mathbf{x}_{b(j,p)}^{(i)}\}_{j,p=1}^{k_b,n_c-1}$, satisfying $l_{b(j,p)}^{(i)} \neq l^{(i)}$ and $l_{b(j_1,p_1)}^{(i)} = l_{b(j_2,p_2)}^{(i)}$ as $p_1 = p_2$ for $j_1, j_2 = 1, 2, \ldots, k_b$. DSA works on dual graphs which are modeled as

$$\begin{cases} \varepsilon_w(\widetilde{\mathbf{C}}_w) = \sum_{i=1}^{n} \left\| \mathbf{x}_i - \sum_{j=1}^{k_w} c_{w(j)}^{(i)} \mathbf{x}_{w(j)}^{(i)} \right\|^2, \\ \varepsilon_b(\widetilde{\mathbf{C}}_b) = \sum_{i=1}^{n} \left\| \mathbf{x}_i - \sum_{j=1}^{k_b} c_{b(j,l*)}^{(i)} \mathbf{x}_{b(j,l*)}^{(i)} \right\|^2, \end{cases} \tag{14.10}$$

where $\widetilde{\mathbf{C}}_w$ and $\widetilde{\mathbf{C}}_b$ are $n \times n$ sparse matrices to encode the graph weights, and the index sets $w(j), b(j, l^*) = 1, 2, \ldots, n$. Note that here l^* indicates the label of the nearest different class.

After we model the graph pair, the objective function for calculating a linear projection of DSA subspace is formulated as follows, which preserves the within-locality affinity while discriminating the between-locality affinity.

$$\begin{cases} \varepsilon_w(\mathbf{P}) = \sum_{i=1}^{n} \left\| \mathbf{P}^T \mathbf{x}_i - \sum_{j=1}^{k_w} c_{w(j)}^{(i)} \mathbf{P}^T \mathbf{x}_{w(j)}^{(i)} \right\|^2, \\ \varepsilon_b(\mathbf{P}) = \sum_{i=1}^{n} \left\| \mathbf{P}^T \mathbf{x}_i - \sum_{j=1}^{k_b} c_{b(j,l*)}^{(i)} \mathbf{P}^T \mathbf{x}_{b(j,l*)}^{(i)} \right\|^2. \end{cases} \tag{14.11}$$

We finally solve the constrained optimization problem by Lagrange optimization, $\widetilde{\mathbf{P}} = \arg\max(\varepsilon_b(\mathbf{P}))$, subject to $\varepsilon_w(\mathbf{P}) = s$.

14.4.2 Correlation Embedding Analysis (CEA) [12]

Based on the Fisher graph criterion, considering the distances (similarity) between sample pairs in the same class is another criterion for discriminative subspace learning. Recent works [7,11–13] suggested the Pearson correlation coefficient as a more effective similarity metric for the real-world pattern recognition applications, in which the Euclidean metric may fail to capture the intrinsic similarities between features with diverse variations. The Correlation Embedding Analysis (CEA) [12] integrates both correlational embedding [11] and discriminating criterion. It is a supervised learning method which potentially enhances the discriminating power with correlation metrics.

Define two correlation affinity graphs $\mathcal{G}_s$ and $\mathcal{G}_d$, both with n nodes. The i-th node corresponds to the datum $\mathbf{x}_i$. For $\mathcal{G}_s$, we only consider each pair of data $\mathbf{x}_i$ and $\mathbf{x}_j$ from the *same* class with $l_i = l_j$. An edge is constructed between nodes i and j if $\mathbf{x}_j$ is among the k_s largest correlation neighbors of $\mathbf{x}_i$. For $\mathcal{G}_d$, we only consider each pair of data $\mathbf{x}_i$ and $\mathbf{x}_j$ from the *different* classes with $l_i \neq l_j$. An edge is constructed between nodes i and j if $\mathbf{x}_j$ is among the k_d largest correlation neighbors of $\mathbf{x}_i$. Here k_s and k_d can be different and chosen with empirical values. Define the $n \times n$ correlation affinity matrices $\mathbf{W}_s$ of graph $\mathcal{G}_s$ and $\mathbf{W}_d$ of graph $\mathcal{G}_d$. The weight of the edge between $\mathbf{x}_i$ and $\mathbf{x}_j$ is set by w_{ij}, with $w_{ij} = 0$ if node i and j are not connected. The parameter w_{ij} is defined afterward. To ensure the affinity in the low-dimensional feature space is with correlation metric, the objective function of CEA is defined as

$$\underset{\mathbf{P}}{\arg\max} \left\{ \varepsilon(\mathbf{P}) = \sum_{i \neq j} \left\| \frac{\mathbf{P}^T \mathbf{x}_i}{\|\mathbf{P}^T \mathbf{x}_i\|} - \frac{\mathbf{P}^T \mathbf{x}_j}{\|\mathbf{P}^T \mathbf{x}_j\|} \right\|^2 \cdot (w_{ij}^{(d)} - w_{ij}^{(s)}) \right\}, \quad (14.12)$$

$$\underset{\mathbf{P}}{\arg\max} \left\{ \varepsilon(\mathbf{P}) = 2 \sum_{i \neq j} \left(1 - \frac{\mathbf{x}_i^T \mathbf{P} \mathbf{P}^T \mathbf{x}_j}{\sqrt{(\mathbf{x}_i^T \mathbf{P} \mathbf{P}^T \mathbf{x}_i)(\mathbf{x}_j^T \mathbf{P} \mathbf{P}^T \mathbf{x}_j)}} \right) \cdot w_{ij}^{(d-s)} \right\}, \quad (14.13)$$

where $w_{ij}^{(d-s)} = w_{ij}^{(d)} - w_{ij}^{(s)}$. This objective function is nonlinear. It essentially couples the neighborhood and class information through the elements of $\mathbf{W}_s$ and $\mathbf{W}_d$. As for generalized feature extraction, we can define the gradient descent rule for optimization by differentiating $\varepsilon(\mathbf{P})$ with respect to matrix $\mathbf{P}$. The (14.13) can be optimized by steepest descent, conjugate gradients, or other fast gradient descent methods. We can only calculate local maxima via iterative optimization techniques for (14.13), since this objective function is not convex. Especially, when the dimension of the data space is too large, the gradient descent may not converge to a good solution. Hence, a good initial solution is critical for achieving a sufficient optimization [12].

14.4.3 Correlation Tensor Analysis (CTA) [13]

When a single sample point is considered, data structure is crucial for feature representation. Since visual features, such as images, contain obvious high-order data patterns, a tensor structure may introduce more powerful properties for feature representation to boost the discriminating power. The Correlation Tensor Analysis (CTA) extends the CEA algorithm by encoding each sample as a second- or higher-order tensor, which learns multiple interrelated subspaces to obtain a low-dimensional data representation. In addition to the same properties of CEA, the CTA also inherits the advantage of tensor representation to be able to avoid the curse-of-dimensionality dilemma, overcome the small sample size problem, and significantly reduce computational costs.

Assume a set of n m-order tensors $\{\mathbf{X}_i \mid \mathbf{X}_i \in \mathbb{R}^{D_1 \times D_2 \times \cdots \times D_m}\}_{i=1}^n$ represents the given high-dimensional data samples, with the corresponding class labels $\{l_i \mid l_i \in \{1, \ldots, n_c\}\}_{i=1}^n$. The objective of CTA is to find a low-dimensional tensor representation $\{\mathbf{Y}_i \mid \mathbf{Y}_i \in \mathbb{R}^{d_1 \times d_2 \times \cdots \times d_m}\}_{i=1}^n$ of $\{\mathbf{X}_i\}_{i=1}^n$, with $\mathbf{Y}_i = \mathbf{X}_i \times_1 \mathbf{U}_1 \times_2 \cdots \times_m \mathbf{U}_m$, where $\mathbf{U}_j \in \mathbb{R}^{D_j \times d_j}$ for $j = 1, \ldots, m$ and $d_j \leqslant D_j$. Again, we define two correlation affinity graphs $\mathcal{G}_s$ and $\mathcal{G}_d$, both with n nodes. Then the objective function of CTA is defined as

$$\arg\max_{\mathbf{U}} \left\{ \varepsilon(\mathbf{U}) = \sum_{i,j=1}^n \left(1 - \frac{\langle \mathbf{Y}_i, \mathbf{Y}_j \rangle}{\sqrt{\langle \mathbf{Y}_i, \mathbf{Y}_i \rangle \langle \mathbf{Y}_j, \mathbf{Y}_j \rangle}} \right) \cdot w_{ij}^{(d-s)} \right\}, \quad (14.14)$$

where $w_{ij}^{(d-s)} = w_{ij}^{(d)} - w_{ij}^{(s)}$. Its right-hand side can be rewritten as

$$\sum_{i,j=1}^n \left(1 - \frac{\langle \mathbf{X}_i \times_q \mathbf{U}_q|_{q=1}^m, \mathbf{X}_j \times_q \mathbf{U}_q|_{q=1}^m \rangle}{\sqrt{\langle \mathbf{X}_i \times_q \mathbf{U}_q|_{q=1}^m, \mathbf{X}_i \times_q \mathbf{U}_q|_{q=1}^m \rangle \langle \mathbf{X}_j \times_q \mathbf{U}_q|_{q=1}^m, \mathbf{X}_j \times_q \mathbf{U}_q|_{q=1}^m \rangle}} \right) \cdot w_{ij}^{(d-s)}. \quad (14.15)$$

As suggested in [18, 30], we define the k-mode unfolding $\mathbf{Z}^{(k)}$ [13] of tensor $\mathbf{X}$. Assume $\mathbf{Z}_i = \mathbf{X}_i \times_1 \mathbf{U}_1 \ldots \times_{k-1} \mathbf{U}_{k-1} \times_{k+1} \mathbf{U}_{k+1} \cdots \times_m \mathbf{U}_m$, we have $\langle \mathbf{Z}_i \times_k \mathbf{U}_k, \mathbf{Z}_i \times_k \mathbf{U}_k \rangle = \langle \mathbf{Z}_i^{(k)T} \mathbf{U}_k, \mathbf{Z}_i^{(k)T} \mathbf{U}_k \rangle$. Then (14.15) can be rewritten as

$$\sum_{i,j=1}^n \left(1 - \frac{\text{Tr}\{\mathbf{U}_k^T \mathbf{Z}_i^{(k)} \mathbf{Z}_j^{(k)T} \mathbf{U}_k\}}{\sqrt{\text{Tr}\{\mathbf{U}_k^T \mathbf{Z}_i^{(k)} \mathbf{Z}_i^{(k)T} \mathbf{U}_k\} \text{Tr}\{\mathbf{U}_k^T \mathbf{Z}_j^{(k)} \mathbf{Z}_j^{(k)T} \mathbf{U}_k\}}} \right) \cdot w_{ij}^{(d-s)}. \quad (14.16)$$

Since there is no closed-form solution for the CTA optimization, we can find the approximate solution with an iterative procedure. We can first initialize $\mathbf{U}_1, \ldots, \mathbf{U}_m$ and assume $\mathbf{U}_1, \ldots, \mathbf{U}_{k-1}, \mathbf{U}_{k+1}, \ldots, \mathbf{U}_m$ are known so that $\mathbf{U}_k$ is solved by fixing the others. Again, following the gradient descent rule, we differentiate $\varepsilon(\mathbf{U})$ with respect to matrix $\mathbf{U}$. A good initial solution is critical for achieving a sufficient optimization [13].

14.5 Comparisons and Experiments

14.5.1 Databases

The CMU PIE (Pose, Illumination, and Expression) face database [25] collects 41368 images of 68 subjects with varying pose, illumination, and expression. The images are manually aligned, cropped, and resized to 20×20 with gray levels. For each subject, we select 168 images covering most varying cases. To further reduce the testing set, 34 samples of each individual are randomly selected (2312 images in total). Random database partitions are done with 5, 10, 15, and 20 images per individual for training, and the remaining data for testing.

The combined Yale Face Database B [15] and extended Yale Face Database B [19] data set contains 5760+16128 single light source images of 38 subjects with 576 viewing conditions (9 poses × 64 illumination conditions). The images are cropped and resized to 32 × 32 with gray levels. To further reduce the testing set, 64 near frontal images under different illuminations per individual are randomly selected. Random database partitions are done with 5, 10, 20, and 30 images per individual for training, and the remaining data for testing.

14.5.2 Theoretical Comparison

The four algorithms – LEA, DSA, CEA, and CTA – are designed from different criteria. LEA, combining learning locality [14] and manifold learning criteria, can be both supervised and unsupervised learning. It can handle continuous labels for supervision, such as face aging and face pose. Since there is no discriminant analysis criterion in the LEA algorithm, its discriminating power can be improved. DSA, CEA, and CTA are all supervised learning. DSA is the combination of learning locality, manifold learning, multiple sample metric, and Fisher graph criteria. CEA is the combination of learning locality, manifold learning, Fisher graph, and correlation metric criteria. CTA is the combination of learning locality, manifold learning, Fisher graph, correlation metric, and high-order data structure criteria. They can handle discrete labels for supervision, such as face identity. Since Fisher criterion is considered, their discriminating power is high.

14.5.3 Experimental Comparison

To compare the performance of the four algorithms, LEA, DSA, CEA, and CTA are applied to face recognition on the PIE and Yale-B databases. The recognition accuracy and corresponding subspace dimension are summarized in Tables 14.1 and 14.2. To be fair in the comparison, LEA and DSA have the same pre-/post-data processing (demean + normalization and correlation affinity classification). This pre-/postprocessing can improve the classification performance, demonstrated in [7]. It can be observed that CEA, CTA, and N-DSA-C all outperform N-LEA-C in all the cases. This result is consistent with the explanation in the above theoretical comparison. On the PIE database, CEA is more powerful than the other methods while on the Yale-B database

Table 14.1 Face recognition accuracy comparison on the PIE database

Method	5 Train. Sample		10 Train. Sample		15 Train. Sample		20 Train. Sample	
	Error	Dim.	Error	Dim.	Error	Dim.	Error	Dim.
N-LEA-C	57.91%	142	26.35%	214	17.03%	209	12.82%	153
N-DSA-C	**34.03%**	169	20.65%	144	13.85%	177	11.13%	212
CEA	35.29%	299	**18.87%**	196	**13.62%**	81	**9.45%**	242
CTA	41.78%	16,16	24.39%	16,16	17.26%	14,14	12.82%	14,14

Table 14.2 Face recognition accuracy comparison on the Yale-B database

Method	5 Train. Sample		10 Train. Sample		20 Train. Sample		30 Train. Sample	
	Error	Dim.	Error	Dim.	Error	Dim.	Error	Dim.
N-LEA-C	20.25%	109	8.87%	156	7.89%	271	8.82%	242
N-DSA-C	20.38%	84	**7.21%**	141	5.74%	166	5.80%	171
CEA	22.61%	103	7.55%	229	6.22%	218	3.56%	202
CTA	**16.99%**	26,26	7.60%	23,23	**4.96%**	23,23	**2.94%**	23,23

CTA performs the best. Since PIE and Yale-B both contain large illumination variations, correlation based methods, CEA and CTA, are more effective and robust. The different performance of the two methods on the two databases may due to the extra pose and expression variations in the PIE database than the Yale-B. The N-DSA-C may outperform the other methods in some cases when training sample size is small, which demonstrates the good performance of the multiple sample metric.

14.5.4 Computational Comparison

Since all the proposed methods are based on the generalized graph embedding framework, they inherit the computational cost from the graph modeling. To build the $n \times n$ graph, where n denotes the number of the training data, we need to build two $n \times n$ weight matrices. These matrices may introduce large computational cost if n is large. In other words, they may suffer from computational difficulty when the sample space is large. One possible way to deal with the problem, for the large database, is to use sample-locality manner [14]. As suggested by our previous work in [14], the query-driven subspace learning framework can significantly relieve the computational cost of the graph embedding methods and further improve the classification accuracy as well. The CEA and CTA may introduce more computational cost in the iteration process. Since the solutions for the two algorithms are not in closed-form, iteration is inevitable. But CTA may also reduce its computational costs owing to the reduced data dimensions in generalized eigen-decomposition.

14.6 Conclusion

In this chapter, a general manifold and subspace learning framework is presented for new algorithm design, based on a review of the state-of-the-art methods for the pattern recognition applications. Four general criteria – manifold learning, Fisher graph, similarity metric, and high-order data structure – are introduced to specify four particular algorithms, namely LEA, DSA, CEA, and CTA. Experimental evaluations in face recognition demonstrate the effectiveness and reveal great potentials for many real-world applications.

References

[1] M. Belkin and P. Niyogi. Laplacian eigenmaps for dimensionality reduction and data representation. *Neural Computation*, 15(6):1373–1396, 2003.

[2] Y. Bengio, J. Paiement, P. Vincent, O. Delalleau, N. Roux, and M. Ouimet. Out-of-sample extensions for LLE, isomap, MDS, eigenmaps, and spectral clustering. In *Proceedings of NIPS*, 2004.

[3] D. Cai, X. He, K. Zhou, J. Han, and H. Bao. Locality sensitive discriminant analysis. In *Proceedings of IJCAI*, pages 708–713, 2007.

[4] H. Chen, H. Chang, and T. Liu. Local discriminant embedding and its variants. In *Proceedings of IEEE Conference on CVPR*, Vol. 2, pages 846–853, 2005.

[5] J.-T. Chien and C.-C. Wu. Discriminant waveletfaces and nearest feature classifiers for face recognition. *IEEE Trans. Pattern Anal. Mach. Intell.*, 24(12):1644–1649, 2002.

[6] D. Donoho and C. Grimes. Hessian eigenmaps: Locally linear embedding techniques for high-dimensional data. *Proceedings of the National Academy of Arts and Sciences*, 100:5591–5596, 2003.

[7] Y. Fu, Unified discriminative subspace learning for multimodality image analysis. Ph.D. Dissertation, Department of Electrical and Computer Engineering, University of Illinois at Urbana-Champaign, 2008.

[8] Y. Fu and T.S. Huang, Locally linear embedded eigenspace analysis, http://www.ifp.uiuc.edu/~yunfu2/papers/LEA-Yun05.pdf, IFP-TR, UIUC, 2005.

[9] Y. Fu and T. Huang. Graph embedded analysis for head pose estimation. In *Proceedings of IEEE Conference on FG*, Southampton, UK, pages 3–8, 2006.

[10] Y. Fu, S. Yan, and T.S. Huang. Classification and feature extraction by simplexization. *IEEE Transactions on Information Forensics and Security*, 3(1):91–100, 2008.

[11] Y. Fu, M. Liu, and T.S. Huang. Conformal embedding analysis with local graph modeling on the unit hypersphere. In *Proceedings of IEEE CVPR Workshop on Component Analysis*, 2007.

[12] Y. Fu, S. Yan, and T.S. Huang. Correlation metric for generalized feature extraction. *IEEE Trans. Pattern Anal. Mach. Intell.*, 30(12):2229–2235, 2008.

[13] Y. Fu and T.S. Huang, Image classification using correlation tensor analysis. *IEEE Trans. Image Processing*, 17(2):226–234, 2008.

[14] Y. Fu, Z. Li, J. Yuan, Y. Wu, and T.S. Huang. Locality versus globality: Query-driven localized linear models for facial image computing. *IEEE Trans. Circuits and Systems for Video Technology*, 18(12):1741–1752, December 2008.

[15] A. Georghiades, P. Belhumeur, and D. Kriegman. From few to many: Illumination cone models for face recognition under variable lighting and pose. *IEEE Trans. Pattern Anal. Mach. Intell.*, 23(6):643–660, June 2001.

[16] X. He and P. Niyogi. Locality preserving projections. In *Proceedings of NIPS*, 2003.

[17] X. He, D. Cai, S. Yan and H. Zhang. Neighborhood preserving embedding. In *Proceedings of IEEE Conference on ICCV*, Vol. 2, pages 1208–1213, 2005.

[18] X. He, D. Cai, and P. Niyogi. Tensor subspace analysis. In *Proceedings of NIPS*, 2005.

[19] K. Lee, J. Ho, and D. Kriegman. Acquiring linear subspaces for face recognition under variable lighting. *IEEE Trans. Pattern Anal. Mach. Intell.*, 27(5):684–698, May 2005.

[20] S.Z. Li and J. Lu. Face recognition using the nearest feature line method. *IEEE Trans. Neural Networks*, 10(2):439–443, 1999.

[21] A. Martinez and A. Kak. PCA versus LDA. *IEEE Trans. Pattern Anal. Mach. Intell.*, 23(2):228–233, 2001.

[22] Y. Pang, L. Zhang, Z. Liu, N. Yu, and H. Li. Neighborhood Preserving Projections (NPP): A novel linear dimension reduction method. In *Proceedings of International Conference on Intelligent Computing*, pages 117–125, 2005.

[23] S. Roweis and L. Saul. Nonlinear dimensionality reduction by locally linear embedding. *Science*, 290:2323–2326, 2000.

[24] L. Saul and S. Roweis. Think globally, fit locally: Unsupervised learning of low dimensional manifolds. *Journal of Machine Learning Research*, 4:119–155, 2003.

[25] T. Sim, S. Baker, and M. Bsat. The CMU pose, illumination, and expression database. *IEEE Trans. Pattern Anal. Mach. Intell.*, 25(12):1615–1618, December 2003.

[26] J. Tenenbaum, V. Silva and J. Langford. A global geometric framework for nonlinear dimensionality reduction. *Science*, 290:2319–2323, 2000.

[27] M. Turk and A. Pentland. Face recognition using eigenfaces. In *Proceedings of IEEE Conference on CVPR*, pages 586–591, 1991.

[28] K. Weinberger and L. Saul. Unsupervised learning of image manifolds by semidefinite programming. In *Proceedings of IEEE Conference on CVPR*, Vol. 2, pages 988–995, 2004.

[29] J. Xia, D.-Y. Yeung, and G. Dai. Local discriminant embedding with tensor representation. In *Proceedings of IEEE International Conference on Image Processing*, pages 929–932, 2006.

[30] S. Yan, D. Xu, B. Zhang, H. Zhang, Q. Yang, and S. Lin. Graph embedding and extensions: A general framework for dimensionality reduction. *IEEE Trans. Pattern Anal. Mach. Intell.*, 29(1):40–51, 2007.

[31] J. Yang, D. Zhang, A.F. Frangi, and J.-Y. Yang. Two-dimensional PCA: A new approach to appearance-based face representation and recognition. *IEEE Trans. Pattern Anal. Mach. Intell.*, 26(1):131–137, January 2004.

[32] J. Ye, R. Janardan, and Q. Li. Two-dimensional linear discriminant analysis. In *Proceedings of NIPS*, 2004.

15

Learning by Generalized Median Concept

Xiaoyi Jiang[1] and Horst Bunke[2]

[1]*Department of Mathematics and Computer Science, University of Münster, D-48149 Münster, Germany; E-mail: xjiang@math.uni-muenster.de*
[2]*Institute of Computer Science and Applied Mathematics, University of Bern, CH-3012 Bern, Switzerland; E-mail: bunke@iam.unibe.ch*

> Three cobblers combined equal the master mind
> – Chinese proverb

Abstract

In this chapter we present a brief overview of the generalized median concept. Given a set of input objects, the generalized median determines a representative from the entire object space within an optimization framework. After introducing the fundamentals, we discuss a variety of concrete generalized median problems, dealing with objects like vectors, contours, strings, graphs, clusterings, and image segmentations. In each case some important algorithms are discussed and the applications are outlined.

Keywords: generalized median, set median, optimization, approximation.

15.1 Generalized Median Concept

The general concept of average, or mean, has turned out to be useful in numerous contexts of science and engineering. In general, we are given a set of noisy samples of the same object and want to infer a representative model. One powerful tool for this purpose is provided by the generalized median concept.

Patrick Shen-Pei Wang (Ed.), Pattern Recognition and Machine Vision – In Honor and Memory of Professor King-Sun Fu, 231–246.

Assume that we are given a set S of objects in some representation space U and a distance function $d(p, q)$ to measure the dissimilarity between any two objects $p, q \in U$. The essential information of the given set of objects is captured by an object $\overline{p} \in U$ that minimizes the sum of distances to all objects from S, i.e.

$$\overline{p} = \arg\min_{p \in U} \sum_{q \in S} d(p, q).$$

Object $\overline{p}$ is called a *generalized median* of S. A related concept is the so-called *set median*, which results from constraining the search to the given set S

$$\hat{p} = \arg\min_{p \in S} \sum_{q \in S} d(p, q).$$

The set median may serve as an approximative solution for the generalized median. Note that neither the generalized median nor the set median is unique in general.

When dealing with real numbers, the set and generalized median correspond to well-known concepts from statistics. If we define the distance function by $d(p, q) = (p-q)^2$, for instance, the generalized median is simply the arithmetic average of the given numbers. For another distance function, $d(p, q) = |p - q|$, the generalized median becomes the usual median of numbers. Both variants represent a powerful technique for image smoothing.

Independent of the object type and the underlying representation space we can always find the set median of N objects by means of $\frac{1}{2}N(N-1)$ pairwise distance computations (although more efficient algorithms have been reported as well). In contrast there is no general approach to computing generalized medians. The reason is that any such algorithm must be of constructive nature and the construction process crucially depends on the structure of the objects under consideration. Additional difficulty is caused by the fact that determining the generalized median is provably of high computational complexity in several cases.

This chapter gives a brief overview of various instances of generalized median problems including vector, contours, strings, graphs, clusterings, and image segmentations. We mention some theoretical considerations with regard to the computational complexity and lower bound of approximative solutions. In addition to theoretical considerations potential applications are also discussed.

15.2 Median Vectors

Extension of scalar median to vector spaces [1, 2] provides a valuable image processing tool for multispectral/color images and optical flow.

The recent work [47] deals with averaging vectors of unit length (hyperspherical data) in $\mathfrak{R}^D$. Four distance functions $d(\mathbf{x}, \mathbf{y})$ are investigated. In addition to three obvious ones (squared distance, arc length, and squared arc length between $\mathbf{x}$ and $\mathbf{y}$), the following one:

$$d(\mathbf{x}, \mathbf{y}) = \sin^2(\mathbf{x}, \mathbf{y}) = 1 - (\mathbf{x} \cdot \mathbf{y})^2$$

leads to an analytic solution being the eigenvector to the largest eigenvalue of $\frac{1}{N} \sum_{i=1}^{N} \mathbf{v}_i^t \mathbf{v}_i$, where v_i, $i = 1, \ldots, N$, are the input (row) vectors of unit length.

In [8, 20] averaging 3D rotations is studied by means of a quaternion representation. Both approaches, however, present rotation-specific solutions which are not general enough for arbitrary vectors.

15.3 Median Strings

The most popular string distance is doubtlessly the Levenshtein edit distance. Let $A = a_1 a_2 \ldots a_n$ and $B = b_1 b_2 \ldots b_m$ be two words over Σ. The Levenshtein edit distance $d(A, B)$ is defined in terms of elementary edit operations which are required to transform A into B. Usually, three different types of edit operations are considered, namely (1) substitution of a symbol $a \in A$ by a symbol $b \in B$, $a \neq b$, (2) insertion of a symbol $a \in \Sigma$ in B, and (3) deletion of a symbol $a \in A$. Symbolically, we write $a \to b$ for a substitution, $\epsilon \to a$ for an insertion, and $a \to \epsilon$ for a deletion. To model the fact that some distortions may be more likely than others, costs of edit operations, $c(a \to b)$, $c(\epsilon \to a)$, and $c(a \to \epsilon)$, are introduced. Let $s = l_1 l_2 \cdots l_k$ be a sequence of edit operations transforming A into B. We define the cost of this sequence by $c(s) = \sum_{i=1}^{k} c(l_i)$. Given two strings A and B, the Levenshtein edit distance is given by

$$d(A, B) = \min\{c(s) \mid s \text{ is a sequence of edit operations transforming } A \text{ into } B\}.$$

The standard algorithm for computing $d(p, q)$ is based on dynamic programming [51]. It provides both the Levenshtein edit distance and the corresponding optimal edit sequence that transforms p into q with minimum cost.

Further string distance functions are known from the literature, for instance, normalized edit distance [41], maximum posterior probability distance [34], feature distance [34], and others [16]. The Levenshtein edit distance is by far the most popular one. Actually, some of the algorithms we discuss below are tightly coupled to this particular distance function.

The computation of generalized median strings is a demanding task. Under the following two conditions it is proved in [21] that computing the generalized median string is NP-hard. Sim and Park [48] proved that the problem is NP-hard for finite alphabet and for a metric distance matrix:

- every edit operation has cost one, i.e., $c(a \rightarrow b) = c(\epsilon \rightarrow a) = c(a \rightarrow \epsilon) = 1$,
- the alphabet is not of fixed size,

Another result comes from computational biology. The optimal evolutionary tree problem there turns out to be equivalent to the problem of computing generalized median strings if the tree structure is a star (a tree with $n + 1$ nodes, n of them being leaves). In [52] it is proved that in this particular case the optimal evolutionary tree problem is NP-hard. The distance function used is problem dependent and does not even satisfy the triangle inequality. All these theoretical results indicate the inherent difficulty in finding generalized median strings.

Given N strings of length $O(n)$ each, an exact algorithm was proposed by Kruskal [36] based on an extension of the dynamic programming algorithm for string edit distance computation [51]. It needs both $O(n^N)$ time and space and thus its applicability is restricted to rather small sets and short strings. If the sum of distances of the (unknown) generalized median string to all input strings is guaranteed to lie under some (relatively small) upper bound, then the computational expense can be reduced significantly, although remaining exponential [38]. In [39] heuristic information about the spatial location of the objects represented by the symbols in a string is used to speed up median string computation. A greedy algorithm is proposed in [6] that requires $O(n^2 N |\Sigma|)$ time. Starting from an empty string, a new symbol is added in each iteration to the partial solution generated so far. All symbols from the alphabet are probed and that one is selected whose resultant new partial solution shows the smallest sum of distances. A slightly refined version of the greedy algorithm with the same time complexity was reported in [37]. The set median or the solution of the greedy algorithm can be used to start an iterative procedure that provides a further improved approximate generalized median string [34, 40]. It is based on a systematic perturbation of the current solution and is repeated until no more improvement is possible. This procedure needs $O(n^3 N |\Sigma|)$ time for each iteration. The algorithm reported in [26] is very different from all methods discussed above. It deals with a dynamic context where one is faced with the situation of steady arrival of new data items represented by strings and proposes an incremental approach without the need of computing the generalized median string from scratch for each new input string.

An important application of generalized median strings is in the field of OCR [3, 38, 39, 43]. Current OCR technology is far from being perfect. One possibility of performance enhancement lies in the combination of multiple classification results. It is well-known that OCR classifiers are sensitive to slight perturbations in their input. As a result, OCR error behavior can vary. The same page rescanned and re-OCR'ed, or OCR'ed by applying multiple classifiers to a single scan of source text may exhibit different sets of errors. Classifier combination aims at combining the outcome of multiple classifiers through voting procedures, with the goal of producing results better than those that could be achieved by any single classifier. Generalized median strings are very useful tools for such a voting procedure.

In conclusion, median strings have been intensively studied. Interested readers are referred to the review [27] for a more detailed discussion.

15.4 Median Contours

The concept of generalized median strings can be applied to compute average contours if contours are represented by strings [23]. This is useful for object prototype learning. In [53] a special class of contours is considered, which start from the top, pass each image row exactly once, and end in the last row of an image. Despite of their simplicity they frequently occur in many applications of image analysis. A dynamic programming algorithm with $O(Nmn)$ time and $O(mn)$ space is designed, where N images of size $m \times n$ are assumed. Recently, this algorithm has been extended to one of expectation-maximization type for handling the case, where the input contours are subject to a varying (unknown) horizontal displacement [7].

15.5 Median Graphs

Graphs are a powerful and universal tool widely used in information processing. In particular, graph representations are extremely useful in image processing and understanding for mapping the initially numeric nature of an image (or images) into symbolic structural representations for subsequent semantic interpretation of the sensed world.

The popular graph edit distance [4] is adapted from the string edit distance and models the structural variation by edit operations reflecting modifications in structure and labeling. A standard set of edit operations is given by insertions, deletions, and substitutions of both nodes and edges. Given two graphs, the source graph g_1 and the target graph g_2, the idea of graph edit distance is to delete some nodes and edges from g_1, relabel (substitute) some of the remaining nodes and edges, and insert some nodes and edges in g_2, such that g_1 is finally transformed into g_2. A sequence of edit operations that transform g_1 into g_2 is called an edit path between g_1 and g_2. Clearly, every edit path between two graphs g_1 and g_2 is a model describing the correspondences found between the graphs' substructures. That is, the nodes of g_1 are either deleted or uniquely substituted with a node in g_2, and analogously, the nodes in g_2 are either inserted or matched with a unique node in g_1. The same applies for the edges. To find the most suitable edit path out of the set of all possible edit paths, one introduces a cost for each edit operation, measuring the strength of the corresponding operation. Then, the edit distance of two graphs is defined by the minimum cost edit path between two graphs. Several other distance functions for graphs have been proposed; see [29, 46] for a discussion.

Algorithms for computing the graph edit distance are typically based on combinatorial search procedures that explore the space of all possible mappings of the nodes and edges of two graphs [4]. A major drawback of those procedures is their computational complexity, which is exponential in the number of nodes of the involved graphs. Consequently, the application of these exact algorithms for edit distance computations is limited to graphs of rather small size in practice.

To render graph edit distance computation less computationally demanding, a number of suboptimal methods have been proposed. In some approaches, the basic idea is to perform a local search to solve the graph matching problem, that is, to optimize local criteria instead of global, or optimal ones [42, 49]. In [31], a linear programming method for computing the edit distance of graphs with unlabeled edges is proposed. The method can be used to derive lower and upper edit distance bounds in polynomial time. Two fast but suboptimal algorithms for graph edit distance computation are proposed in [42]. The authors propose simple variants of a standard edit distance algorithm that make the computation substantially faster. In [9] another suboptimal method has been proposed. The basic idea is to decompose graphs into sets of subgraphs. These subgraphs consist of a node and its adjacent nodes and edges. The graph matching problem is then reduced to the problem of finding a match between the sets of subgraphs. In [45] a method somewhat similar to the method described in [9] is proposed. However, while the optimal correspondence between local substructures is found by dynamic programming in [9], a bipartite matching procedure is employed in [45]. A detailed discussion of graph distance computation can be found in [29, 46].

The idea of median graphs was recently introduced in [24]. Determining the generalized median graph[1] is computationally complex. On the one hand, the computation time is clearly exponential in the size of input graphs. On the other hand, it is also exponential in terms of the number of input graphs. The reason for this behavior is that already for the special case of strings, the required time is exponential in the number of input strings. As a consequence, search-based exact algorithms are able to compute generalized median graphs for very small graphs only. The recent work [12] uses a particular cost function that permits the definition of the graph distance in terms of the maximum common subgraph. A prediction function is proposed for the backtracking algorithm so that the size of the search space is substantially reduced, avoiding the evaluation of a great amount of states and still obtaining the exact median graph.

In general, however, we are forced to resort to approximate solutions that can be found in reasonable time. In [24] a genetic solution is proposed for this purpose. A comparison of the genetic algorithm against combinatorial search is described in [5]. A subsequent paper [22] focuses on building the median graph through a two step process: finding a set of likely nodes for the median using a clustering procedure and assigning labels to edges.

Recently, graph embedding [44] has been studied, giving a means of mapping graphs into a vector space using the graph edit distance. By doing so we can combine advantages from both domains: We keep the representational power of graphs while being able to operate in a vector space. This idea has been applied to approximately compute the median graph [11, 13]. After embedding the input graphs, the median of the resulting vectors can be easily computed in the vector space. Then, a method has been designed which

---

[1] Note that in graph theory the term median graph is also used to define a special class of graphs. According to [33], a median graph is a connected graph such that, for every triple of nodes u, v, w, there is a unique x lying on a geodesic (i.e., shortest path) between each pair of u, v, w. By contrast, in our work, a median graph is the representative of a given set of graphs.

permits to go from the vector domain back to the graph domain in order to finally obtain an approximative median graph.

Generalized median graphs have been applied to learning prototype models of graphs. In addition, graph clustering based on k-means benefits from generalized graph for computing the cluster centers [14].

15.6 Median Clustering

Clustering is the assignment of objects to groups, called clusters, so that objects from the same cluster are more similar to each other than objects from different clusters. Median clustering, also called consensus clustering, has emerged as an important elaboration of the classical clustering problem. It refers to the situation in which a number of different (input) clusterings have been obtained for a particular dataset and it is desired to find a single (consensus) clustering which is a better fit in some sense than the existing clusterings.

Multiple clusterings of the same data arise in many situations. Different attributes of large data sets lead to different clusterings of the entities. For example, high-throughout experiments in molecular biology and genetics often yield data, which provide a lot of useful information about different aspects of the same entities, in this case genes or proteins. Due to varying experimental conditions, different clusterings of the genes may be obtained. In addition to the individual outcome of each experiment, combining the data across multiple experiments could potentially reveal different aspects of the genomic system and even its systemic properties. A second instance of multiple clusterings results from situations where multiple runs of the same non-deterministic clustering algorithms yield multiple clusterings of the same dataset. Non-deterministic clustering algorithms, e.g. k-means, are sensitive to the choice of the initial seed clusters; running k-means with different seeds may yield very different results. Median clustering is thus the problem of reconciling clustering information about the same dataset coming from different sources or from different runs of the same algorithm.

There exist a large number of distance functions for measuring the dissimilarity of two clusterings [28]. Any of them can be used to define a concrete median clustering problem. Sometimes, similarity measures such as mutual information and its variants are preferred instead of dissimilarity measures. In this case the minimization problem in the definition of generalized median has to be replaced by a maximization.

The median clustering problem implicitly appears as early as the late 18th century; see [15] for the history. In 1986 it was proved [35] that the median clustering problem is NP-complete. The NP-completeness holds in general, yet it is not known whether it is NP-complete for any particular N (number of clusterings). The case of $N = 1, 2$ is trivial since one input clustering is the median clustering [15]. Nothing is known for $N \geq 3$.

Exactly solving large instances of the median clustering problem is obviously intractable. A greed algorithm is described in [50], which uses the set median to start. Then, for each object, the current label is changed to each of the other $k - 1$ possible labels (k is the number of clusters) and the optimization function is re-evaluated. If it decreases, the object's label is changed to the best new value and the algorithm proceeds to the next object. When all objects have been checked for possible improvements, a sweep is completed. If at least one label was changed in a sweep, a new sweep is initiated. The algorithm terminates when a full sweep does not change any labels, thereby indicating that a local optimum is reached. In [50] three additional algorithms are proposed which are far more efficient than the greedy approach and deliver a similar quality of results. These algorithms have been developed from intuitive heuristics rather than from the vantage point of a direct optimization. The evidence accumulation algorithm in [17] follows this line. Taking the co-occurrences of pairs of objects in the same cluster as votes for their association, the N partitions of n objects are mapped into a $n \times n$ co-association matrix: $C(i, j) = n_{ij}/N$; where n_{ij} is the number of times the object pair (i, j) is assigned to the same cluster among the N partitions. The co-association matrix C is then fed to a hierarchical clustering and the final result is regarded as the median clustering. In this work the optimization function for generalized median (defined by means of normalized mutual information) is not involved in the median computation at all, but instead serves as a figure of merit for performance evaluation only. In practice, the median clustering problem is apparently easy to approximate by heuristics. But guidance is necessary as to which heuristic to use. In [18] the authors compare the performance of a number of heuristics on both simulated and real datasets.

15.7 Median Segmentation

Despite of decades of intensive research, unsupervised image segmentation remains a difficult task. Recently, researchers started to investigate combination of multiple segmentations. We may consider an image segmentation as a clustering of pixels and apply some clustering combination algorithm for the segmentation combination purpose. The work reported in [32] is such an example.

In [30] a greedy algorithm finds the matching between the regions from the input segmentations which build the basis for the combination. However, it assumes that all input segmentations contain the same number of regions (same as in [32]).

A more general approach without this limitation is described in [54]. Its basis is the random walker algorithm for image segmentation [19]: Given a small number of K seeds (groups of pixels with user-defined labels), the algorithm labels unseeded pixels by resolving the probability that a random walker starting from each unseeded pixel will first reach each of the seeds. A final segmentation is derived by selecting for each unseeded pixel the most probable seed destination for the random walker. The algorithm can produce a segmentation of high quality provided suitable seeds are placed manually. There exists a natural link of the median segmentation problem at hand to the random walker based image segmentation. The input segmentations provide strong hints about where to *automatically* place some seeds. Given such seed regions one is then faced with the same situation as image segmentation with manually specified seeds and can thus apply the random walker algorithm [19] to achieve a high quality final segmentation. To develop a median segmentation algorithm three components are needed: Generating a graph to work with, extracting seed regions, and computing a final combined segmentation result using the random walker algorithm; see [54] for details.

All the algorithms discussed above have been developed from intuitive heuristics rather than from the vantage point of a direct optimization. Indeed, only [54] mentions the median segmentation optimization function based on normalized mutual information. It is used to select the best segmentation from a set of combination segmentations with different number K of regions. In some sense this approach can be regarded as an approximation of generalized median segmentation by investigating the subspace of U (all possible segmentations of an image), which consists of the combination segmentations for the considered range of K.

The work [54] is motivated by two scenarios. Segmentation algorithms mostly have some parameters and their optimal setting is a non-trivial task. The authors propose to implicitly explore the parameter space (*without* the need of ground truth segmentation). A reasonable parameter subspace (i.e. a lower and upper bound for each parameter) is assumed to be known and is sampled into a finite number N of parameter settings. Then, the segmentation procedure is run for all these N parameter settings and a final combined segmentation of the N segmentations is computed. The rationale behind this approach is that the combination segmentation tends to be a good one within the explored parameter subspace. An extensive experimental study [55] clearly demonstrated the advantages of this combination approach. In addition to alleviating the parameter problem it is also possible to combine the results of different segmenters to benefit from their strengths.

15.8 Optimal Lower Bound in Metric Space

The discussion above reveals that the computation of generalized median is frequently of high time complexity and we thus need to resort to approximative approaches. In this case, however, the question of accuracy of the approximative generalized median $\tilde{p}$ arises. In [25] a lower bound is proposed to answer this question. An approximate computation method gives us a solution $\tilde{p}$ such that

$$\text{SOD}(\tilde{p}) = \sum_{q \in S} d(\tilde{p}, q) \geq \sum_{q \in S} d(\overline{p}, q) = \text{SOD}(\overline{p})$$

where SOD stands for sum of distances and $\overline{p}$ represents the (unknown) true generalized median. The quality of $\tilde{p}$ can be measured by the difference $\text{SOD}(\tilde{p}) - \text{SOD}(\overline{p})$. Since $\overline{p}$ and $\text{SOD}(\overline{p})$ are unknown in general, we resort to a lower bound $\Gamma \leq \text{SOD}(\overline{p})$ and measure the quality of $\tilde{p}$ by $\text{SOD}(\tilde{p}) - \Gamma$. Note that the relationship

$$0 \leq \Gamma \leq \text{SOD}(\overline{p}) \leq \text{SOD}(\tilde{p})$$

holds. Obviously, $\Gamma = 0$ is a trivial, and also useless, lower bound. We thus require Γ to be as close to $\text{SOD}(\overline{p})$ as possible.

The distance function $d(p, q)$ is assumed to be a metric. Given the input $S = \{p_1, p_2, \ldots, p_N\}$, the generalized median $\overline{p}$ is characterized by:

minimize $\text{SOD}(\overline{p}) = d(\overline{p}, p_1) + d(\overline{p}, p_2) + \cdots + d(\overline{g}, p_N)$ subject to

$$\forall i, j \in \{1, 2, \ldots, N\}, \ i \neq j, \quad \begin{cases} d(\overline{p}, p_i) + d(\overline{p}, p_j) \geq d(p_i, p_j) \\ d(\overline{p}, p_i) + d(p_i, p_j) \geq d(\overline{p}, p_j) \\ d(\overline{p}, p_j) + d(p_i, p_j) \geq d(\overline{p}, p_i) \end{cases}$$

$$\forall i \in \{1, 2, \ldots, N\}, \ d(\overline{p}, p_i) \geq 0$$

Note that the constraints except the last set of inequalities are derived from the triangular inequality of the metric $d(p, q)$. By defining N variables x_i, $i = 1, 2, \ldots, N$, we replace $d(\overline{p}, p_i)$ by x_i and obtain the linear program LP:

minimize $x_1 + x_2 + \cdots + x_N$ subject to

$$\forall i, j \in \{1, 2, \ldots, N\}, \ i \neq j, \ \begin{cases} x_i + x_j \geq d(p_i, p_j) \\ x_i + d(p_i, p_j) \geq x_j \\ x_j + d(p_i, p_j) \geq x_i \end{cases}$$

$$\forall i \in \{1, 2, \ldots, N\}, \ x_i \geq 0$$

If we denote the solution of LP by Γ, then the true generalized median $\overline{p}$ satisfies $\Gamma \leq \mathrm{SOD}(\overline{p})$, i.e. Γ is a lower bound for $\mathrm{SOD}(\overline{p})$.

For a fixed N value, any set S of N objects specifies $K = N(N-1)/2$ distances $d(p, q)$, $p, q \in S$, and can be considered as a point in the K-dimensional real space $\mathfrak{R}^K$. Due to the triangular inequality required by a metric, all possible sets of N objects only occupy a subspace $\mathfrak{R}_*^K \subset \mathfrak{R}^K$. Abstractly, any lower bound is thus a function $f : \mathfrak{R}_*^K \to \mathfrak{R}$. The lower bound Γ derived above is such a function. Does a lower bound exist that is tighter than Γ? This optimality question is interesting from both a theoretical and a practical point of view. In [25] it could be shown that there exists no lower bound that is tighter than Γ.

The lower bound above presumes a metric distance function $d(p, q)$. The triangle inequality of a metric excludes cases in which $d(p, r)$ and $d(r, q)$ are both small, but $d(p, q)$ is very large. In practice, however, there may exist distance functions which do not satisfy the triangle inequality. The work [10] extends the concept of metrics to so-called quasi-metrics with a relaxed triangle inequality. Instead of the strict triangle inequality, the relation:

$$d(p, r) + d(r, q) \geq \frac{d(p, q)}{1 + \varepsilon}$$

is required, where ε is a small nonnegative constant. As long as ε is not very large, the relaxed triangle inequality still retains the human intuition of similarity. Note that the strict triangle inequality is a special case with $\varepsilon = 0$. The lower bound above can be easily extended to quasi-metric distance functions.

15.9 Conclusion

This brief overview has demonstrated the universality of the generalized median concept. Various instances have been presented. Several applications have been discussed and indicate the substantial potential of generalized median.

References

[1] J. Astola, P. Haavisto and Y. Neuvo. Vector median filters. *Proceedings of the IEEE*, 78(4):678-689, 1990.

[2] F. Bartolini, V. Cappellini, C. Colombo and A. Mecocci. Enhancement of local optical flow techniques. in *Proceedings of 4th International Workshop on Time Varying Image Processing and Moving Object Recognition*, Florence, Italy, 1993.

[3] R. Bertolami and H. Bunke. Hidden Markov model based ensemble methods for offline handwritten text line recognition. *Pattern Recognition*, 41(11):3452–3460, 2008.

[4] H. Bunke and G. Allermann. Inexact graph matching for structural pattern recognition. *Pattern Recognition Letters*, 1:245–253, 1983.

[5] H. Bunke, A. Münger and X. Jiang. Combinatorial search versus genetic algorithms: A case study based on the generalized mean graph problem. *Pattern Recognition Letters*, 20:1271-1277, 1999.

[6] F. Casacuberta and M.D. de Antoni. A greedy algorithm for computing approximate median strings. in *Proceedings of National Symposium on Pattern Recognition and Image Analysis*, Barcelona, Spain, pages 193–198, 1997.

[7] D.-C. Cheng, A.Schmidt-Trucksäss, S.-H. Liu and X. Jiang. Improved arterial inner wall detection using generalized median computation. In *Computer Analysis of Images and Patterns*, X. Jiang and N. Petkov (Eds.), LNCS, Vol. 5702, pages 622–630. Springer, 2009.

[8] W.D. Curtis, A.L. Janin and K. Zikan. A note on averaging rotations. in *IEEE Virtual Reality Annual International Symposium*, pages 377–385, 1993.

[9] M.A. Eshera and K.S. Fu. A graph distance measure for image analysis. *IEEE Transactions on Systems, Man, and Cybernetics (Part B)*, 14(3):398–408, 1984.

[10] R. Fagin and L. Stockmeyer. Relaxing the triangle inequality in pattern matching. *Int. Journal on Computer Vision*, 28(3):219–231, 1998.

[11] M. Ferrer, E. Valveny, F. Serratosa, K. Riesen, and H. Bunke. An approximate algorithm for median graph computation using graph embedding. in *Proceedings of International Conference on Pattern Recognition*, 2008.

[12] M. Ferrer, E. Valvenya and F. Serratosa. Median graph: A new exact algorithm using a distance based on the maximum common subgraph. *Pattern Recognition Letters*, 30(5):579–588, 2009.

[13] M. Ferrer, D. Karatzas, E. Valveny and H. Bunke. A recursive embedding approach to median graph computation. In *Graph-Based Representations in Pattern Recognition*, A. Torsello et al. (Eds.), LNCS, Vol. 5534, pages 113–123. Springer, 2009.

[14] M. Ferrer, E. Valveny, F. Serratosa, I. Bardaji and H. Bunke. Graph-based k-means clustering: A comparison of the set versus the generalized median graph. In *Computer*

Analysis of Images and Patterns, X. Jiang and N. Petkov (Eds.), LNCS, Vol. 5702, pages 342–350. Springer, 2009.

[15] V. Filkov and S. Skiena. Integrating microarray data by consensus clustering. *International Journal on Artificial Intelligence Tools*, 13(4):863-880, 2004.

[16] A.L.N. Fred and J.M.N. Leitão. A comparative study of string dissimilarity measures in structural clustering. In *Proceedings of International Conference on Document Analysis and Recognition*, pages 385–394, 1998.

[17] A.L.N. Fred and A.K. Jain. Combining multiple clusterings using evidence accumulation. *IEEE Trans. on Pattern Analysis and Machine Intelligence*, 27(6):835–850, 2005.

[18] A. Goder and V. Filkov. Consensus clustering algorithms: Comparison and refinement. In *Proceedings of Workshop on Algorithm Engineering and Experiments*, San Francisco, pages 109–117, 2008.

[19] L. Grady. Random walks for image segmentation. *IEEE Transactions on Pattern Analysis and Machine Intelligence*, 28(11):1768–1783, 2006.

[20] C. Gramkow. On averaging rotations. *Int. Journal of Computer Vision*, 42(1/2):7–16, 2001.

[21] C. de la Higuera and F. Casacuberta. Topology of strings: Median string is NP-complete. *Theoretical Computer Science*, 230(1/2):39–48, 2000.

[22] A. Hlaoui and S. Wang. Median graph computation for graph clustering. *Soft Computing*, 10(1):47–53, 2006.

[23] X. Jiang, L. Schiffmann, and H. Bunke. Computation of median shapes. In *Proceedings of 4th. Asian Conference on Computer Vision*, Taipei, pages 300–305, 2000.

[24] X. Jiang, A. Münger and H. Bunke. On median graphs: Properties, algorithms, and applications. *IEEE Trans. on Pattern Analysis and Machine Intelligence*, 23(10):1144–1151, 2001.

[25] X. Jiang and H. Bunke. Optimal lower bound for generalized median problems in metric space. In *Structural, Syntactic, and Statistical Pattern Recognition*, T. Caelli et al. (Eds.), pages 143–151. Springer, 2002.

[26] X. Jiang, K. Abegglen, H. Bunke and J. Csirik. Dynamic computation of generalized median strings. *Pattern Analysis and Applications*, 6(3):185–193, 2003.

[27] X. Jiang, H. Bunke and J. Csirik. Median strings: A review. In *Data Mining in Time Series Databases*, M. Last et al. (Eds.), pages 173–192. World Scientic, 2004.

[28] X. Jiang, C. Marti, C. Irniger and H. Bunke. Distance measures for image segmentation evaluation. *EURASIP Journal on Advances in Signal Processing*, 1–10, 2006.

[29] X. Jiang and H. Bunke. Graph matching. In *Case-Based Reasoning on Images and Signals*, P. Perner (Ed.), pages 149–173. Springer-Verlag, 2008.

[30] Y. Jiang and Z.-H. Zhou. SOM ensemble-based image segmentation. *Neural Processing Letters*, 20(3):171–178, 2004.

[31] D. Justice and A. Hero. A binary linear programming formulation of the graph edit distance. *IEEE Trans. on Pattern Analysis and Machine Intelligence*, 28(8):1200-1214, 2006.

[32] J. Keuchel and D. Küttel. Efficient combination of probabilistic sampling approximations for robust image segmentation. In *Proceedings of DAGM-Symposium*, pages 41-50, 2006.

[33] S. Klavzar and H.M. Mulder. Median graphs: Characterizations, location theory and related structures. *J. Combinatorial Math. and Combinatorial Computing*, 30:103–127, 1999.

[34] T. Kohonen. Median strings. *Pattern Recognition Letters*, 3:309–313, 1985.

[35] M. Krivanek and J. Moravek. Hard problems in hierarchical tree clustering. *Acta Informatica*, 23:311–323, 1986.

[36] J.B. Kruskal. An overview of sequence comparison: Time warps, string edits, and macromolecules. *SIAM Reviews*, 25(2):201–237, 1983.

[37] F. Kruzslicz. Improved greedy algorithm for computing approximate median strings. *Acta Cybernetica*, 14:331–339, 1999.

[38] D. Lopresti and J. Zhou. Using consensus sequence voting to correct OCR errors. *Computer Vision and Image Understanding*, 67(1):39–47, 1997.

[39] V. Marti and H. Bunke. Use of positional information in sequence alignment for multiple classifier combination. In *Multiple Classifier Combination*, J. Kittler and F. Roli (Eds.), pages 388–398. Springer-Verlag, 2001.

[40] C.D. Martinez-Hinarejos, A. Juan and F. Casacuberta. Use of median string for classification. In *Proceedings of International Conference on Pattern Recognition*, pages 907–910, 2000.

[41] A. Marzal and E. Vidal. Computation of normalized edit distance and applications. *IEEE Trans. on Pattern Analysis and Machine Intelligence*, 15(9):926–932, 1993.

[42] M. Neuhaus and H. Bunke. *Bridging the Gap between Graph Edit Distance and Kernel Machines*. World Scientific, 2007.

[43] S.V. Rice, J. Kanai, and T.A. Narker. An algorithm for matching OCR-generated text string. In *Document Image Analysis*, H. Bunke et al. (Eds.), pages 263–272. World Scientific, 1994.

[44] K. Riesen and H. Bunke. A graph kernel based on vector space embedding. *International Journal of Pattern Recognition and Artificial Intelligence*, 23(6):1053–1081, 2009.

[45] K. Riesen and H. Bunke. Approximate graph edit distance computation by means of bipartite graph matching. *Image and Vision Computing*, 27(4):950–959, 2009.

[46] K. Riesen, X. Jiang and H. Bunke. Exact and inexact graph matching – Methodology and applications. In *Managing and Mining Graph Data*, C. Aggarwal and H. Wang (Eds.), pages 215–246. Springer, 2009.

[47] K. Rothaus, X. Jiang and M. Lambers. Comparison of methods for hyperspherical data averging and parameter estimation. In *Proceedings of International Conference on Pattern Recognition*, Vol. 3, pages 395–399, 2006.

[48] J.S. Sim and K. Park. The consensus string problem for a metric is NP-complete. *Journal of Discrete Algorithms*, 1(1):111–117, 2001.

[49] S. Sorlin and C. Solnon. Reactive tabu search for measuring graph similarity. In *Proceedings 5th International Workshop on Graph-based Representations in Pattern Recognition*, L. Brun and M. Vento (Eds.), LNCS, Vol. 3434, pages 172–182. Springer, 2005.

[50] A. Strehl and J. Ghosh. Cluster ensembles – A knowledge reuse framework for combining multiple partitions. *Journal of Machine Learning Research*, 3:583–617, 2002.

[51] R.A. Wagner and M.J. Fischer. The string-to-string correction problem. *Journal of the ACM*, 21(1):168–173, 1974.

[52] L. Wang and T. Jiang. On the complexity of multiple sequence alignment. *Journal of Computational Biology*, 1(4):337–348, 1994.

[53] P. Wattuya and X. Jiang. A class of generalized median contour problem with exact solution. In *Proceedings of SSPR*, Hong Kong, LNCS, Vol. 4109, pages 109–117. Springer, 2006.

[54] P. Wattuya, K. Rothaus, J.-S. Praßni and X. Jiang. A random walker based approach to combining multiple segmentations. In *Proceedings of International Conference on Pattern Recognition*, Tampa, Florida, 2008.

[55] P. Wattuya and X. Jiang. Ensemble combination for solving the parameter selection problem in image segmentation. In *Proceedings of SSPR*, Orlando, FL, LNCS, Vol. 5342, pages 392–401. Springer, 2008.

16

Ensembles Methods for Machine Learning

H. Allende[1,2], C. Moraga[3], R. Ñanculef[2] and R. Salas[2,4]

[1]*Facultad de Ingeniería y Ciencia, Universidad Adolfo Ibáñez, Viña del Mar, Chile;*
E-mail: hallende@uai.cl
[2]*Departamento de Informática, Universidad Técnica Federico Santa María,*
Casilla 110-V, Valparaíso, Chile; E-mail: {hallende, jnancu, rsalas}@inf.utfsm.cl
[3]*European Centre for Soft Computing, 33600 Mieres, Spain, and*
Technical University of Dortmund, 44221 Dortmund, Germany;
E-mail: mail@claudio-moraga.eu
[4]*Departamento de Ingeniería Biomédica, Facultad de Ciencias, Universidad de*
Valparaíso, Chile; E-mail: rodrigo.salas@uv.cl

Abstract

Ensemble methods have gained considerable attention from machine learning
and soft computing communities in the last years. There are several practical
and theoretical reasons, mainly statistical reasons, why an ensemble may be
preferred. A set of learners with similar training performance may have different generalization performance, when exposed to sparse data, large volume
of data or data fusion. The basic idea of an ensemble learning algorithm is to
build a predictive model by combining a set of learning hypotheses instead
of laboriously designing the complete map between inputs and responses in a
single step. Ensemble based systems have shown to produce favorable results
compared to those of single-expert system for a broad range of applications
such as financial, medical and social models, network security, web mining or
bioinformatics, to name a few. The focus of this chapter will be on ensembles
for classification and regression estimation. Design, implementation and application will be the main topics of the chapter. Specifically, conditions under
which ensemble based systems may be more beneficial than their single ma-

Patrick Shen-Pei Wang (Ed.), Pattern Recognition and Machine Vision – In Honor
and Memory of Professor King-Sun Fu, 247–261.

chine, algorithms for generating individual components of ensemble systems and various procedures through which the individual classifiers can be combined. Finally, future research directions will be included, like incremental learning, machine fusion, and other areas in which ensemble of machines have shown great promise.

Keywords: ensemble learning, diversity, multiple classifiers machine, classifier combination.

16.1 Ensembles: Basic Concepts

Probably the earliest idea of an ensemble may be traced back to a proposal of the Marquis of Condorcet in 1784 (see e.g. [3, 33]) saying that "if each voter has a probability p of being correct and the probability of a majority of voters being correct is P, then $p > 0.5$ implies that $P > p$. In the limit, for all $p > 0.5$, as the number of voters approaches infinity, P approaches 1". In the context of an ensemble of neural networks, for instance, the requirement that $p > 0.5$ is not difficult to satisfy.

When a classical multilayer feedforward neural network is prepared, the adjustment of its parameters is based on a training set, the elements of which are expected to be representatives of an unknown distribution. A neural network is said to have been successfully trained if it not only makes small errors with respect to the training set, but if it "discovers and learns" the unknown distribution. This is called the generalization capability of the network and it is measured against the error with respect to a test set not used during the training. Now consider the following scenario: several neural networks using possibly different samples (of a common data set), are trained with different training algorithms considering different activation functions. If all these neural networks are well trained they will achieve a good generalization, but highly probably they will make different generalization errors. The word diversity has been coined to denote this property [22]. A proper aggregation of these networks will lead to an ensemble, which will have a more robust performance than its individual components, the more different are the generalization errors, since a compensating effect may be induced. The idea of combining under diversity is not a development exclusive for neural networks. People working in forecast were already combining forecasts much earlier than the time – 1986 – when applications of neural networks started to grow [16]. Furthermore, in everyday life people are familiar with the principle behind combining under diversity (without using this name). For instance a patient will probably like to hear the opinion of several physicians before accepting a surgical intervention; before buying some home appliance, a person will probably like to know the opinion of other people that have the same device, etc.

In the following sections we will analyze the questions of generation, and target-oriented control of diversity to obtain effective ensembles.

16.2 Ensemble Methods

Ensemble algorithms have gained considerable attention from the machine learning and soft-computing communities in the last years [6, 37]. The basic idea is to build a learning hypothesis by combining a set of simple functions instead of carefully designing the complete map between X and Y in one single step. Well-known ensemble methods include Bagging [5], AdaBoost [13], Mixture of Experts [18], Stacking [54], Negative Correlation [25], and their many variations. These algorithms have demonstrated to be a flexible way to improve the generalization performance of a base learning algorithm in different tasks including classification [21], regression estimation [6, 10], novelty detection [40] and clustering [47].

As mentioned earlier, diversity in the generalization is of key importance for the design of ensembles. In principle all following alternatives contribute to this goal: (1) different initialization of the weights and biases of the component neural networks, (2) use of different training algorithms to adjust weights and biases, (3) use of different architectures for the component neural networks, (4) use of different training sets, and (5) enhancement of the negative correlation of errors in the components.

If for a given problem the cardinality of the available data set were large, it would be simple to obtain representative but disjoint data subsets to train the different neural networks building an ensemble. This certainly would provide a good diversity. In most cases, however, the data set is not that big. In that case bootstrapping [11] can be an effective method. In statistics, bootstrapping is a general purpose approach to statistical inference, falling within a broader class of resampling methods. It is used for estimating the sampling distribution of an estimator by resampling with replacement from the original sample. This has given origin to the method disclosed in 1995 by Breiman and known as Bagging [5] (short for "bootstrapping aggregating"), to successfully design ensembles. The available training data is resampled with replacement to generate additional training samples (called replicates), which are used to train the component neural network. Each component is trained with a replicate, independently of the results of other trainings. These will then be aggregated according to the task to be solved: (weighted) average in the case of regression (approximation, prediction) or (weighted) voting in the case of classification. Although very simple, this algorithm exhibits results in many cases superior to more elaborated algorithms. Recent works have been conducted in order to study the nature of the benefits provided by this and other sampling techniques. In [36] it is shown that for well behaved loss functions, certain data partition schemes can provide a generalization ability of the same order as Tikhonov regularization [49], a technique widely used to avoid overfitting (i.e. learning without generalization) [42]. The key observation is that after sampling, a training point does not influence all the hypotheses in the ensemble and hence the composite predictor results more stable than the individual ones. Similar results can be obtained for random sampling schemes if the probability that a point belongs to a given sample can be bounded. Using uniform resampling as in Bagging, the probability that a pattern appears in the resulting sample is exactly $1 - (1 - 1/n)^n$ which converges to ~ 0.632 as $n \to \infty$. This means that the expected fraction of hypotheses that are affected by a given training point is around 0.6. The ability of resampling to limit the effect of a training point on the resulting model has been carefully measured in several experiments [15]. Empirical evidence suggests that Bagging equalizes the influence of training points in such a way that points highly influential (the so called leverage points) are down-weighted. Since in most situations leverage points are badly influential, Bagging can improve generalization by making robust an unstable base learner. From this point of view, resampling has an effect similar to robust M-estimators in statistics, where the influence of sample points is bounded using appropriate loss functions. Since in uniform resampling all the points in the sample have the same probability of being selected, it seems counterintuitive that Bagging has the ability to selectively reduce the influence of leverage points. The explanation is in the nature of leverage points itself. Leverage points are usually isolated in the feature space while the others are located in more dense regions of the input space. To remove the influence of a leverage point it is enough to eliminate this point from the sample but to remove the influence of a non-leverage point we must in general remove a group of observations. Now, the probability that a group of size k be completely ignored by Bagging is $(1 - k/n)^n$ which decays exponentially with k. For $k = 2$ for example $(1 - k/n)^n \sim 0.14$ while $(1 - 1/n)^n \sim 0.37$. This reasoning is actually consistent with the formal definitions of stability of a learning algorithm presented for example in [36] and [4]. The importance of Bagging is not only that is able to produce successful ensembles, but that it has been analyzed [5] under which conditions the performance of an ensemble will be better than that of its components.

Another important algorithm for the design of ensembles is AdaBoost [13], introduced by Freund and Shapire in 1997, where the name is an abbreviation of "adaptive boosting". This algorithm has a theoretical background based on the "PAC" learning model [51]. The authors of this model were the first to pose the question of whether a weak learning algorithm that is slightly better than random guessing can be "boosted" into a strong learning algorithm. The classic AdaBoost takes as an input a training set $Z = \{(x_1, y_1), \ldots, (x_n, y_n)\}$ where each x_i is a variable that belongs to $X \subset \Re^d$ and each label y_i is in some characterizing set Y. Assume for now a classification scenario that $Y = \{1, -1\}$, i.e., the algorithm will be considered from the point of view of classification, as the authors did in their original proposal. AdaBoost calls a weak or base learning algorithm repeatedly in a sequence of stages $t = 1 \ldots T$. The main idea of AdaBoost is to maintain a sampling distribution over the training set. Let $D_t(i)$ be the sampling weight assigned to the example i at the t-th iteration. At the beginning of the algorithm the distribution is uniform, that is, the distribution $D_1(i) = 1/n$ for all i. At each round of the algorithm however, the weights of the incorrectly classified examples are increased, so that the following step the weak learner is forced to focus on the "hard" examples of the training set. The job of each weak or base learner is to find a hypothesis $h_t : X \to \{-1, 1\}$ appropriate for the distribution D_t. The goodness of the obtained hypothesis can be quantified as the weighted error:

$$\epsilon_t = \Pr_{i \sim D_t}[h_t(x_i) \neq y_i] = \sum_{i:h_t(x_i) \neq y_i} D_t(i).$$

Notice that the error is measured with respect to the distribution D_t on which the weak learner was trained. Once AdaBoost has computed a weak hypothesis h_t, it measures the importance that the algorithm assigns to h_t with the parameter

$$\alpha_t = \frac{1}{2} \ln \left(\frac{1 - \epsilon_t}{\epsilon_t} \right).$$

After choosing α_t the next step is to update the distribution D_t with the following rule,

$$D_{t+1}(i) = \frac{1}{Z_t} D_t(i) \exp(-\alpha_t y_i h_t(x_i)),$$

where Z_t is a normalization factor (to keep D_{t+1} a distribution). The effect of this rule is the following: when an example is misclassified its sampling weight for the next round is increased, and the opposite occurs when the classification is correct. This updating rule makes the algorithm focus on the "hard" examples, instead of on the correctly classified examples. After a sequence of T iterations have been carried out, the final classification hypothesis H is computed as

$$H_T(x) = \text{sign}\left(\sum_{t=1}^{T} \alpha_t h_t(x)\right).$$

The original AdaBoost [13] was developed for classification, but an adaptation for regression was introduced by Drucker in 1997 [10]. AdaBoost is inherently sequential, however a suggestion for parallelizing may be found in [29]; asymmetric boosting is discussed in [28], meanwhile quadratic boosting is addressed in [35]. Robust AdaBoost is introduced in [1].

A third family of algorithms to design ensembles by improving diversity, specially for regression problems, is based on "Negative Correlation" NC (a short term used to denote the idea of enhancing the negative correlation among the generalization error made by different components of an ensemble). In the case of a single neural network realizing a function $f_{NN} : \mathfrak{R}^n \rightarrow \mathfrak{R}$, it is known that the mean square error equals the square of the bias plus the variance (see e.g. [14]). In the case of ensembles, Ueda and Nakano proved in 1996 [50] that a similar decomposition is possible, but besides the square of the bias and the variance, also the covariance among components of the ensemble has to be considered. This is important, since only the covariance may be negative and thus it may contribute to decrease the square error of the ensemble. Motivated by these results, Rosen [39] suggested expressing the error of a learner as the sum of the square error and a weighted penalty function p to "decorrelate" the errors of the components of an ensemble with respect to the expected output. If $x \in \mathfrak{R}^n$ denotes an input to the ensemble, $y \in \mathfrak{R}$ denotes the expected output, and $f_i(x)$ represents the output of the i-th learner, then the core of the penalty function of Rosen is given by

$$p_i = \lambda(f_i(x) - y)\sum_{j=1}^{i-1}(f_j(x) - y),$$

where λ controls the amount of penalty to be considered. Liu continued this idea [25, 26] and developed a method, which he called (for the first time)

"Negative Correlation", where he included a way to train all components of the ensemble simultaneously, meanwhile Rosen considered a sequential training. Brown disclosed a new penalty function [6] in the line of the "ambiguity decomposition" introduced earlier by Kroch and Vedelsby [20]. Brown's penalty function may be given a representation close to that of Rosen and Liu, but also a compact one as the following:

$$p_i = \lambda(f_i(x) - f_{\text{ens}}(x)) \sum_{i \neq j} (f_j(x) - f_{\text{ens}}(x)) = -\lambda(f_j(x) - f_{\text{ens}}(x))^2,$$

where $f_{\text{ens}}(x)$ denotes the output produced by the ensemble. With this, the error ϵ_i of the i-th learner is given by

$$\epsilon_i = 1/2(f_i(x) - y)^2 - \lambda(f_i(x) - f_{\text{ens}}(x))^2.$$

This error function consists of the (normalized) square deviation with respect to the target and the (weighted) square deviation with respect to the output of the ensemble. The first part represents the accuracy of the component meanwhile the second, represents the diversity. The scaling factor λ regulates the tradeoff between these two features. Brown showed [6] that λ should be asymptotically bounded by 1, but for a small number of components λ is larger than 1. In the case of an ensemble with only 2 learners, λ equals 2. In a very impressive example using the breast cancer benchmark Brown showed [6] that if the ensemble has only a few components, a fine adjustment of λ may thoroughly improve its performance.

All former methods discussed above are oriented to produce and possibly control the diversity to obtain an ensemble, whose performance is better than that of its components. A possibly different approach to use diversity was introduced by Wolpert under the name "Stacking" (an abbreviation of "Stacked Generalization"). Stacking represents a novel concept to improve the performance of an ensemble by correcting the errors [54]. This algorithm addresses the issue of classifier bias with respect to a training set, by using this information in the learning process to improve the classification task. The main philosophy behind this algorithm is to use a new classifier to correct the errors of a previous one (hence the classifiers are "stacked" on top of one another). The idea is that if an input-output pattern (x, y) is excluded of the training set of the current hypothesis h_i, after the training has finished for h_i, the y output can still be used to correct the model's error.

16.3 Modular Approaches

Modular Systems were first inspired by the divide an conquer paradigm for algorithm design in computer science, where a hard problem is divided in smaller simpler parts that are easy to solve and the results are combined to obtain the whole solution (according to Knuth, von Neumann introduced the earlier technique as an application of this approach; see [19] for more details about this paradigm). Later, the modularity in human brain function and structure inspired the construction of modular machine learning models (see [18]).

Recent advances in neuropsychology and neurobiology have experimentally proven the structural and functional modularity of the brain, and this capability is a key property of the efficient and intelligent behaviour in human and animal beings. Structural engineering design, learning efficiency and performance, and model complexity issues are important factors that motivate the modular design of machine learning techniques. One key property of modular systems is their rapid and efficient learning, but furthermore, the modular structure enables a system to better generalize its learned behaviour to new instances [17]. Azam [2] highlights some capabilities of modular systems as follows: (1) model complexity reduction due to the specialized modules that are simpler and smaller than their monolithic counterpart solution; (2) fault tolerance when some module fails; (3) the scalability of the system when additional information needs to be stored in a new module; (4) computational efficiency by dividing the task into a set of simpler tasks reducing both computational effort and time; (5) better learning capacity of complex tasks; (6) economy of learning by reusing or adapting the modules needed for an specific new task.

Modular Ensembles Methods are characterized by their ability of inducing a hard or soft partition of the input space. Each partition is accomplished by creating basic learners that best model a specific region where they become experts. The outputs of the learners are mediated by an integrated unit to obtain the aggregated output of the system. Moreover, the integrating unit selects which modules should learn which training patterns. Under these conditions, the expertise of the basic learners is obtained by means of a competitive and cooperative learning, where each basic learner competes to map a region and at the same time the integration unit mediates this competition.

Jacobs et al. proposed a technique known as Mixture of Experts (ME) [18] as a generalization of the statistical mixture models. The ME architecture consists of K basic learners called experts and a gating network. The experts solve a function approximation problem over local regions of the input space, so the architecture needs a mechanism to identify for each input which experts are more adequate to model the desire output, work that is accomplished by the gating network. The gating network implements a soft partition of the input space into regions corresponding to experts; the outputs of the experts are weighted as dictated by the gating networks to obtain a combined output.

16.4 Future Work: Dealing with Complex Environments

Advances in computing, communication technologies, storage methods and user-interaction paradigms have resulted in many large scale, highly dynamic and decentralized data environments which naturally calls for a new generation of knowledge extraction tools. Fraud detection in electronic commerce, content annotation on social networks or home surveillance based on sensor networks, are examples of applications that challenge the view of a classifier as a monolithic and static component but pose the requirement of a model constantly adapting to new observations, capable to extract and integrate information from different sources and dealing with high volumes of maybe heterogeneous data. The flexibility and low implementation cost characteristic of ensemble methods make them a feasible and attractive solution to deal with the constraints underlying these domains.

Data arriving continuously in time is called *streaming data*. Probably the simplest approach to process a data stream consists in periodically rebuilding the model from time to time putting together all the old and new observations and retraining a standard learner. However, a data stream is usually in vast volume, changing dynamically and possibly arriving to a very high rate which prevents the use of this approach in practice. The learner is required in contrast to *incrementally* update the current model by detecting novel information without cumulating all the data. A straightforward way to implement incremental ensembles is to use base members already supporting incremental learning. For example in [45] each member of a fixed size set of incremental decision trees is continuously trained with the new observations. Old models can be replaced with "fresh" ones when the error goes beyond a limit. An advantage of this approach is scalability, however the ensemble does not result in strong advantages with respect to the use of a single incremental decision tree. Another approach consists in different component members to learn new increments of data. For example in [52] a new classifier is triggered when the number of new observations reaches a given limit and trained on this data. A disadvantage of this approach is that when new observations do not contain new knowledge several redundant classifiers are generated. Dealing with this issue several approaches have focused on filtering useful incoming observations and training new component members from them. In [46] the way for detecting new knowledge is that the observations be misclassified by the current model. Most algorithms are however based on an AdaBoost-inspired resampling, weighting the observations according to the error of the current model [38, 43]. Different strategies to combine the obtained set of classifiers have been explored, but since resampling induces differential pieces of knowledge; most of them are not fixed but are dynamically adjusted depending on the instance to be predicted. The performance as well as the scalability of these methods in real streaming environments depend on the way in which observations are weighted, the way in which weights are dynamically adjusted and the periods at which new classifiers are incorporated to the ensemble. Probably a best strategy would combine characteristics of online component learners and resampling based generation of new classifiers, but up to our knowledge, efficient implementations of this alternative remain quite unexplored. An additional problem of dealing with data streams appears when the patterns underlying the observations change over time, phenomenon known as *concept drift*. In a content annotation application, for example, the features defining a given category can become no longer valid due to a change in preferences of the users. In this case, our algorithm should be capable not only to integrate new knowledge but also to refine, correct and possible eliminate previously acquired knowledge which is no longer valid. Most ensemble based algorithms to deal with data streams mentioned before have considered the problem of drifting observations but the problem is far away from being solved, regarding mainly the accurate detection of a concept drift scenario, the efficiency/scalability of the model in terms of the number of classifiers cumulated along the learning process, and from the theoretical side the conditions under which learning is possible in drifting scenarios.

Another dimension of current data rich environments is the decentralized way in which data sources are available. Energy, time, cost, security and privacy concerns prevent the centralization of data in a single highly controlled computation node, as required by traditional algorithms. New learning systems need to provide exact or approximate solutions to such ideal but infeasible centralized solution, minimizing the communication costs, guaranteeing robustness to changes in the availability of the datasets and being scalable in the size of the network and the dataset, both of which could be large scale. Ensemble systems are emerging as a promising solution to this problem. In [30] simulations are presented to show that an ensemble of decision trees or k-nearest neighbors classifiers trained with only partial information of the complete dataset can approximate reasonably well a centralized solution even without communicating the data sources in the training process. Similar evidence is presented in the work of [8]. In [23] a synchronous parallelization of AdaBoost to learn from distributed data is proposed, in which the sampling distribution weights are computed considering the results obtained from the different data sources. Although the algorithm is exact, this requires a high level of synchronization and the final network load can be prohibitive if convergence is slow. Taking into account the topology of the network on which data sources resides [53] proposes to consider in contrast only the information contained in a neighborhood of each node, obtaining good approximations to an equivalent base learner trained in a centralized way. Recent works like [44] have also studied the effect of a vertical partitioning of datasets, where the data sources can store values for different subsets of features of a given example.

Certainly in this emerging scene of ensemble methods being applied to decentralized environments several questions remains yet open for new research. For example, up to our knowledge, the effect of unbalanced data sources on the final performance of the model as well as convergence times for synchronous procedures has not been covered. Research in distributed data mining with other techniques like (single) support vector machines [27] has shown that local computation times and communication overheads are monotonically deteriorated when distribution through the network is unbalanced. Another question to address is when the feasibility of ensemble approaches to successfully work in these environments depends on the sensibility of base learners through data partition, either horizontally or vertically. Finally, the effect of the (logical) network topology (through which a node is able to interact with others) on the tradeoff between local efficiency and the cooperation overhead needs to be studied and considered in the design.

16.5 Closing Remarks

Research in the area of ensembles is a very dynamic field, which, among other things, has motivated a fruitful dialog between people interested in neural networks and statisticians. Since the appearance of Bagging and AdaBoost, many variations have been published customizing these algorithms for particular classes of applications (see e.g. www.boosting.org). Locality for diversity and for Negative Correlation has been discussed [31, 32], to improve efficiency keeping the effectiveness. Even though, a great number of ensembles use feedforward neural networks with sigmoidal activation, positive results based on RFB-networks [7] as well as SVM [24] have been reported. Paying attention to special features of real-world problems, work has been done in adaptation to changing environment [21] and in incremental learning [12, 38]. Other areas of Soft Computing have also been considered for the design of ensembles: A genetic algorithm has been used to obtain a highly diverse set of accurate trained neural networks [34] and methods of fuzzy logic have been used to build/design combination rules [9,48]. Boosting weak learners to become strong learners is not an exclusive topic of research in neural networks and has a long history in other areas, as mentioned earlier. All ensembles mentioned above combine the output signals of the basic learners to generate the final output. This may be considered as "data fusion". A different, but close related approach is the one that focuses on the fusion of the (architecture of the) basic learners – "machine fusion" – to obtain one final machine with improved performance [41].

Acknowledgements

This work was supported in part by the research grants: Fondecyt 1070220, DGIP-UTFSM, Chile, and in part by the Foundation for the Advancement of Soft Computing, Mieres, Spain. Thanks go to our colleages at INCA Laboratory and the Departament on Informatics of the Federico Santa María Technical University for the wonderful intellectual discussion. Special thanks are due to Héctor Allende-Cid for his valuable contribution and collaboration during the preparation of this chapter.

References

[1] H. Allende-Cid, R. Salas, R. Ñanculef, and H. Allende. Robust alternating AdaBoost. In *Progress in Pattern Recognition, Image Analysis and Applications*, Lecture Notes in Computer Science, Vol. 4756, pages 427–436. Springer, 2008.

[2] F. Azam. Biologically inspired modular neural networks. PhD Thesis, Virginia Polytechnic Institute and State University, 2000.

[3] P.J. Boland. Majority systems and the Condorcet jury theorem. *Statistician*, 38(3):181–189, 1989.

[4] O. Bousquet and A. Elisseeff. Stability and generalization. *Journal of Machine Learning*, 2(1):499–526, 2002.

[5] L. Breiman. Bagging predictors. *Machine Learning*, 24(2):123–140, 1996.

[6] G. Brown and J. Wyatt. The use of the ambiguity decomposition in neural network ensemble learning methods. In T. Fawcett and N. Mishra (Eds.), *Proceedings of 20th International Conference on Machine Learning (ICML'03)*, pages 67–74, 2003.

[7] G. Brown, J. Wyatt, and Y. Bengio. Managing diversity in regression ensembles. *Journal of Machine Learning Research*, 6:2005, 2005.

[8] N. Chawla, L. Hall, K. Bowyer, and P. Kegelmeyer. Learning ensembles from bites: A scalable and accurate approach. *Journal of Machine Learning Research*, 5:421–451, 2004.

[9] S.B. Cho and J.H. Kim. Combining multiple neural networks by a fuzzy integral for robust classification. *IEEE Transactions on Systems, Man, and Cybernetics*, 25(2):380–384, 1995.

[10] H. Drucker. Improving regressors using boosting techniques. In *Machine Learning: Proceedings of the Fourteenth International Conference*, pages 107–115, 1997.

[11] B. Efron and R. Tibshirani. *An Introduction to the Bootstrap*. Chapman & Hall, 1993.

[12] Z. Erdem, R. Polikar, F. Gurgen, and N. Yumusak. Ensemble of SVMs for incremental learning. In *Multiple Classifier Systems*, Lecture Notes in Computer Science, Vol. 3541, pages 246–256. Springer-Verlag, 2005.

[13] Y. Freund and R.E. Schapire. A decision-theoretic generalization of on-line learning and an application to boosting. In *EuroCOLT'95: Proceedings of the Second European Conference on Computational Learning Theory*, London, UK, pages 23–37. Springer-Verlag, 1995.

[14] S. Geman, E. Bienenstock, and R. Doursat. Neural networks and the bias/variance dilemma. *Neural Comput.*, 4(1):1–58, 1992.

[15] Y. Grandvalet-Heudiasyc and Y. Grandvalet. Bagging down-weights leverage points. In *IJCNN*, Vol. IV, pages 505–510. IEEE, 2000.

[16] C.W.G Granger. Combining forecasts – Twenty years later. *Jr. of Forecasting*, 8(1):167–173, 1989.

[17] B.L.M. Happel and J.M.J. Murre. The design and evolution of modular neural network architectures. *Neural Networks*, 7:985–1004, 1994.

[18] M.I. Jordan and R.A. Jacobs. Hierarchical mixtures of experts and the EM algorithm. Technical Report AIM-1440, Massachusetts Institute of Technology, 1993.

[19] D. Knuth. *The Art of Computer Programming, Volume 3: Sorting and Searching*. Addison-Wesley, 1998.

[20] A. Krogh and J. Vedelsby. Neural network ensembles, cross validation, and active learning. In *Advances in Neural Information Processing Systems*, pages 231–238. MIT Press, 1995.

[21] L.I. Kuncheva. Classifier ensembles for changing environments. In *In Multiple Classifier Systems*, pages 1–15. Springer, 2004.

[22] L.I. Kuncheva and C.J. Whitaker. Using diversity with three variants of boosting: Aggressive, conservative and inverse. In *Proceedings of International Workshop on Multiple Classifier Systems*, Lecture Notes in Computer Science, Vol. 2364, pages 81–90. Springer, 2002.

[23] A. Lazarevic and Z. Obradovic. Boosting algorithms for parallel and distributed learning. *Distriuted Parallel Databases*, 11(2):203–229, 2002.

[24] C.A.M. Lima, A.L.V. Coelho, and F.J. Von Zuben. Hybridizing mixtures of experts with support vector machines: Investigation into nonlinear dynamic systems identification. *Inf. Sci.*, 177(10):2049–2074, 2007.

[25] Y. Liu and X. Yao. Ensemble learning via negative correlation. *Neural Networks*, 12:1399–1404, 1999.

[26] Y. Liu and X. Yao. Simultaneous training of negatively correlated neural networks in an ensemble. *IEEE Transactions on Systems, Man, and Cybernetics, Part B: Cybernetics*, 29:716–725, 1999.

[27] Y. Lu, V. Roychowdhury, and L. Vandenberghe. Distributed parallel support vector machines in strongly connected networks. *IEEE Transactions on Neural Networks*, 19(7):1167–1178, July 2008.

[28] H. Masnadi-Shirazi and N. Vasconcelos. Asymmetric boosting. In *ICML'07: Proceedings of the 24th International Conference on Machine Learning*, pages 609–619. ACM, 2007.

[29] S. Merler, B. Caprile, and C. Furlanello. Parallelizing AdaBoost by weights dynamics. *Comput. Stat. Data Anal.*, 51(5):2487–2498, 2007.

[30] C. Moretti, K. Steinhaeuser, D. Thain, and N.V. Chawla. Scaling up classifiers to cloud computers. In *Proceedings of the ICDM 2008*, pages 472–481. IEEE Computer Society, 2008.

[31] R. Ñanculef, C. Valle, H. Allende, and C. Moraga. Ensemble learning with local diversity. In S. Kollias, A. Stafylopatis, W. Duch, and E. Oja (Eds.), *Artificial Neural Networks – ICANN 2006*, Lecture Notes in Computer Science, Vol. 4131, pages 264–273. Springer, 2006.

[32] R. Ñanculef, C. Valle, H. Allende, and C. Moraga. Local negative correlation with resampling. In E. Corchado, H. Yin, V. Botti, and C. Fyfe (Eds.), *Intelligent Data Engineering and Automated Learning – IDEAL 2006*, Lecture Notes in Computer Science, Vol. 4224, pages 570–577. Springer, 2006.

[33] S.I. Nitzan and J. Paroush. *Collective Decision Making*. Cambridge University Press, 1985.

[34] D. W. Opitz, J.W. Shavlik, and O. Shavlik. Actively searching for an effective neural-network ensemble. *Connection Science*, 8:337–353, 1996.

[35] T.V. Pham and A.W.M. Smeulders. Quadratic boosting. *Pattern Recognition*, 41(1):331–341, 2008.

[36] T. Poggio, R. Rifkin, S. Mukherjee, and A. Rakhlin. Bagging regularizes. Technical Report, AI Memo 2002-003, CBCL Memo 214, MIT AI Lab, 2002.

[37] R. Polikar. Bootstrap – Inspired techniques in computation intelligence. *IEEE Signal Processing Magazine*, 24(4):59–72, July 2007.

[38] R. Polikar, L. Udpa, S. Udpa, and V. Honavar. Learn++: An incremental learning algorithm for supervised neural networks. *IEEE Transactions on System, Man and Cybernetics*, 31:497–508, 2001.

[39] B. Rosen. Ensemble learning using decorrelated neural networks. *Connection Science*, 8:373–384, 1996.

[40] G. Rtsch, S. Mika, B. Schlkopf, and K.R. Mller. Constructing boosting algorithms from svms: An application to one-class classification. *IEEE Transactions on Pattern Analysis and Machine Intelligence*, 24(9):1184–1199, 2002.

[41] C. Saavedra, R. Salas, H. Allende, and C. Moraga. Fusion of topology preserving neural networks. In *Proceedings of HAIS 2009*, Lecture Notes in Artificial Intelligence, Vol. 5572, pages 517–524. Springer-Verlag, 2009.

[42] B. Schölkopf and A. J. Smola. *Learning with Kernels: Support Vector Machines, Regularization, Optimization, and Beyond*. MIT Press, 2001.

[43] M. Scholz and R. Klinkenberg. Boosting classifiers for drifting concepts. *Intelligent Data Analysis*, 11(1):3–28, 2007.

[44] D. Skillicorn and S. McConnell. Distributed prediction from vertically partitioned data. *Journal of Parallel Distributed Computing*, 68(1):16–36, 2008.

[45] K.O. Stanley. Learning concept drift with a committee of decision trees. Technical Report AI-03-302, Department of Computer Sciences, University of Texas at Austin, 2003.

[46] N. Street and Y. Kim. A streaming ensemble algorithm (sea) for large-scale classification. In *Proceedings of KDD 2001*, pages 377–382. ACM, 2001.

[47] A. Strehl, J. Ghosh, and C. Cardie. Cluster ensembles – A knowledge reuse framework for combining multiple partitions. *Journal of Machine Learning Research*, 3:583–617, 2002.

[48] C. Sung-Bae and J.H. Kim. Multiple network fusion using fuzzy logic. *IEEE Transactions on Neural Networks*, 6(2):497–501, March 1995.

[49] A. Tikhonov and V. Arsenin. *Solutions of Ill-Posed Problems*. Winston, 1977.

[50] N. Ueda and R. Nakano. Generalization error of ensemble estimators. In *IEEE International Conference on Neural Networks*, Vol. 1, pages 90–95, June 1996.

[51] L.G. Valiant. A theory of the learnable. *Commun. ACM*, 27(11):1134–1142, 1984.

[52] H. Wang, W. Fan, P. Yu, and J. Han. Mining concept-drifting data streams using ensemble classifiers. In *Proceedings of the ACM SIGKDD 2003*, pages 226–235. ACM, 2003.

[53] S. Wang and C. Zhang. Collaborative learning by boosting in distributed environments. In *Proceedings of the ICPR 2008*, pages 1–4, 2008.

[54] D.H. Wolpert. Stacked generalization. *Neural Networks*, 5:241–259, 1992.

17

Modern Human-Machine Modelling Using Information, Control, and Complexity Theory

D.W. Repperger[1], C.A. Phillips[2] and J.M. Flach[3]

[1]711 Human Performance Wing, Air Force Research Laboratory, Wright-Patterson Air Force Base, Dayton, Ohio 45433, USA; E-mail: daniel.repperger@wpafb.af.mil
[2]Department of Biomedical, Industrial, and Human Factors Engineering, Wright State University, Dayton, Ohio 45435, USA
[3]Department of Psychology, Wright State University, Dayton, OH 45435, USA

Abstract

The Cybernetics viewpoint for modeling and analyzing a human-machine system within the context of control and information theory is first pursued. The Fitts' Law paradigm, one of the most highly embraced mathematical developments employed to model human movement, is first analyzed from both a control theory and information theory perspective. Additionally, the modern field of complexity theory presents a third viewpoint on this classical law in the study of human-machine system interaction. Both across and within the fields of complexity, information and control theory are compared and contrasted discovering links between the constituent areas involving a fundamental law of human motor response.

Keywords: Fitts' Law, control and communication models, complexity.

Patrick Shen-Pei Wang (Ed.), Pattern Recognition and Machine Vision – In Honor and Memory of Professor King-Sun Fu, 263–273.

17.1 Introduction

17.1.1 Cybernetics from N. Wiener to Modern Times

The area of Cybernetics initiated in the early days of Professor Norbert Wiener [11] to the present time has evolved into modern models of human-machine interactions based on concepts from control and information theory. The basic premise that Cybernetics is interdisciplinary and concerned with regulation and feedback was rooted in its early definition. Combining control and information theory models to characterize human response is not a trivial process. The control system's role in cybernetics is due to the need to regulate systems and to provide feedback, when it may be beneficial. Physical systems exhibit dynamics which attempt to optimize certain variables in the environment subject to constraints under conditions of uncertainty. Professor King-Sun Fu [2] had revolutionized thinking from traditional means by viewing pattern recognition and other problems within and across multiple and eclectic contexts. A theoretical framework to examine human-machine interaction across the fields of control and information theory also requires showing how one model may relate to another.

Collectively, in the area of human-machine systems, the authors of this chapter have considered [1, 3–5, 7–9] both control and information theory models of human-machine interaction and performance. In addition, the link between the control action and information-theoretic variables [3,7,8] for humans performing certain motion tasks can be objectively determined through the Fitts' Law paradigm, which is described in the sequel. Such methods offer not only a theoretical means to understand the cybernetic relations of humans with machines, but also how to empirically validate such interactions in a quantitative manner. This approach to modelling and analysing performance in human-machine systems is consistent with the early concepts of Professors N. Wiener [11] and King-Sung Fu [2] to provide an objective means of quantifying and understanding the control theory and information theory interactions of biological systems with their external environments.

Fitts' Law is now investigated under the purview of several different technical areas. The basic motor tracking task is first described.

17.2 The Basics of Fitts' Law – Information Theory Model

Figure 17.1 illustrates a physical rendering of Fitts' Law to be analyzed herein to model human motor response.

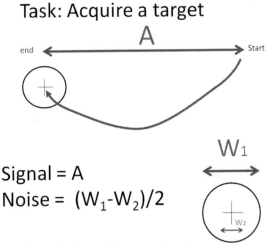

Task: Acquire a target

Signal = A

Noise = $(W_1 - W_2)/2$

Goals: Minimum time, maximum accuracy.

Figure 17.1 Fitts' Law as a reciprocal tapping task

The human will start at the right with a stylus object of diameter W_2. To complete the task, the goal is to move the stylus to the left and end up inside the circle of diameter W_1 for a certain minimum amount of time. The amplitude of movement is A, which can be considered analogous to a signal $S(t)$. The noise (randomness or slop in the system) in the processes is related to

$$\text{noise or uncertainty} = \eta(t) = \frac{1}{2}(W_1 - W_2) \qquad (17.1)$$

The goal is to complete the task in minimum time but with sufficient accuracy so as to not overshoot the circle of diameter W_1. To recapitulate, the two important variables are

$S(t)$ = signal variable with amplitude A

$\eta(t)$ = noise variable defined in Equation (17.1)

The data resulting from Figure 17.1 are now summarized in Figure 17.2. Figure 17.2 presents a plot of the data collected from a scenario such as in Figure 17.1, if different values of the amplitude A and circle sizes W_1 and W_2 are varied. The straight line is obtained (with high correlation) when plotting on the y axis the acquisition time in seconds versus task difficulty on the x

Information Theoretic Property:

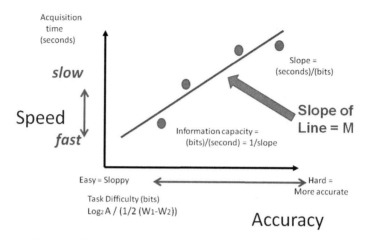

Figure 17.2 Fitts' Law data for information capacity calculation

axis. The units on the x axis are bits where

$$x \text{ axis units} = \text{bits} = \log_2 \left[\frac{A}{\frac{1}{2}(W_1 - W_2)} \right] \tag{17.2}$$

The more difficult task (high accuracy) is plotted to the far right in Figure 17.2. To the left are the less difficult tasks (lower accuracy). A speed accuracy tradeoff results (fast acquisition times occur for easy (less accurate tasks)). The slope M of the line has units of seconds/bit. However the reciprocal slope of the line has units of bits/second which is analogous to the capacity in an information theory channel.

If M is the slope of the line in Figure 17.2, then channel capacity can be written as

channel capacity $= 1/M$ (slope of the line in Figure 17.2)   (17.3)

The traditional information channel viewpoint is described next.

Following the classical concept of an information channel [10], in Figure 17.3 the input signal is $S(t)$. The channel or environment adds noise $\eta(t)$. The signal $S(t) + \eta(t)$ is obtained at the receiver. From Shannon [10], the maximum capacity through the system can be determined:

$$\text{channel capacity} = \text{BW} \log_2 \left(1 + \frac{S}{N} \right) \tag{17.4}$$

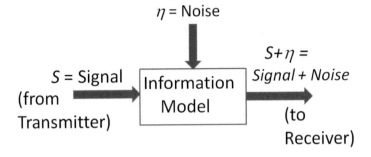

Shannon Information Channel

Figure 17.3 Information theory model

It is noted from [10] that S and N have units of power (not amplitude), and the bandwidth (BW) has units of 1/seconds. Channel capacity has units of bits/second. Bits are actually a dimensionless quantity but signify a log scale conversion.

17.3 Fitts' Law and Control Theory

Fitts' Law can be derived from a control theory perspective and two such derivations are presented here. The purpose is to understand the link between the limb or arm motion, the time for execution, the relationship between human-machine bandwidth, and the pertinent control theory parameters as well as the effect of uncertainty on the limb's response. Two cases will be considered. Case 1 will deal with the situation of no time delay; case 2 will deal with the situation of a nonzero time delay, which is not insignificant.

17.3.1 Case 1 – No Time Delay Approximation

For a position tracking system in Figure 17.4 the reference signal is $R(s)$ (using Laplace transforms) with the output being $C(s)$. The error signal is $E(s)$. The forward loop gain is approximated by the human-machine transfer function [6] as follows:

$$\text{human-machine transfer function} \approx \frac{\omega_c}{s} \qquad (17.5)$$

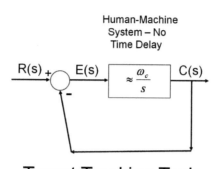

Target Tracking Task

Figure 17.4 Human-machine system with no time delay

where ω_c is termed the crossover frequency. The transfer function between $E(s)$ and $R(s)$ can be written as

$$\frac{E(s)}{R(s)} = \frac{1}{1 + \frac{\omega_c}{s}} \qquad (17.6)$$

Thus from Equation (17.6), the human-machine bandwidth is ω_c radians/second. If $R(s)$ is a step input (Laplace transform of 1/s) then it is easy to show that

$$E(s) = \frac{1}{s + \omega_c} \qquad (17.7)$$

with time domain solution of the error signal

$$e(t) = e(0)\varepsilon^{-\omega_c t} \qquad (17.8)$$

where $\varepsilon = 2.71828+$. But $e(0) = A$ and solving for the time to complete the acquisition of the target can be written as

$$t = -\frac{1}{\omega_c} \log_\varepsilon \frac{e(t)}{A} \qquad (17.9)$$

At the terminal time t_f when the task is complete, the error signal is inside the circle with diameter W_1. Denote as the tolerance of completion of the task:

$$e(t_f) = \frac{1}{2}W := \frac{1}{2}[W_1 - W_2] \qquad (17.10)$$

If the initial task starts at time $t_0 = 0$, then the total time to complete the task is

$$t_f - t_0 = t_f = -\frac{1}{\omega_c}\left[\log_\varepsilon\left(\frac{\frac{1}{2}W}{A}\right) - 0\right] \qquad (17.11)$$

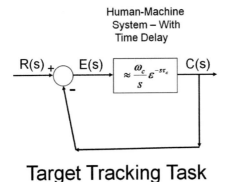

Target Tracking Task

Figure 17.5 Human-machine system with time delay

which is in a form similar to Fitts' Law:

$$t_f = \frac{1}{\omega_c} \log_\varepsilon \left(\frac{2A}{W} \right)$$ (17.12)

Thus the slope of the Fitts' Law curve is proportional to $(1/\omega_c)$ and the capacity of the information channel (reciprocal slope) is proportional to ω_c. Hence the bandwidth of the human-machine system (ω_c) is proportional to the information capacity. Systems with large bandwidth (large ω_c values) have high information capacity, etc. The case of a significant time delay in the loop can also be handled analytically.

17.3.2 Case 2 – Time Delay in the Human-Machine System

In [8, appendix c] the entire derivation was obtained for the situation in Figure 17.5 where the time delay (τ_c) was not insignificant in the human-machine system.

Since brevity must be the style here, only the points to be made are the following presumptions when considering the case of nonzero time delay:

(a) A first order Pade' approximation had to be utilized for the time delay.
(b) A dominant eigenvalue was found to contribute to 90% of the solution.
(c) The exponential form of Equation (17.12) still holds for the case of nonzero time delay. Thus the approximation in Equation (17.12) is valid and Fitts' Law is related to ω_c as stated previously. Hence the effect of time delay does not seem to be a major issue.

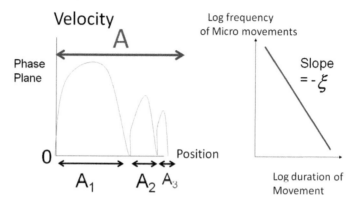

Figure 17.6 Relations to complexity theory and power laws

17.4 Complexity Theory and Fitts' Law

This last investigation will consider a complexity viewpoint on Fitts' Law and relate it to the previously defined quantities. From Figure 17.6, a phase plane is plotted (velocity versus position) of the arm trajectory motion from Figure 17.1. There appears to be a self similarity property in the micro structure of movements where the ratios

$$\frac{A_2}{A_1} = \frac{A_3}{A_2} = \xi, \ldots \tag{17.13}$$

The fractional ratio ξ can be found from the sum of the micro movements which should total to A, the total distance to traverse, i.e.

$$A = A_1 + A_2 + A_3 + A_4 + A_5 + \cdots \tag{17.14}$$

or

$$A = A_1 + \xi A_1 + \xi^2 A_1 + \xi^3 A_1 + \xi^4 A_1 + \cdots \tag{17.15}$$

or

$$A = A_1[1 + \xi + \xi^2 + \xi^3 + \xi^4 + \cdots] \tag{17.16}$$

But

$$[1 + \xi + \xi^2 + \xi^3 + \xi^4 + \cdots] = \frac{1}{1 - \xi} \tag{17.17}$$

Thus

$$A = A_1 \frac{1}{1 - \xi} \tag{17.18}$$

The presumption is that the first sub-movement is within the diameter of the larger circle in Figure 17.1 yielding

$$A - A_1 \approx W_1 \tag{17.19}$$

Thus A_1 can be approximated via

$$A_1 \approx A - W_1 \tag{17.20}$$

But

$$A - A_1 = W_1 \Rightarrow \frac{W_1}{A_1} = \frac{\xi}{1 - \xi} \tag{17.21}$$

which is a measure of the normalized noise in the system. Hence the slope of the power law curve in Figure 17.6 $(-\xi)$ is proportional to the noise in the information channel model. From all these results some key links between the technical disciplines of control theory, information theory, and complexity theory can be summarized.

17.5 Unifying the Different Fields

So far some of the following conclusions may be drawn:

(a) From a control theory perspective, the human-machine bandwidth (ω_c) in Figure 17.4 is directly proportional to the information channel capacity (reciprocal slope $1/M$ in Figure 17.2).

(b) From a complexity perspective, the large negative slope in Figure 17.6 (larger values of ξ) is related to a noisy system or operation on the left-hand side of Figure 17.2 where sloppy tasks occur. This is consistent with ballistic motion of the arm movement.

(c) A resource model (limiting information capacity rate) may be constructed, constraining the slope of the straight line in Figure 17.2. This gives rise to a maximum reciprocal slope of $1/M$ bits/second for a limited channel capacity. In Figure 17.4 this mitigates the overall bandwidth of the human-machine system (lowering the ω_c value). Finally, in Figure 17.6, the structure of the micro movements adjust accordingly to operate in the phase plane to match up the level of uncertainty with the slope of $(-\xi)$ in Figure 17.6 and the relationship in Equation (17.21).

17.6 Some Extensions of This Research

To extend this research, additional studies will add value in future understandings and interrelating the three fields of information theory, control theory and complexity theory. Some suggestions include:

1. Experimental studies involving Fitts' Law could verify the macro and micro structure of the trajectories in the phase plane of Figure 17.6. By adjusting different experimental parameters (A, W_1, and W_2) to see if the shape of the micro movements can be manipulated. The fractal dimension (slope in Figure 17.6) can be varied and the performance results plotted and analyzed in the control and information theory frameworks to see if they are concurrently affected.

2. The high correlation of the line to the data in Figure 17.2 over a wide range of the experimental parameters (A, W_1, and W_2) is well known and provides credence to Fitts' Law in multiple settings. Similar analysis should also be performed for the parameters from the complexity analysis (ξ in Figure 17.6) or the control parameter (ω_c in Figure 17.4) should be tested to demonstrate the consistency that Fitts' Law has shown to the information theory model. These validations should be tested experimentally.

17.7 Conclusions

Three ways to view the ubiquitous Fitts' Law are presented. Comparisons are made within and across the various disciplines. It is shown that Fitts' Law provides a platform to better understand the relationships between control and communications theory. What also has been newly added in this work is relating complexity theory to the aforementioned areas when developing a better understanding of Fitts' Law.

References

[1] J.M. Flach, M.A. Guisinger, and A.B. Robison. Fitts' Law: Nonlinear dynamics and positive entropy. *Ecological Psychology*, 8(4):281–325, 1996.

[2] K.S. Fu. *Syntactic Methods in Pattern Recognition*. Academic Press, New York, 1974.

[3] R.J. Jagacinski and J.M. Flach. *Control Theory for Humans-Quantitative Approaches to Modeling Performance*. Lawrence Erlbaum Associates, 2002.

[4] R.J. Jagacinski, D.W. Repperger, M. Moran, S. Ward, and B. Glass. Fitts' Law and the microstructure of rapid discrete movements. *Journal of Experimental Psychology: Human Perception and Performance*, 6:309–320, 1980.

[5] R.J. Jagacinski, D.W. Repperger, S. Ward, and M.S. Moran, A test of Fitts' Law and moving targets. Human Factors, 22(22):225–233, 1980.

[6] D.T. McRuer and K.S. Krendel. The man-machine concept. In *Proceedings IRE*, pages 1117–1123, May 1962.

[7] C.A. Phillips. Human Factors Engineering: Human Systems Design and Modeling. Wiley, New York, 2000.

[8] D.W. Repperger, C.A. Phillips, and T. Chelette. Study of spatially induced 'virtual force' with an information theoretic investigation of human performance. *IEEE Transactions on Systems, Man, and Cybernetics*, 25(10):1392–1404, October 1995.

[9] D.W. Repperger, C.A. Phillips, J. Berlin, A. Neidhard-Doll, and M. Haas. Human-machine haptic interface design using stochastic resonance methods. *IEEE Transactions on Systems, Man, and Cybernetics, Part A, Humans and Systems*, 35(4): 574–582, July 2005.

[10] C.E. Shannon and N. Weaver. *The Mathematical Theory of Communication*. University of Illinois Press, Urbana, 1949.

[11] N. Wiener. *Cybernetics or Control and Communication in the Animal and Machine*. Houghton Mifflin Co., Boston, 1950.

PART IV

SYSTEMS AND
BASIC TECHNOLOGIES

18

Multi-View People Counting System – Pedestrian Representation

Jung-Ming Wang[1], Wan-Ya Liao[2], Sei-Wang Chen[2]
and Chiou-Shann Fuh[1]

[1]*Department of Computer Science and Information Engineering, National Taiwan University, Taiwan, ROC; E-mail: {d97030, fuh}@csie.ntu.edu.tw*
[2]*Department of Computer Science and Information Engineering, National Taiwan Normal University, Taipei, Taiwan, ROC; E-mail schen@csie.ntnu.edu.tw*

Abstract

In this chapter, a multi-view people counting system is presented. This system uses as the input data the video sequences acquired by a camcorder. The camcorder can be mounted anywhere (e.g., below a ceiling, on a side wall, or at a corner) with any viewing direction. In order to manage various appearances of people, a multi-view representation of pedestrian is introduced. This representation is characterized by a unit sphere, called the viewsphere. The viewsphere is composed of a number of nested spherical layers all centered at the core of the viewsphere. Each layer forms a 2D manifold of viewing directions (viewpoints), which are uniformly distributed over the layer. Pedestrian views generated according to the viewpoints of the viewsphere are clustered into so-called aspects. The pedestrian views within an aspect possess the same characteristics of silhouette.

Keywords: multi-view representation of pedestrian, sequential Monte Carlo method, static parameters, dynamic parameters, viewsphere.

Patrick Shen-Pei Wang (Ed.), Pattern Recognition and Machine Vision – In Honor and Memory of Professor King-Sun Fu, 277–292.

18.1 Introduction

People counting system play an important role in a variety of applications regarding security, management and commerce. Considering a skyscaper, it is important for the security division of the building to know both the total number of people in the building and the number of people on each floor of the building. Such information becomes vital once an emergency, such as fire, explosion, and toxic gas, occurs in the building. Strategies for effectively evacuating and rescuing people from the spot of emergency heavily rely on the information of people count. Likewise, for the places, such as the public areas of transportation stations, stadiums, museums, and malls as well as the restricted areas of government buildingsm military camps, and construction sites, where the control of people count is essential, automatic people counters provide a reliable and persistent tool for governing the number of people in the regions.

People counters have also been considered to count the passengers getting in and out of transit carriages, such as buses and tains. The data provided by the counters can be used to schedule proper times and time intervals of carriage dispatch. Furthermore, the boarding and alighting behaviors at each station can be investigated based on the information of passenger count collected at the station. Accordingly, adequate utilities as well as facilities can be suggested for different stations.

Typically, three stages are involved in a people counter: (a) people detection, (b) people tracking, and (c) counting. The factors that can influence the implementation technique of a people counter include: (a) the scene to be considered, (b) the position and orientation of the camera, and (c) the goal of the application. The difficulties include occlusion, overlapping, and merge-split and shadow effect.

In detecting step, many detection methods, such as model matching [2], temporal differencing [8], and background subtraction [9], have been developed. After obtaining the foreground, we have to separate that if some human figures are connected with each other. Using a contour (ellipse or rectangular) to represent a pedestrian figure is a common idea [12], but it is too rough to detect while they are occluded. Level set [16] and snake model [3] can be applied to model a human's silhouette dynamically [20], but they are too precise to tolerate the imperfect observation. Detecting each part of a human body and combing them according to a predefined architecture can solve the above problems [11, 15], but there are many constrain required to define the human architecture and the combinations may require many computation time to prove.

In the tracking step, we want to obtain the moving trajectory of each pedestrian. This can be done by matching the pedestrian silhouette between sequential frames [18], or changing the model's state to achieve the current state [20]. Because the detecting step may not be reliable, some prediction or filtering models are applied here to compensate that. Kalman filter [13] is a common filtering while the target has a stable moving. Since a pedestrian often moves at will, this method would have some problem in tracking people. Some updating versions of Kalman filter [4] have been proposed to solve the nonlinear system that could be approximated by a Gaussian distribution. Particle filter [7], unlike Kalman filter, is a non-parametric filtering model without predefining the distribution of the prior and posterior knowledge. Recently, some researches have shown that it is robust to track people [12].

After detecting and tracking, the people number can be obtained by counting the trajectory number. This counting can be applied in two kinds of area. In the first one, the area is closed and we want to count the people number in it. Such area often has an entrance that we can monitor that and it often requires to define a cross line for counting [1]. In an open area, since we cannot define inside or outside part, only the passing number can be counted. The passing number means the number of the people that have been detected [13], which will be the same as the trajectory number in the scene.

18.2 Multi-view Representation of Pedestrian

A multi-view representation of pedestrian is characterized by a unit sphere, which is called the viewsphere. The viewsphere is composed of a number of nested spherical layers all centered at the core of the viewsphere (Figure 18.1). Each layer forms a 2D manifold of viewing directions (viewpoints), which are uniformly distributed over the layer. A viewpoint located at longigudinal degree φ, latitudinal degree θ on layer d is specified by (φ, θ, d).

Let us consider one camera on the layer d. If the width of the observation object is w, we will have an observation angle γ to cover that. Those parameters would satisfy the following equation:

$$w = d\gamma, \quad d = w\gamma.$$

Figure 18.1 Viewsphere is composed of a number of nested spherical layers

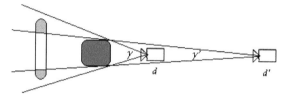

Figure 18.2 Viewing ranges of the cameras with different distances to the monitoring objects

While we move the camera to the other layer d' under the same φ and θ angles, we may see the other object larger and behind the observation object. Take for example Figure 18.2, where the background object will be viewed in the camera on the layer d' while the camera on the layer d cannot see that. If we want a significant change in the scene, there will be significant difference, $\Delta\gamma$, between γ and γ'. We have $d'/d = \gamma/(\gamma' - \Delta\gamma)$. Since the observation object is small in the most case, $\Delta\gamma$ would be large comparing with γ in our case. That means that we do not need to design too many layers in our model (there is only one layer in our research to reduce the following processing).

The above parameters can be regarded as the external parameters of the camera. In addition to those parameters, some local parameters, pan, tilt, and rotation angles, at each viewing point should be considered as well. Let we denote these angles as *viewing angle*. These parameters, however, influence the object position and shape distortion in the image plane. Take Figure 18.3, for example, where Image planes I_1 and I_2 have different viewing angles, so that they may have the same object shapes except for the different foreshortening result.

Suppose the reflecting light of the object project to the image plane at the position p_l with the distance l to the image center. We have the viewing angle a defined as

$$\alpha = \tan^{-1}\frac{l}{f},$$

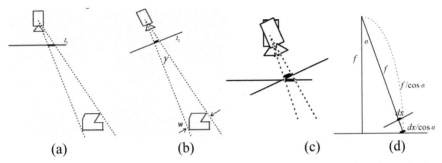

Figure 18.3 Image planes I_1 and I_2 have different viewing angles. (a)–(b) The projecting result on the image plane I_1 and I_2. (c)–(d) The relationship between the projecting results

where f is the focal length of the camera. Let dx be the projecting size on the center of the image plane. The same content projecting on the image position p_l will have size $dx / \cos \alpha$, since the focal length is $f \cos \alpha$. Due to the rotate angle of the plane, the projecting result will be $dx / \cos^2 \alpha$.

Besides, model view may have different scale size, σ, and roataion angle, τ, because of image resolution, pedestrian distance to the camera, and camera setting. Finally, we have all parameters, $(\varphi, \theta, d, l, f, \sigma, \tau)$, in our model view. In a static camera, θ, f, and τ are static. Considering the specific position on the image plane, l also is static. If we could obtain the static parameter values before the system operating, we can reduce the searching space to the dynamic parameters, (φ, d, σ).

Our system consists of three components, which are training step, running step, and memory states. Image sequences with unoccluded pedestrians are applied to measure the static parameter values. This process, called as training step, is applied to define the parameter values using global search. After this process, the pedestrian views are clustered into so-called aspects. The pedestrian views within an aspect possess the possible silhouette on a specific image point. In the running step, the parameter space is much smaller than the original design, and we solve the dynamic parameters using sequential Monte Carlo (SMC) method. Finally, memory states are designed here to memory the distribution of the pedestrian state in the secene. They will support the information of the possible state of the pedestrian for detection and tracking in the previous two steps.

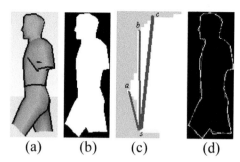

Figure 18.4 (a) Model shapes and (b) its binary image; (c) the process of detecting the line segment and (d) the polygonization result

18.3 Measurement of the Static Parameters

In each viewing direction, we have a model view as shown in Figure 18.4a. Since a pedestrian shape has different contents while comparing with each other, we use the boundary of the shape to measure their similarity with the model shape. The model shape is extracted from the model view and represented using a binary image as shown in Figure 18.4b. After the foreground shape being extracted, it is compared with all of the model shapes to obtain the values of the parameters.

Comparing each boundary point between the foreground and the model shape can help to measure their matching degree. However, this method is too detail and may cause the comparing result too sensitive. Beside, we may need to compare with various poses of the model, so the search space is really large $(\varphi, \theta, d, l, f, \sigma, \tau)$. Most of the camera is perspective camera with little viewing angle α. The distoration $dx / \cos^2 \alpha$ would be too small to be ignored, and l and f can be ignored from the parameters.

Even l and f are ignored; such searching space is still intractable. We simplify the shape using a polygonization method. The boundary of the polygonization consists of some line segments which are represented as a graph using scale and rotation invariant feature values. This representation will reduce the searching space to fewer parameters space, (φ, θ, d). Assume the camera have a significant distance to the pedestrian, d can be set as one to reduce the searching space.

Our polygonization method is applied from the boundary point on the top left of a shape. Along the boundary under clockwise, the orientation from the start point to the current boundary point is calculated and recorded. While the difference between the maximum and minimum orientation values is greater than a threshold, one line segment is assigned from the start point to the current one, and set the current one as the start point of the next line. Figure 18.4c shows the process. In this figure, *s* means start point, and we will get the orientations along the boundary. When the maximum orientation, *a*, and the minimum orientation, *c*, have a significant difference, we connect *s* to *c* as one line segment of the polygonization. Figure 18.4d shows the example of the polygonization result of a model shape.

Foreground objects are extracted using the method purposed in [19]. This method has the advantage of the complete shape comparing with the traditional background subtraction method. In this step, non-occluded pedestrians in the monitoring scene are used for training, where the training means that we want to locate the static parameters in this step. After the foreground object being extracted, the above polygonization method is applied and line segments also are defined as mentioned before.

Line segments of a shape are represented as a full connected graph $G = (V, E)$, where V is the set of the nodes showing the line segments, and E is set of the edges showing the relation between line segments. After representing line segments as graph, the matching process between the pedestrian model and foreground figures can be considered as graph matching problem. Both graphs may have different node noumbers, so this matching is an inexact matching. Our solution not only can apply to partial matching but also give the maching degree value.

We represent a graph G as a set of matrices $A_k, k = 1, 2, \ldots, m$, where m is the number of kinds of features. The elements of matrix A_k are the feature values calculated according for the k-th kind of feature for all nodes. For unary features, we construct a diagonal matrix whose element (i, i) contains the feature value of the i-th node. The binary feature values form a symmetric matrix with element (i, j) representing the binary feature between the i-th and j-th nodes.

Line segments for the pedestrian figure can also be represented using matrices A'_k. The correspondence between the feature points in different images can be obtained by solving the following equation to obtain the permutation matrix, P:

$$P = \min_P \left(\sum_{k=1}^m \| A_k - P A'_k P^T \| \right),$$

where P can be solved by the method proposed in [18], and $\| \cdot \|$ means some norm and can be computed by the square root of the sum of square of elements.

The above method, however, can be applied to A_k whose elements have been normalized. Here we modify this method by applying new measurement function to adapt to the variant types of feature values. P is solved using the following two steps: weighted matrix construction and optimal assignment. Each element $W(i, j)$ of the weighted matrix is computed by

$$W(i, j) = \frac{\sum_{k=1}^{m} \max_{s} \left[\min_{t} |A_k(i, s) - A_k'(j, t)| \right] + \max_{s} \left[\min_{t} |A_k(s, i) - A_k'(t, j)| \right]}{\max[A_k(i, .), A'(j, .), A_k(., i), A_k'(., j)]},$$

where $A_k(.)$ and $A_k'(.)$ mean the element in A_k and A_k' respectively. The calculation result means the degree of the correspondence between node i and node j.

The graph with fewer nodes is assigned some null nodes to equal their node numbers. In the matrices, feature values of the null nodes and its corresponding edges are set as nulls. Null values are ignored in constructing the weighted matrix. After constructing the weighted matrix, we assign the optimal value for each element of P. Some Hopfield model can be applied here, for example, the Hopfield memory in the neural network. In this application, we use Hungarian algorithm [10] because of its polynomial processing time.

Binary feature values are assigned in calculating for size and rotation invariant. In this research, we use four feature values: difference of the orientations, ratio of the segment lengths, relative distance between line segments, and the angle from the center of the shape to the centers of the line segments. After this computation, each node will match one of the nodes in the other graph. The redundant nodes will match a null node or one redundant node of the other graph.

Matching degree is than computed according to the minimized term, $\sum_{k=1}^{m} \| A_k - P A_k' P^T \|$ in defining P. This term combines several value types so that cannot show the real degree value. In our application, one of the feature values, diference of the orientations, is assigned to compute. The model views with higher degree values are extracted as the comparing results (Figure 18.5). After obtaining the candidate model views, we compare their boundary points with the foreground object under various scale sizes and

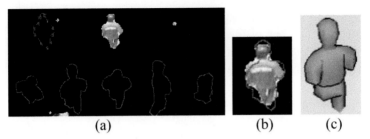

(a)   (b)   (c)

Figure 18.5 The bottom row of (a) are the candidates of the model view and (b) is the best comparing result; (c) is the corresponding model view

rotation angles:

$$p^e(.) = \frac{1}{k} \sum_{i=1}^{k} g_i \times G\big(D(v_i, C(v_i))\big), \qquad (18.1)$$

where $\{v_1, v_2, \ldots, v_k\}$ are the boundary points of the model, g_i is the corresponding weight, $G(.)$ is a Gaussian function, $D(.)$ is the distance two points, and $C(.)$ is the closet boundary point of the foreground object on the normal vector of v_i. The model view with the greatest $p^e(.)$ value (Figure 18.5b) is the most likely model at this image position. In most case, the shape of the human head is more reliable than the body, so we give the boundary points of the model head greater g_i values. This will help to locate the human shape more precisely (Figure 18.5c).

After training some pedestrian shape on this monitoring range, the distribution of the parameter values, $(\varphi, \theta, s, \tau)$, can be constructed. Among them, (θ, τ) are static parameters and their values are defined using the expected value. In addition, their distribution will be updated in running step while the human states have been detected. The values of (φ, s) will be computed in running step, because they are dynamic parameters.

18.4 Sequential Monte Carlo Method

In our application, we want to know the human behavior (hidden states) according to the monitoring image sequence (observations). Since the actual human behavior can not be known, its distribution of the state given the passing observations, $z_{0:t} = (z_0, \ldots, z_t)$, is defined as $p(s_t \mid z_{0:t})$, where s_t is the human behavior shooting at time t. In this system, each image frame is computed to the distribution of the human behavior $p(s_t \mid z_{0:t})$, and the behavior s_t with a highest probability is determined as the pedestrian detecting result. The number of the behaviors along time, $s_{0:t} = (s_0, \ldots, s_t)$, is regarded as the detection result of one pedestrian.

Using the Bayesian rule, this prior probability can be updated to

$$p(s_t \mid z_{0:t}) = \frac{p(z_t \mid z_{0:t-1}, s_t)p(s_t \mid z_{0:t-1})}{p(z_t \mid z_{0:t-1})}, \tag{18.2}$$

where $p(z_t \mid z_{0:t-1})$ is the predictive distribution of z_t given the past observation $z_{0:t-1}$, and it is a normalizing term in most case. Assume that $p(z_t \mid z_{0:t-1}, s_t)$ depends only on s_t through a predefined measurement model $p(z_t \mid s_t)$. Equation (18.2) can be rewritten as

$$p(s_t \mid z_{0:t}) = \alpha p(z_t \mid s_t)p(s_t \mid z_{0:t-1}), \tag{18.3}$$

where α is a constant.

Now, suppose s_t is Markovian, then its evolution can be described through a transition model, $p(s_t \mid s_{t-1})$. Based on this model, $p(s_t \mid z_{0:t-1})$ can be calculated using the Chapman–Kolmogorov equation:

$$p(s_t \mid z_{0:t-1}) = \int p(s_t \mid s_{t-1})p(s_{t-1} \mid z_{0:t-1})ds_{t-1}. \tag{18.4}$$

Equations (18.3) and (18.4) show that we can obtain $p(s_t \mid z_{0:t})$ recursively if we have the following requirements: the measurement model $p(z_t \mid s_t)$, the transition model $p(s_t \mid s_{t-1})$ and the initial distribution $p(s_0 \mid z_0)$.

The main problem is how to represent the distribution of the transition model $p(s_t \mid s_{t-1})$ and the unknown state given the observation $p(s_0 \mid z_0)$. This problem also induces the problem of the calculation of Equation (18.3). Particle filter is a Monte Carlo method that uses m particles, $s^{(i)}, i = 1 \ldots m$, and their corresponding weights, $w^{(i)}$, to simulate the distribution $p(s_t \mid z_{0:t})$. This simulation also can be applied to Equation (18.4):

$$p(s_t \mid z_{0:t-1}) = \sum_{i=1}^{m} p(s_t \mid s_{t-1}^{(i)})p(s_{t-1}^{(i)} \mid z_{0:t-1}). \tag{18.5}$$

The distribution $p(s_t \mid z_{0:t-1})$ also can be simulated using the these particles if we have a new transition equation $s_t^{(i)} = f(s_{t-1}^{(i)}, u_{t-1})$ matching the transition model $p(s_t \mid s_{t-1})$, where u is a noise sequence with zero mean. The distribution of the propagating result

$$f(s_{t-1}^{(i)}, u_{t-1})p(s_{t-1}^{(i)} \mid z_{0:t-1}), \quad i = 1, \ldots, m, \tag{18.6}$$

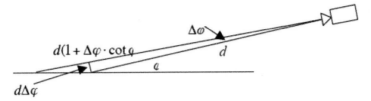

Figure 18.6 The relationship between the scale and the object distance

will be the same as $p(s_t \mid z_{0:t-1})$ if the particle number is infinite. For more details on the particle filter, the reader is referred to [7].

For each pedestrian moving in the monitoring range, we detect his behavior and denote the detecting result as $s_{0:t}$. State s_t is calculated to the MAP of $p(s_t \mid z_{0:t})$ using the mean shift method [6]. The distribution of $p(s_t \mid z_{0:t})$ can be obtained if we have the following requirements. The first requirement, initial state s_0, is defined according to the foreground object extracted using the method proposed in [19]. The second requirement, measurement model, is computed by comparing the current image with our predefined human model. The final requirement, transition model, would be defined by some prior knowledge, and updated according to the detecting result. People number is then calculated to the number of the moving trajectory $s_{0:t}$).

18.5 Measurement of the Dynamic Parameters

At the beginning, a model view requires parameters $(\varphi, \theta, d, l, f, \sigma, \tau)$ to locate the foreground pedestrians. After the training step, they are reduced to (φ, σ) in running step. Scale value, σ, has different range according to φ value. Take Figure 18.6 for example. Let d be the distance from the camera to the pedestrian and φ is the tilt angle. If we limit the viewing direction of the model view in $\Delta\varphi$ angle, the distance from the pedestrian to the camera will be value from $d(1 - \Delta\varphi \cdot \cot\varphi)$ to $d(1 + \Delta\varphi \cdot \cot\varphi)$. The range of the scale value would be between $\sigma(1 - \Delta\varphi \cdot \cot\varphi)$ to $\sigma(1 + \Delta\varphi \cdot \cot\varphi)$, where σ is the scale value trained in the training step. In most case, $\Delta\varphi \cdot \cot\varphi$ is so small that we can ignore that, except that φ is very close to zero. Under the previous assumptions and training, we just need to define φ and σ (if φ is close to zero) values in system running.

Initial distribution $p(s_0 \mid z_0)$ is the base of the sequential Bayesian estimation. Here we give particle set S_t using Markov Chain Monte Carlo (MCMC) method [22]. MCMC method has been proved [21] that it is more efficient than the particle filter-based method when the dimensionality of the search space is high [17]. The state of a particle is set as the following features: shape, intensity histogram, and moving velocity. Among these features, the shape should be designed to match a pedestrian's silhouette. Here we use the multi-view representation to model this shape. The second feature, color histogram is constructed using the values of the pixels within the model view. The velocity includes the moving direction and moving distance that can be computed while we have two successive states.

Since people would be occlued with each other, their shape cannot be complete most of the times. Assume that their head and shoulder always be shown in the monitoring range. We use the models with head and shoulder parts to measure the $p(z_t \mid s_t)$ value, which will be given in a later section. In an image, the region not covered by any state is denoted as R, and the feature values of the first particle, $s^{(1)}$, is set on the boundary of R. The static features of the model view are extracted at the corresponding position. Scale σ and viewing angel φ are given as the expected value learned in the training step.

We sample a candidate state s' according to $s^{(i-1)}$ from the proposal distribution $q(s^{(i)} \mid s^{(i-1)})$, and then set $s^{(i)}$ as s' with the probability

$$p = \min\left\{1, \frac{p(s' \mid z_t)q(s^{(i-1)} \mid s')}{p(s^{(i-1)} \mid z_t)q(s' \mid s^{(i-1)})}\right\};$$

otherwise, set $s^{(i)}$ as $s^{(i-1)}$. Our proposal distribution is combined with random and data-driven proposal probability. Some feature values are proposed randomly, such as longigudinal degree and scale value, while others are proposed in data-driven, such as position value. In our application, $q(s' \mid s)$ is computed by $q(s' \mid s) = Ke^{-(x'-x)/2\sigma^2}$, where $\mathbf{x}'$ and $\mathbf{x}$ are the position of s' and s respectively, and K is a normalized term. To let the sampling more efficient, the positions of the samples are limited on the boundary of the foreground figures extracted by the method proposed in [19].

While applying the Bayesian rule to $p(s \mid z_t)$, we have $p(s \mid z_t) \propto p(z_t \mid s)p(s)$. Assume $p(s)$ is constant, we can computer $p(z_t \mid s)$ as $p(s \mid z_t)$. That also is the measurement model in sequential Monte Carlo model discussed in previous section. Our measurement model consists of two measurements, one is edge likelihood p^e and the other is color likelihood p^c. That is

$$p(z_t \mid s) = p^e(z_t \mid s) \times p^c(z_t \mid s),$$

where p^e is defined in Equation (18.1), and now we define p^c as below.

Human's face and hair are considered as our color information. At first, we detect the skin color [17] and hair color (black) to separate the input image into two binary images, I^s and I^h. The p^c is defined as

$$p^c(z_t \mid s) = p^c(I_t^s \mid s) \times p^c(I_t^h \mid s),$$

where $p^c(I \mid s)$ is defined according to the pixel number. As the s is given, we count the number of skin pixel falling on the face of the model, and denote this number as N_d^s. The number of non-skin pixel is denoted as N_f^s, and the number of hair pixel falling on the face of the model is denoted as N_n^s. Similarly, we calculate N_d^h, N_f^h, and N_n^h for the hair part of the model. Then we can define

$$p^c(I_t^{\{s,h\}} \mid s) = \frac{N_d^{\{s,h\}}}{(N_f^{\{s,h\}} + N_n^{\{s,h\}})}. \tag{18.7}$$

Unfortunately, Equation (18.7) will let the smaller model have larger measurement value, so we must refer to the pedestrian size in the scene. According to the position $\mathbf{x}$ and the scale σ values in the state s, we define a search region to calculate the skin area N_r^s and and hair N_r^h. Finally, the $p^c(I \mid s)$ value is defined as

$$p^c(I_t^{\{s,h\}} \mid s) = \frac{N_d^{\{s,h\}}}{(N_f^{\{s,h\}} + N_n^{\{s,h\}})} \times \min\left(\frac{N_s^{\{s,h\}}}{N_r^{\{s,h\}}}, \frac{N_r^{\{s,h\}}}{N_s^{\{s,h\}}}\right).$$

After sampling, the distribution of $\{s^{(i)}, i = 1 \ldots N\}$ can show the initial distribution $p(s_0 \mid z_0)$ (Figure 18.7a). However, there will be more than one peak in this initial distribution if there are more than one pedestrian in the foregound region. Mean shift with flat kernel [6] is applied here to locate the peak values. The radius of the kernel is designed as the half of the model length. After locating one cluster, we remove those samples closing to the center for the cluster, and do the mean shift algorithm again to locate other clusters.

For each cluster, we can compute an expected state, $\bar{s}$, and its corresponding measurement $p(z \mid \bar{s})$. A cluster with high $p(z \mid \bar{s})$ value is regarded as a human head, or we will stop to locate the other cluster. Figure 18.7c shows the detecting result. After we detect the human head, each human head is given a set of particles according to the distribution of the sample states. This set of particles model the distribution of $p(s_0 \mid z_0)$.

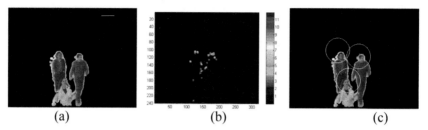

Figure 18.7 (a) The original foreground image, (b) the distribution constructed by MCMC method, (c) using mean shift to locate the pedestrians

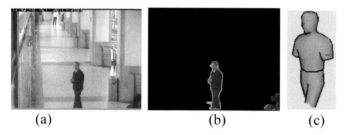

Figure 18.8 (a) The foreground object, (b) the corresponding model view

18.6 Experiments

We test our algorithm on the CAVIAR data set [5]. Each image is reduced to the size 320×240. Because foreground object detection and model comparing are time consumed, the processing to compute the static parameter values is 3.5 minutes on an Intel Core 2 Duo 2 GHz machine. Fortunately, it needs to compute only once after setting the monitoring system, and we can do it off-line. Figure 18.8 shows model view corresponding to the detection result.

The dynamic parameter values is computed on an Intel Core 2 Quad CPU 2.66 GHz machine. The size of the image is 320×240. The processing time is 4 seconds. Figure 18.9 shows some initialization results. Our algorithm can robust to many situations: (a) independent pedestrian, (b) connected pedestrians, and (c) occluded pedestrians.

18.7 Conclusion

Counting people cannot avoid detecting and tracking people. In this paper, we propose a people counting system based on the particle filter applied to track people. When the particle filter is applied, the initial state, measure-

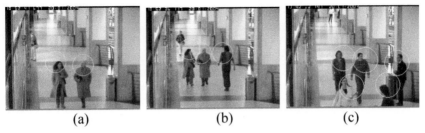

Figure 18.9 The test result of (a) independent pedestrian, (b) connected pedestrians, and (c) occlued pedestrians

ment model and transition model need to be defined before hand. We give a processing method to define the initial state automatically and to construct the prior knowledge for measurement and transition model. Among those requirements, we define the initial state using our multi-view representation of the pedestrian with MCMC algorithm. The final calculation result is the probability of the human state; mean shift is applied here to locate the peak value instead of the expectation value.

In the future works, particle filter will be applied to track the pedestrians along time. The tracking results will show the human's trajectories, and we can obtain the people number by counting the trajectory number.

References

[1] A. Albiol, I. Mora, and V. Naranjo. Real-time high density people counter using morphological tools. *IEEE Trans. on Intelligent Transportation Systems*, 2(4):204–218, 2001.

[2] A. Broggi, M. Bertozzi, A. Fascioli, and M. Sechi. Shape-based pedestrian detection. In *Proceedings of the IEEE Intelligent Vehicles Symposium*, Dearborn, pages 215–220, 2000.

[3] F. Buccolieri, C. Distante, and A. Leone. Human posture recognition using active contours and radial basis function neural network. In *Proceedings IEEE Conference on Advanced Vide and Signal Based Surveillance*, pages 213–218, 2005.

[4] O. Cappe, S.J. Godsill, and E. Moulines. An overview of existing methods and recent advances in sequential Monte Carlo. *Proceedings of the IEEE*, 95(5):899–924, 2007

[5] The CAVIAR Data Set, http://homepages.inf.ed.ac.uk/rbf/CAVIAR/, 2008.

[6] Y. Cheng. Mean shift, mode seeking, and clustering. *IEEE Trans. on Pattern Analysis and Machine Intelligence*, 17(8):790–799, 1995.

[7] N.J. Gordon, D.J. Salmond, and A.F.M. Smith. Novel approach to nonlinear/non-Gaussian Bayesian state estimation. *IEE Proc.-F Radar and Signal Processing*, 140(2):107–113, 1993.

[8] S. Jehan-Besson, M. Barlaud, and G. Aubert. A 3-step algorithm using region-based active contours for video objects detection. *Journal on Applied Signal Processing*, 2002(1):572–581, 2002.

[9] J.-W. Kim, K.-S. Choi, B.-D. Choi, and S.-J. Ko. Real-time vision-based people counting system for the security door. In *Proceedings International Technical Conference on Circuits/Systems Computers and Communications*, pages 1416–1419, 2002.

[10] H.W. Kuhn. The Hungarian method for the assignment problem. *Naval Research Logistics Quarterly*, 2:83–97, 1955.

[11] M.-W. Lee and I. Cohen. A model-based approach for estimating human 3D poses in static images. *IEEE Trans. on Pattern Analysis and Machine Intelligence*, 28(6):905–916, 2006

[12] E. Maggio, F. Smerladi, and A. Cavallaro. Adaptive multifeature tracking in a particle filtering framework. *IEEE Trans. on Circuits and Systems for Video Technology*, 17(10):1348–1359, 2007

[13] O. Masoud and N.P. Papanikolopoulos. A novel method for tracking and counting pedestrians in real-time using a single camera. *IEEE Trans. on Vehicular Technology*, 50(5):1267–1278, 2001.

[14] E. Poon and D.J. Fleet. Hybrid Monte Carlo filtering: Edge-based people tracking. In *Proceedings Workshop on Motion and Video Computing*, Orlando, pages 151–158, 2002

[15] D. Ramanan, D.A. Forsyth, and A. Zisserman. Tracking people by learning their appearance. *IEEE Trans. on Pattern Analysis and Machine Intelligence*, 29(1):65–81, 2007

[16] J.A. Sethian, *Level Set Methods and Fast Marching Methods: Evolving Interfaces in Computational Geometry, Fluid Mechanics, Computer Vision, and Materials Science*, 2nd edn. Cambridge University Press, 1999.

[17] J.-M. Wang, H.-W. Lin, C.-Y. Fang, and S.-W. Chen. Detecting driver's eyes during driving. In *Proceedings of the 18th IPPR Conference on CVGIP*, Taipei, Taiwan, 2005.

[18] J.-M. Wang, S.-W. Chen, S. Cherng, and C.-S. Fuh. People counting using fisheye camera. In *Proceedings of the IPPR Conference on CVGIP*, Mauli, Taiwan, 2007.

[19] J.-M. Wang, S. Cherng, C.-S. Fuh, and S.-W. Chen. Foreground object detection using two successive images. In *Proceedings of IEEE International Conference on Advanced Video and Signal-based Surveillance*, Santa Fe, pages 301–306, 2008.

[20] A. Yilmaz, X. Li, and M. Shah. Contour-based object tracking with occlusion handling in video acquired using mobile cameras. *IEEE Trans. on Pattern Analysis and Machine Intelligence*, 26(11):1521–1536, 2004.

[21] T. Zhao and R. Nevatia. Tracking multiple humans in crowded environment. In *Proceedings Computer Vision and Pattern Recognition*, Washington, Vol. 2, pages 406–413, 2004.

[22] T. Zhao, R. Nevatia, and B. Wu. Segmentation and tracking of multiple humans in crowded environments. *IEEE Trans. on Pattern Analysis and Machine Intelligence*, 30(7):1198–1211, 2008.

19

Automatic Cheque Processing System

Xiaoxiao Niu, Ching Y. Suen and Tien D. Bui

Centre for Pattern Recognition and Machine Intelligence, Concordia University, Suite EV003.403, 1455 De Maisonneuve Blv. West, Montreal, Quebec, Canada H3G 1M8; E-mail: {x_n, suen, bui}@encs.concordia.ca

Abstract

This chapter introduces the implementation of a cheque processing system developed at CENPARMI (Centre for Pattern Recognition and Machine Intelligence). To recognize a cheque image, there are three major recognition targets that need to be achieved, consisting of the courtesy amount, the legal amount and the date. The implementation begins with the extraction of these target regions based on the detection of baselines. Next, the extracted courtesy amount is split into isolated digits, the legal amount sentence is separated into words, and the date field is divided into Year, Month and Day zones. Then, the relevant recognizers are invoked to process digits, cursive words, and a mixture of numbers and words. Although recognizers have different functions, certain components can be shared. We describe the algorithms and strategies used in the implementation of each module and present some experimental results.

Keywords: check processing, item extraction, courtesy amount recognition, legal amount recognition, date recognition.

19.1 Introduction

Automatic bank cheque processing has been intensively studied for many years, since hundreds of millions of cheques are processed all over the world

Patrick Shen-Pei Wang (Ed.), Pattern Recognition and Machine Vision – In Honor and Memory of Professor King-Sun Fu, 293–318.

ever year. With the help of efficient cheque reading systems, manual efforts in the typing and verification of cheques can be largely minimized. However, it is not a trivial task to develop a reliable cheque recognition system, and many challenges have to be dealt with, e.g. removal of complex backgrounds and noise, extraction of handwritten information, separation of touching and overlapping characters and words, requirement of high performance for recognizers, and the verification between courtesy and legal amounts. To meet these challenges, a great deal of efforts focusing on various aspects have been developed. A comprehensive overview of bank cheque processing can be found in [10]. Moreover, many implementations of cheque reading systems in different languages in many countries have been proposed by research groups, for example, U.S.A. [4, 12], Canada [24], French [7, 9], Italy [1], Switzerland [11], Brazil [16], China [23], Israel [18], etc.

In spite of the intensive studies in the research fields, several commercial cheque processing systems have been successfully designed and widely used in banks, such as A2iA ChekReader [7] from the A2iA company, Check-Plus [4] from the Parascript company, OrboCAR [18] from the Orbograph company, SoftCAR+ [26] from the Unisys company, etc. Here, we briefly introduce some representative commercial systems. The A2iA CheckReader [7] system starts with the courtesy amount recognition by combining four Neural Networks. If the courtesy amount recognition result has a high confidence value, the input cheque image is accepted. Otherwise, the legal amount recognizer is involved, and the legal amount and courtesy amount results are combined with a log-linear integrator to make the final decision. The CheckPlus [4] system performs the courtesy amount recognition based on the procedure of matching input subgraphs to graphs of symbol prototypes. This system independently recognizes the legal amount by matching the input word feature to all lexicon entry representations. After the cross validation of recognition results between legal and courtesy amounts, the final result is achieved.

For both the commercial and research purposes, a prototype cheque processing system has been implemented at CENPARMI for the automatic reading and recognition of legal amounts, courtesy amounts and date amounts on Canadian bank cheques. Given a scanned grey scale cheque image, interesting handwriting regions are firstly located by the detection of baselines which are later eliminated through a morphological operation. The lost information resulting from such an operation is restored by using morphological and topological methods. Next, after binarization and cleaning as preprocessing steps, each extracted field is passed to the corresponding recognition engine. The courtesy amount is separated into isolated numerals by a digit splitter. These numerals are recognized by a combination of three Neural Network digit classifiers for recognizing the isolated digits, and a Convolutional Neural Network designed for recognizing the touching pairs "00" and "000". The legal amount sentence, after the slant correction, is separated into individual words by a feedback segmentation method. The word recognizer is the combination of Multilayer Perceptron classifiers, based on the implicit segmentation, and the Hidden Markov Model, based on the explicit segmentation. In processing the handwritten date, the date image is divided into three parts: Year, Month and Day fields. The Year field is firstly to be extracted. To split the Month and Day fields, a knowledge-based model is used to solve easier cases by detecting the obvious separator and then leaves the more ambiguous cases to the Multi-hypotheses model. During the separation process, the category of each field (Day and Month) is also determined so that relevant classifiers can be applied to recognize numbers or a mixture of numerals and words.

The details for each model will be described in the following sections, with some experimental results. Future suggestions are presented in the end. This is an extension to the work of [24] and many research members from CENPARMI [2, 5, 8, 13, 14, 17, 21, 27, 29–31] have contributed to it.

19.2 Automatic Extraction of Items

The first and most basic step of an automatic cheque reading system is the item extraction. It includes the extraction of interesting zones, such as courtesy amounts, legal amounts and data fields, etc. After those items have been obtained, the recognition engine is able to process them, respectively. Nowadays, the color image is easily acquired by scanning, which tremendously requires increased memory storage and preprocessing time. Therefore, in our cheque item extraction module, grey-level images are used and analyzed directly. The three main steps of handwriting data extraction include: detection and elimination of baselines, restoration of the lost image information, and post processing to remove noises, such as printed characters, "$" signs, long dashes, and other intruding strokes.

The regions of items on standard Canadian cheque images are determined by a set of baselines. For a typical Canadian cheque as shown in Figure 19.1, the legal amount is written in the middle of the cheque on a baseline which is half a size longer than the width of the image. This baseline is called "legal amount baseline". Above it, a parallel line with almost the same length is

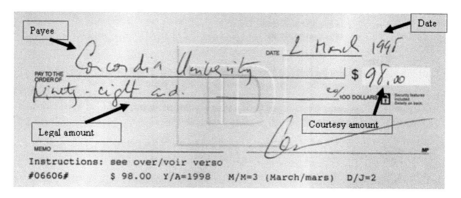

Figure 19.1 A typical Canadian cheque image

used for filling in the name of a payee. It is named "payee baseline". In the right corner above the payee baseline, one short line for writing the date is called "date baseline". But sometimes, there are two printed lines for the date entered. We assume that the two baselines coincide in this case. The courtesy amount on the righthand side of the image may be placed above a short baseline, which is called "courtesy baseline". Another instance of the courtesy amount is that it is surrounded with a rectangular box (Figure 19.1). No matter how the regions are designed for the courtesy amount to be filled, a printed "$" mark always appears inside the courtesy amount field and works as an indicator for the beginning of a handwritten date. Hence, to capture these handwriting fields, we adhere to the rules used for the analysis of the layout of cheque images. The rules are:

- Legal amounts are written between "payee baseline" and "legal amount baseline".
- Dates are filled on the upper right of the "payee baseline".
- Courtesy amounts are placed around the right hand of the region of the "payee baseline".

The baseline detection method will be described below.

19.2.1 Baseline Detection

The baselines are detected by using a mathematical morphology operation method. The morphology operation works as a detector of certain baseline features which are described by our defined structuring elements. By applying

(a) before the line removal

(b) after the line removal

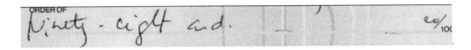

(c) after information restoration

Figure 19.2 Legal amount images

a series of closing, subtraction and summation operations [29] on the grey-level cheque image, baselines are derived.

After baselines have been determined, we can locate the search areas which contain legal amounts, courtesy amounts and dates. It should be noted that for the location of the courtesy amount, if there exists no courtesy baseline, then a bounding box is used instead. To solve such a problem, we propose another searching method. The method detects the machine printed dollar sign "$", which is used as the indication of the beginning of the courtesy amount. However, in real applications, it is difficult to detect the "$" sign because of the crossing and touching made by the handwritten numeral strokes. Therefore, different courtesy amount extraction procedures have been developed to search the righthand part of the cheque image. The main idea of courtesy amount extraction is that we try to find the connected components of pixels, based on the assumption that the grey value of the handwritten strokes is lower than a specific threshold.

19.2.2 Lost Information Restoration

We eliminate baselines after they have been detected, but it results in a loss of information in samples where handwritten strokes may intersect with a baseline. Figure 19.2 shows an example before (Figure 19.2(a)) and after the line elimination (Figure 19.2(b)). As a remedy, morphological closing operations followed by topological processing have been implemented to save the lost information.

The morphological operation is conducted around the location of each known baseline for efficiency purposes. The grey image, after baselines have been eliminated, is restored by applying closing operations with three defined structuring elements [19]. The purpose of such operations is to propagate the intensity difference between the baseline and the intersecting handwritten strokes. However, when the handwritten strokes have the same grey level as the intersected baseline, the morphological closing operation could fail to preserve the lost information of the handwritten data. In that case, a topological process is invoked to detect and fill the gaps according to certain rules [17]. Figure 19.2(c) illustrates the image after this restoration procedure.

19.2.3 Post Processing

Before sending each item to the next stage, it is necessary to do the post processing on each field. The purpose is to enhance the quality of the images and to facilitate the following recognition procedures. Two main tasks are involved, that is, binarization and noise removal. The local threshold binarization method is implemented by considering both the threshold of the image's grey-level and the threshold of the image's local contrast. As for noise removal, we can determine which of the objects that intersect with the border of the located region should be included, or eliminated and cropped. Those objects need to be cleaned, such as the printed bounding box in the courtesy amount field, the words "order of" in the legal amount field , and the "$" sign in the courtesy amount zone. Other examples include smallish noises when each length is below a certain threshold, the long dashes inside regions, etc. Figure 19.3 illustrates the legal amount (Figure 19.3(a)), the courtesy amount (Figure 19.3(b)) and the date ((Figure 19.3(c)) after post processing. Once post processing has been completed, each zone with its cleaned data is submitted to the corresponding recognizer engine for further processing.

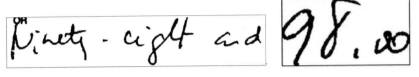

(a) the legal amount   (b) the courtesy amount

(c) the date

Figure 19.3 Images after post processing

19.3 Item Processing

After handwritten items have been retrieved from the cheque image, each item is processed by an independent specialist module. Those modules include: courtesy amount processing, legal amount processing and date processing. In this section, the implementations of each of the three modules are discussed.

19.3.1 Courtesy Amount Recognition

The courtesy amount recognition is the basic procedure in a cheque reading system. In the earlier cheque systems, they have been designed by relaying only on the courtesy amount recognition, based on the fact that bank employees prefer to read the numbers for quick processing, and they only examine the legal amount for verification purposes when the corresponding courtesy amount can hardly be recognized. Therefore, the performance of the courtesy amount engine has great influence on the success of the entire cheque system. To build an efficient courtesy amount recognition system, four critical tasks should be considered: (1) removing spurious connected components and noises, (2) grouping relevant component neighbors, (3) splitting touching digits, and (4) designing a high performance numeral recognizer with a high rejection on garbage (parts of a digit and stroke noises).

Our courtesy amount recognition system works to solve those critical problems in the following way: The connected components are found firstly from a binary image by tracing its contour. For those components, with a perimeter below a threshold, they are thought of as noises and removed. Then, the remaining components are sorted from left to right, and recognized into one of 14 classes (digits, dashes, periods, commas and others) by a context-free classification. If some connected components are recognized without a reliable confidence value, they may seem like potential broken parts of a digit. Hence, we group the neighborhood components according to some criteria. For instance, a dash in the upper part of the image is grouped with its neighboring component on the left ("five-hats") and a stroke "1" is combined with the upper image component on its left ("4"). If the group components can not be recognized by the digit recognizer, we try to re-segment those group components by a digit splitter. When one segmentation path has been found with a high confidence value, the results are concatenated from left to right. Otherwise, the courtesy amount is rejected.

The courtesy amount recognizer has been designed for both punctuation (dashes, periods, and commas) recognition and digit (individual or string) recognition. For punctuation symbol recognition, the symbols can be easily detected through their shape features. Three shape features on the contour component are used: length of the contour, straightness of the component and slopes of it. As for the numeral string recognition, the common idea is to segment them firstly and then recognize them. The detailed procedures of the digit string recognition are given next.

19.3.1.1 Digit Segmentation

Our digit splitter, dividing numeral strings into individual digits, is based on the hypotheses-then-verification methodology. The splitter attempts to separate and recognize the leftmost numeral from a connected component as a single digit. It tries to achieve recognition of the leftmost digit with a certain confidence value, within at most five cuts. If successful, the digit splitter recursively separates the leftmost numeral from the remaining string. Once the rest of the connected component yields confident recognitions, results are concatenated from left to right. If recognition fails during any splitting step, the connected component is re-segmented by adjusting certain parameters based on the previous recognition results [2]. The re-segmentation process terminates either when an accepted segmentation path has been found or all the combinations of adjusted parameters have been unsuccessfully exhausted.

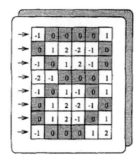

Figure 19.4 The horizontal pixel distance feature array for a character "8" [22]

19.3.1.2 Isolated Digit Recognition

Once the isolated digits are successfully segmented by the digit splitter, the recognition is performed with the combination of three Neural Networks, which use different sets of features extracted from each digit. The feature extraction method from binary images was first proposed by Strathy [22], which is simple and has proven to be efficient and powerful in real applications. It is called Pixel Distance Features (PDF). Two features for each pixel are calculated on the image, in the horizontal direction and in the vertical direction, respectively. The feature value of one pixel indicates both the direction and the distance information to its nearest black pixel on the same scan line. The direction is represented by one of three signs. A negative sign indicates that the pixel is located on the left of its nearest black pixel; positive means that it is on the right; and zero denotes the black pixel itself. As for the distance information, by moving away from the black pixel, the absolute PDF value increases and achieves its maximum value at the midpoint. Figure 19.4 shows the pixel distance features on the horizontal line. The vertical PDFs are analogous to the horizontal PDFs.

We have combined three standard backpropogation Neural Networks with similar structures to perform the isolated digit recognition task. The combination is conducted by summing up three lists of several candidates with their probabilities into a single candidate list. The three Neural Networks trained with PDFs extracted from different pre-processed binary images include:

- Neural network A: Horizontal and vertical PDFs are obtained from the thinned input image, averaged into the size of 12×14, and normalized into $[-1, 1]$. Network A has 336 inputs, 70 hidden units and 10 output units.

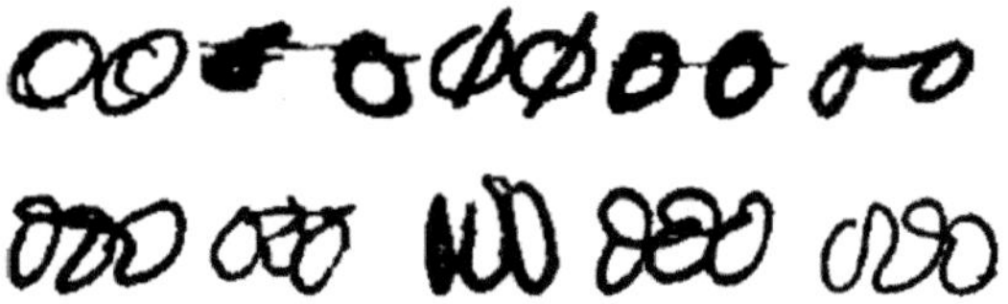

Figure 19.5 Some examples of touching numeral strings "00" and "000"

- Neural network B: The input image is scaled to a size of 16×16 firstly, and then PDFs are extracted and normalized. Network B has 256 inputs, 560 hidden units and 10 output units.
- Neural network C: The only difference with Network C lies in that the input image is rotated 45 degrees before it is scaled as in Network B. The structure is the same as Network B.

Each network was trained on the 18,468 isolated training samples and the training rate is higher than 99.8%. The recognition rate on 2213 testing samples reaches up to 99.19% by combining those three networks [24].

19.3.1.3 Recognizer for Touching Strings "00" and "000"

There is a great dependence between the digit recognition and the digit segmentation processes. If the digit splitter results in the incorrect cutting path, even with the perfect recognizer, we still can not get the correct answer. Moreover, it is impossible to achieve perfect segmentation results even by sophisticated algorithms. Based on the analysis of the CENPARMI cheque data, connected numerals like "00" and "000" appear more often than other touching pair strings, and propose big challenges to the segmentation processing. Some image examples are shown in Figure 19.5. Therefore, it is efficient to integrate a specific recognizer for "00" and "000" without segmenting them. A recognizer for these touching strings has been designed in order to release the burden of the segmentation procedure and to improve the performance of our courtesy amount recognition engine.

A convolutional neural network (CNN) has been built to recognize the overlapping numeral strings "00" and "000". CNN is one kind of a backpropagation network that can extract topological characteristics automatically from a raw image by avoiding the hand-designed feature extraction. In addition, CNN tolerates some degree of shift, scale, and distortion variance. More details about the architecture of CNN in our application can be found in [3]. Training images, with 3107 strings of "00" and 141 strings of "000", were normalized to the size of 29×29 before they were sent to the CNN.

Table 19.1 The performance of our courtesy amount system on real cheques

Module	Courtesy number	Error number	Recognition rate (%)
Courtesy amount	9648	3337	74.3

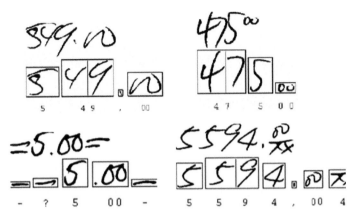

Figure 19.6 Examples of courtesy amount recognition

19.3.1.4 Experimental Results of Courtesy Amounts

To evaluate the performance of our courtesy amount system consisting of the isolated digit recognizer and the CNN string recognizer, we have tested our engine on 12,985 real cheques from the CENPARMI database. Table 19.1 shows the recognition results. Whenever any digit in a numeral string was misrecognized, an error occurred. The results were tested on a large number of samples with zero-rejection. Some examples of courtesy amount images with their recognition results are given in Figure 19.6. After a careful examination of recognition errors generated by our courtesy amount engine, we noticed that errors belong to the following categories: (1) item extraction errors; (2) non-digit recognition errors (e.g. "xx" in Figure 19.6); (3) segmentation errors; (4) recognition errors. Future research could focus on dealing with such errors.

19.3.2 Legal Amount Recognition

The processing of legal amounts is a big challenge due to the great variability in cursive handwriting styles. For instance, there may be sloppy written and touching characters within a word, irregular transition gaps among individual words, etc. Thus, it is difficult to achieve a perfect segmentation performance on segmenting sentences into words, particularly on the segmentation of words into characters. All of the errors in the segmentation stage will accumulate and contribute to errors in the final decisions. Therefore, to avoid the difficulty in designing a sophisticated segmentation stage, we perform recognition on the word level and use the combination of the implicit and explicit segmentation methods for the recognition task. Our legal amount processing includes preprocessing, sentence-into-word segmentation, and cursive word recognition. These will be discussed in the next few subsections.

19.3.2.1 Preprocessing

The main tasks in the preprocessing stage are the slant detection and slant correction. In the slant detection process, we try to normalize the slant of the handwriting to the vertical position. The slant of legal amounts is calculated by looking for the greatest position derived from all of the computed slanted histograms [8]. Once the best slant angle has been detected, we correct the slant of the legal amount sentence by using a shear transformation. Figure 19.7 shows some image examples before and after the slant correction process.

19.3.2.2 Sentence-into-Word Segmentation

Our segmentation approach is based on a feedback segmentation strategy [31], which uses spatial distance information between components as well as a recognition engine. There are two steps. In the pre-segmentation step, we initially segment the legal amount sentence by calculating the Euclidean distance between the connected components. When the gap distance exceeds a certain threshold, we treat it as the inter-word gap. The threshold is chosen based on the idea that we want to minimize the over-segmented (when one word is cut into more than one piece) cases as much as possible. Next, a re-segmentation approach is invoked to handle those under-segmented (containing more than one word) cases. The under-segmented images are detected by evaluating the length information firstly. Then, a set of segmentation options with cutting path hypotheses is generated. Finally, a trained Neural Network classifier chooses the best path with the highest confidence value, based on the assumption that the recognition confidence value of a word is higher than the confidence value of a part of a word.

19.3.2.3 Word Classification

An MLP-HMM hybrid recognizer has been implemented for recognizing each separated cursive word. The main idea of this methodology is that distinct classifiers can complement each other and boost the performance,

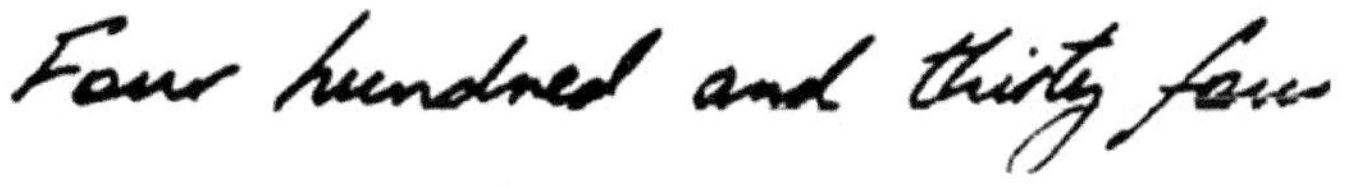

(a) original image A

(b) after slant correction A

(c) original image B

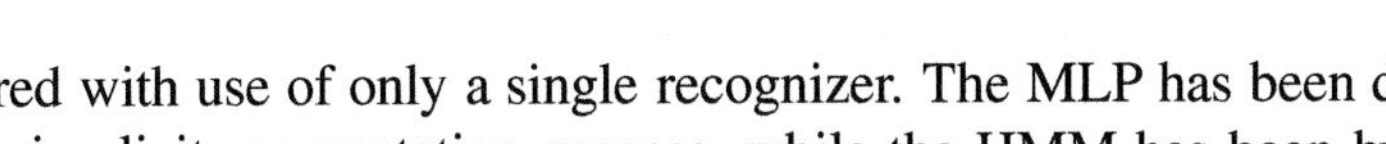

(d) after slant correction B

Figure 19.7 Slant corrections of legal amounts

compared with use of only a single recognizer. The MLP has been designed with an implicit segmentation process, while the HMM has been built with an explicit segmentation process. In the end, the combination task uses a weighted multiplication strategy. Each model implementation is described below.

19.3.2.3.1 The MLP model   The implementation of the MLP (Multilayer Perceptron) model for the legal amount recognition, as shown in Figure 19.8, is a fusion of two MLPs at the architectural level. Each of the two MLPs is trained with different feature sets. Then, a new MLP is designed by using the outputs of the neurons in the two hidden layers as the new input features.

There are two feature sets used to train the two MLP model: one set includes mesh features, chain features, crossing features and distance features;

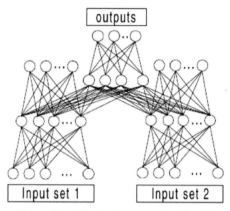

Figure 19.8 The fusion of two MLPs [15]

the other contains only gradient features. These basic features are derived from a binary legal amount image. They are defined as follows:

- Mesh feature: The number of black pixels in each of the subdivided local regions from the image.
- Chain feature: The number of same directional pixels in each of the subdivided local regions from the contour image. Four directions are considered here.
- Distance feature: The distance from the boundary rectangle of the word image to the first black pixel of the word image.
- Gradient feature: The number of pixels that have the same gradient angle, in each of the subdivided local regions from the image.

19.3.2.3.2 The HMM model The HMM model proceeds with explicit segmentation of the cursive word into graphemes, and thus complements the MLP word recognition model. Here, the grapheme represents part of a word image, which looks like a character, but sometimes just contains part of a character or more than one character. One word-level HMM is built by cascading the character-level HMMs. It is designed to have the ability to solve the over-segmented or under-segmented cases. The topology of such a word-level HMM is illustrated in Figure 19.9. Here, α_{ij} is a single state transition, representing one grapheme. Three concatenated states compose one character to cope with the under-segmentation cases. α'_{ij} is a null transition for the letter skipped due to the under-segmentation. $b_{ij}(o_t)$ denotes the output probability for the given observation of symbol sequence O. Since

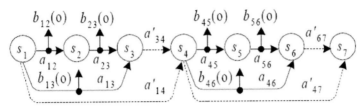

Figure 19.9 Word level HMM with two cascaded character level HMMs [14]

the lexicon for the legal amount recognition is very small, e.g. 32 classes for the English cheques, we have built one HMM word model for each word class. The detailed algorithms to calculate the re-estimated parameters and the posterior probability can be found in [14].

The features used to obtain the observation sequence for training the HMM model are extracted from graphemes in the contour image. They consist of shape features, direction code distribution features, curvature features and moment features [14]. Shape features represent topological characteristics of a grapheme. They include ascenders, descenders, loops and dots. Which are detected based on reference lines. Direction code distribution features are generated in the same way as the chain features, which were described in the MLP model. Curvature features are calculated as the ratios of the Euclidean distance, between two end points of each grapheme and the number of pixels connecting them. Three moment features are extracted from the contour of the grapheme. All of the feature values need to be converted to the observation symbol sequence before transferring them to the HMMs.

Once the MLP and the HMM models have been trained with their own feature sets, the combination of these classifiers is implemented. We will introduce the combination method in the following part.

19.3.2.3.3 Combination of MLP and HMM A weighted multiplication scheme has been applied, to combine the implicit segmentation-based MLP model and the explicit segmentation-based HMM model. Before the combination, each output from two classifiers is normalized to an equal measurement level. The weighted multiplication becomes a new probability, computed by multiplying two probabilities from each model with weighting factors. The formula is defined as follows:

$$P(c_i|H, M) = P(c_i|H)^{w1} \times P(c_i|M)^{w2}$$

where $P(c_i|H)$ means a conditional probability for one word computed from the HMM model; $P(c_i|M)$ represents a posterior probability for the same word given by the MLP model; and $P(c_i|H, M)$ is the combination probability for the word result. The weighting factors $w1$ and $w2$ are derived from each classifier performance, respectively. As a result, a ranked list of legal amount word candidates is obtained with a decreasing order of probability values after the combination process. Then they are passed to the parser for the post processing.

We have evaluated the MLP-HMM word recognition engine. Experiments were carried out on the legal amount words, extracted from 2500 handwritten English cheques in the CENPARMI database. Nearly 800 writers participated. Our MLP-HMM hybrid classifier was trained on 5224 training words in 32 classes, and the 2483 testing words were tested. The MLP-HMM hybrid model has a recognition rate of 92.2% for the Top 1 word candidate and of 99.2% for the Top five word candidates. Due to different databases used in the literature, it is difficult to compare our recognition results with others. However, the recognition rate with 92.2% for 32 classes is a very promising result among other reported works.

19.3.3 Date Recognition

There is great variability and uncertainty of handwritten dates written on standard Canadian bank cheques. For instance, the machine-printed century symbol (i.e., "19" or "20" either appears or does not appear on the year field of the date. Cursive words or numerals can be mixed when written as the month field. The abbreviated month forms (i.e., Jan., Aug., Oct.) are often used. Numerals presented on the day field are sometimes followed by suffixes (i.e., "st", "nd", "rd", or "th"), either as superscripts or at the same horizontal level. Besides these variations, both punctuations such as period ("."), comma (","), slash ("/") and hyphen ("-") and big gaps are often used as identifiers as the end of each date field. When the cheque is in French, the article "Le" is sometimes handwritten at the beginning of the date image. Figure 19.10 shows some examples of various date types.

It is difficult to develop an effective processing system to recognize the entire date image all at once, since there is no prior knowledge of the format or the category of a date image. Therefore, a segmentation-based method has been developed in our date recognition system. The main procedures of our date recognition system [28] consist of: the segmentation of the date image into three fields: year, month, and day; the identification of the nature of the month, i.e., whether it is written in alphabetic letters or numerals, and the recognition of each field by the appropriate classifier. Numeral strings are

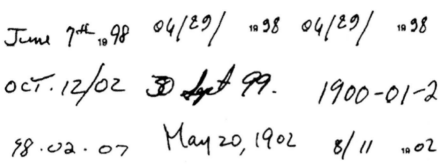

Figure 19.10 Examples of date images from our cheque database

sent to the digit recognizer with the same recognizer used for the courtesy amount, while the months in the alphabetic format are passed to the cursive month word recognizer, which is also used in the legal amount process, and is trained on both English and French month words.

In order to improve the reliability of our date recognition system, a verification model has been designed in the post processing stage. The aim is to let the system not only make a rejection but also to correct some errors. However, some ambiguously written dates, such as 08-11-1998, which is ambiguous due to the difficulty in determining the Day and Month specifically, cannot be interpreted by the system and are left to the user to make the final decision.

We will discuss the segmentation of the date field, the recognition, and the verification in the following sections.

19.3.3.1 Date Image Segmentation

The objective of the date segmentation step is to distinguish between the Year, the Day, and the Month fields from the date image. The process includes the following steps:

1. Extract the Year field from the date image firstly;
2. Divide the Day and the Month by detecting the obvious punctuation;
3. If no punctuation exists, the gap separator between the Day and the Month field continues to be searched;
4. If ambiguous Day & Month sub-images cannot be separated by step 3, they are sent to a multi-hypotheses module, which tries to detect the best cutting option.

The details of the methods for separating the Year, the Day and the Month are introduced in the following sections.

19.3.3.2 Year Detection

The first step in our date segmentation is to separate the Year field from the Day & Month fields. Based on our cheque database analysis, 80% of date images have the standard format, with the machine-printed "19" (or "20") as an indicator of the beginning of the Year field. The remaining of the date images, without machine-printed century symbols in front of the Year zone, are categorized as free format dates. For the standard format, we locate the Year field by detecting the printed numerals. When the detection for the standard format failed, the free format processing is applied and selects a reasonable Year field by comparing two candidates from two ends of the date image.

19.3.3.3 Day and Month Separation

After the Year field extraction, the sub-image containing the Day & Month fields should be segmented and identified. Two tasks should be solved within this step: (1) the segmentation of Day and Month by locating punctuations and gaps; (2) the determination of the nature for each field, e.g. whether it is written with alphabetic letters or with numerals. To solve these problems, we have integrated two models, Knowledge-Based and Multi-Hypotheses Models. Easy cases are performed by using the knowledge-based module, while ambiguous cases are sent to the multi-hypotheses module for future evaluation. In the following sections, we will describe both the knowledge-based module and the multi-hypotheses module.

19.3.3.3.1 Knowledge-Based Module   There are three handwriting styles that correspond to the Day & Month fields, based on the analysis of the writing styles of date images from our cheque database. They are: NSN, NSA, and ASN. Here, S denotes a separator (a punctuation or a gap), N represents a numeral string (Day or Month field), and A describes an alphabet string (Month field). Therefore, our Knowledge-Based segmentation strategy incorporates this knowledge information. The separator is detected in two steps: the punctuation is searched, firstly; if no punctuation can be found or confirmed, then we start to locate the gap transitions between Month and Day fields, secondly. Next, we will discuss the details of how to detect the separator symbols: punctuation and gaps.

A. *Punctuation Detection*

The four punctuations most frequently used in the cheque date are slash "/", hyphen "-", comma ",", and period ",". They are primarily found by extracting the shape and spatial features [6]. The former describes shape characteristics of the connected component itself, such as: a long length and a slope for the slash; a flat stroke for the hyphen; a small size for the comma and period, etc. The latter explains the symbol's relative location in the Day & Month sub-image. For example, slashes usually extend from the top to the bottom of the entire image; hyphens are placed in the middle; and commas and periods appear at relatively low positions.

If shape and spatial features of an unknown punctuation are significant and are recognized with a high confidence value, then the type of punctuation can be indicated immediately. Otherwise, the primary punctuation candidate should be verified by a confirmation process, which utilizes the knowledge of the writing style of Day and Month. Strong relationships between punctuations and the nature of writing styles on cheque images are:

- Slashes and hyphens are usually located in the middle of two numeral strings, or at least with one numeral string at one side.
- Periods and commas are mostly used in the pattern ASN, where the Month is written in alphabetic letters.

Based on these observations, condition-action grammar rules are generated to conduct the confirmation process for uncertain punctuations. Table 19.2 shows some condition-action rules for confirming a separator candidate. We get the pattern module by using the combination of multiple NNs (Neural Networks), which differentiates between the N (numeral string) and the A (alphabet string). If the condition separator candidate is waiting for verification, the corresponding action is taken. For example, when a "-" candidate is found through shape and spatial features, and both neighbors of this separator are numeral strings with high confidence values returned by the multiple NN classifiers, the separator is confirmed as a hyphen. Sometimes, a punctuation such as the slash "/" is easily confused with the handwritten digit "1" or the letter "l". In this case, more features like distances from the separator candidate to its neighbors should be checked as additional contextual characteristics in order to confirm one pattern. However, some ambiguous separators cannot be solved in the knowledge-based model, so the multi-hypotheses process is activated to handle such uncommon cases.

B. *Gap Detection*
When either no punctuation is detected or punctuation candidates are passed to the multi-hypotheses module, the gap searching process starts by locating the gap between the Day & Month fields. In the analysis of the Day & Month

Table 19.2 Examples of condition-action rules

Pattern	Condition	Action
NSN	-	Hyphen candidate is confirmed
NSN	/	Slash candidate is confirmed if additional distance features can be satisfied
ASN	.(,)	Period or comma candidate is confirmed
NSA	.(,)	Multi-hypotheses module is activated
…	…	…

sub-images in our cheque database, we have found that 98% of gaps exist at the transition point between numerals (Day) and alphabetic letters (Month). However, to find a correct gap is not an easy task since peoples' free writing styles varied. Thus, to cope with this problem, we selected a few gap candidates, and then used a confirmation step to figure out the best candidate option which had the highest probability as the final gap separator. If no gap could be detected initially, then multi-hypotheses lists based on gap candidates were generated and sent to the multi-hypotheses module for future verification. We will introduce the multi-hypotheses module in the next part. Here, we will firstly discuss how to find the gap separator.

The gap detection process also adopts a candidate-then-confirmation strategy. Each two gap candidates with the biggest distances are selected from both the horizontal bounding box distance algorithm [20] and the maximum distance method (which scans all lines within the middle zone of a sub-image) [24], respectively. Thus, four gap candidates are generated. If the maximum gap based on the horizontal bounding box has a high confidence value (e.g. it is much wider than the second one), it is identified as the gap separator. Otherwise, a confirmation is called, where one gap candidate is selected from the horizontal bounding box algorithm; the other gap candidate is chosen from the maximum distance method; and the final gap separator is detected from these two candidates.

The basic idea for the selection method in the confirmation step mentioned above is to cluster sub-images on each side of a gap candidate to see whether one part is a numeral string or an alphabetic string. The differentiation process between numeral string and an alphabetic string is performed by the distance_to_numeral measurement [25]. For example, in Figure 19.11, there are three sub-images (Im1, Im2, and Im3) separated by gap1 and gap2. After clustering, Im1 and Im2 are grouped together, and gap2 is detected.

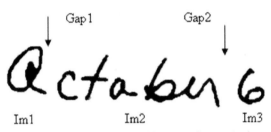

Figure 19.11 Gap detected by clustering method

19.3.3.3.2 Multi-Hypotheses Module The multi-hypotheses module is implemented to solve uncertain separator types passed from the knowledge-based segmentation module. A set of segmentation options, where each option represents a different way of segmenting the Day and the Month fields, contains a separator candidate and both neighbors of Day/Month. For each hypothesis, the confidence values for three writing patterns are calculated as the following weighted sums:

NSN ConfidenceType1 = w1* DigitConfidenceLeft + w1* DigitConfidenceRight + w3 * SeparatorConfidence

NSA ConfidenceType2 = w1* DigitConfidenceLeft + w2* WordConfidenceRight+ w3 * SeparatorConfidence

ASN ConfidenceType3 = w2* WordConfidenceLeft + w1* DigitConfidenceRight+ w3 * SeparatorConfidence

where DigitConfidenceLeft and DigitConfidenceRight are confidence values from a digit recognizer, for the left and right sides of a separator candidate, respectively. WordConfidenceLeft and WordConfidenceRight are confidence values from a cursive word recognizer. The SeparatorConfidence is the confidence value for the separator candidate generated from the knowledge-based segmentation stage. The weights w1,w2, and w3 are the equation parameters that are determined during the training step for both the digit recognizer and the cursive word recognizer of the date [27].

For each hypothesis in the list, the maximum ConfidenceType is recorded. By comparing all the maximum ConfidenceType values, the one with the highest value is usually selected, and the corresponding separator and the pattern type are finally determined for the input Day & Month sub-image.

Once the date images have been segmented into three subimages, they are submitted to the recognizers.

19.3.3.4 Date Image Recognition and Verification

After each segmentation part (Day, Month and Year) and the writing type of the Month (alphabetic letters versus numerals) have been derived, we use the appropriate trained recognizer to recognize the actual numerals and month words, respectively. The digit recognizer for the courtesy amount can be used to process the numeral date field. The cursive word recognizer used for the Legal Amount can be applied to recognize the Month's alphabetic part, but first this classifier has to be trained on the lexicon of month words for the date.

Finally, a verification module has been designed to perform a reliable date processing procedure. Two tasks are involved in the verification process: (1) detecting the possible errors from the segmentation stage and trying to correct the errors; and (2) accepting the valid and reliable recognition results and rejecting others. In addition, three rejection criteria are applied as follows, to guarantee the reliability of date processing:

- Low rejection: when no separator can be found by the segmentation step.
- Medium and high rejection: when the recognition results for each field represent valid parts of a date and the average confidence value from the three fields is above a certain threshold.

The overall performances of our date recognition module on the English and French databases have been achieved from 36% to 62% based on different rejection strategies. The errors mainly come from segmentation, word misrecognition and the numeral misrecognition. In order to improve the overall performance, one possible way is to design the specific recognizers based on the inherent restrictions of the date instead of using recognizers for courtesy amounts and legal amounts directly without modifications.

19.4 Summary and Future Work

In this chapter, we have introduced the cheque processing system developed at CENPARMI since the 1990s. To design such an automatic system, many complex issues have been addressed and most of them have been solved through various algorithms and strategies. The main modules in our system include: extracting target regions with image preprocessing; separating isolated digits from touching numeral strings, cutting the legal amount sentence into individual words, and dividing the Year, Month and Day fields through the punctuation and gap detection; designing the digit recognizer and the word recognizer by the combination of classical classifier models; and post processing the recognition hypotheses to make the final decision.

For such a comprehensive system, there have been many challenges. Some solutions include applying the morphological and topological operations, but they can only be used in certain cases. The item extraction becomes difficult when the background has a similar grey-level to the foreground of handwritten strokes. Noise needs to be further removed, such as the printed symbols, intruders from outside fields, and backgrounds that remain after binarization. Some parts are missed due to the restriction of the searching region, such as parts of words that have been written under the legal amount baseline, and the end part of a written date which exceeds our designed searching box. When these problems occur, partial cheque information is lost. Therefore, it is hard to figure out the exact item information on the input image when the part item information is missing, even with the perfect digit or word recognizers.

Many problems have come from the segmentation process. In our system, segmentation processing and recognition processing highly depend on each other. If the uncorrected segmentation parts are returned with reliable confidences from the recognizer, then the segmentation process terminates and generates a wrong cutting path. Therefore, to minimize segmentation errors, we need more investigations of other segmentation algorithms and further improvements to the recognition and rejection performances of each recognizer.

The cent parts in both courtesy amounts and legal amounts can also cause problems in each processing field. Some special layouts in the cent part on courtesy amounts, such as "XX/XX", "XX/00", "00/100", have been poorly analyzed by our current courtesy amount recognition engine. The same situation happens with the cent parts on legal amounts, which are written in the numeral format. Therefore, both a cent part extractor and a cent digit recognizer need to be developed and incorporated.

To overcome problems arised due to the extraction and segmentation procedures, legal amounts and courtesy amounts need to be cross-verified. The validation process involves combining the legal amount recognition candidates with the courtesy amount recognition candidates, and ranking the final candidate list. Many cheque reading systems [4, 7] have shown that the verification analysis of both fields significantly improves the recognition performance and reduces the error rate. In our system, the verification model is currently under development.

All of these problems come from the underlying humans' handwriting styles, which include variability and cursive styles. Therefore, more efforts are needed to reach a satisfactory solution for such complex issues.

Acknowledgements

The authors would like to acknowledge research funding from the National Sciences and Engineering Research Council of Canada, and the efforts by Ms. Shira Katz for editing this article.

References

[1] G. Dimauro, S. Impedovo, G. Pirlo, and A. Salzo. Automatic bankcheck processing: A new engineered system. *International Journal of Pattern Recognition and Artificial Intelligence*, 11(4):467–503, 1997.

[2] W. Ding. Courtesy amount recognition using a feedback-based segmentation algorithm. Master Thesis, Computer Science, Concordia University, Montreal, September 2008.

[3] W. Ding, C.Y. Suen, and A. Krzyzak. A new courtesy amount recognition module of a check reading system. In *Proceedings of International Conference on Pattern Recognition*, Tampa, Florida, pages 1–4, 2008.

[4] G. Dzuba, A. Filatov, D. Gershuny, I. Kil, and V. Nikitin. Check amount recognition based on the cross validation of courtesy and legal amount fields. *International Journal of Pattern Recognition Artificial Intelligence*, 11(4):639–655, 1997.

[5] R. Fan. Recognition of dates handwritten on cheques. Master Thesis, Computer Science, Concordia University, Montreal, September 1998.

[6] R. Fan, L. Lam, and C.Y. Suen. Processing of date information on cheques. In A.C. Downton and C. Impedovo (Eds.), *Progress in Handwriting Recognition*, pages 473–479. World Scientific, Singapore, 1997.

[7] N. Gorski, V. Anisimov, E. Augustin, O. Baret, and S. Maximov. Industrial bank check processing: the A2iA checkreaderTM. *International Journal on Document Analysis and Recognition*, 3:196–206, 2001.

[8] D. Guillevic. Unconstrained handwriting recognition applied to the processing of bank cheques. PhD Thesis, Computer Science, Concordia University, Montreal, September 1995.

[9] L. Heutte, P. Pereira, O. Bougeois, J. Moreau, B. Plessis, and P. Courtellemont. Multi-bank check recognition system: Consideration on the numeral amount recognition module. *International Journal of Pattern Recognition and Artificial Intelligence*, 11(4):595–617, 1997.

[10] S. Impedovo, P.S.P. Wang, and H. Bunke. *Automatic Bankcheck Processing*. World Scientific, Singapore, 1997.

[11] G. Kaufmann and H. Bunke. Automated reading of cheque amounts. *Pattern Analysis Appication*, 3:132–141, 2000.

[12] G. Kim and V. Govindaraju. A lexicon driven approach to handwritten word recognition for real-time applications. *IEEE Trans. Pattern Analysis and Machine Intelligence*, 19:366–379, 1997.

[13] I.C. Kim, K.M. Kim, and C.Y. Suen. Word separation in handwritten legal amounts on bank cheques based on spatial gap distances. In *Proceedings of 17th International Conference on Industrial and Engineering Applications of Artificial Intelligence and Expert Systems*, Ottawa, Canada, pages 453–462, May 2004.

[14] J.H. Kim, K.K. Kim, and C.Y. Suen. An HMM-MLP hybrid model for cursive script recognition. *Pattern Analysis and Application*, 3(4):314–324, 2000.

[15] J.H. Kim, K.K. Kim, and C.Y. Suen. Hybrid schemes of homogeneous and heterogeneous classifiers for cursive word recognition. In *Proceedings of 7th International Workshop on Frontiers in Handwriting Recognition*, Amsterdam, the Netherlands, pages 433–442, 2000.

[16] L. Lee, M. Lizarraga, N. Gomes, and A. Koerich. A prototype for brazilian bankcheck recognition. In S. Impedovo, P.S.P. Wang, and H. Bunke (Eds.), *Automatic Bankcheck Processing*, pages 549–569. World Scientific, Singapore, 1997.

[17] K. Liu, C.Y. Suen, M. Cheriet, J.N. Said, C. Nadal, and Y.Y. Tang. Automatic extraction of baselines and data from check images. *International Journal of Pattern Recognition and Artificial Intelligence on Automatic Bankcheck Processing*, 11(4):675–697, 1997.

[18] Orbograph Company. Orbocar engine utilization strategies for CAR/LAR recognition solutions. Technical Report, Orbograph Company, 2007.

[19] J.N. Said, M. Cheriet, and C. Y. Suen. Dynamical morphological processing: A fast method for base line extraction. In *Proceedings of 13th International Conference Pattern Recognition*, Vienna, Austria, pages 8–12, 1996.

[20] G. Seni and E. Cohen. External word segmentation of off-line handwritten text lines. *Pattern Recognition*, 27(1):41–52, 1994.

[21] N.W. Strathy. A method for segmentation of touching handwritten numerals. Computer Science, Concordia University, Montreal, September 1993.

[22] N.W. Strathy and C.Y. Suen. A new system for reading handwritten zip codes. In *Proceedings of the 3rd International Conference on Document Analysis and Recognition*, pages 74–77, Montreal, Canada, 1995.

[23] H. Su, B. Zhao, F. Ma, S. Wang, and S. Xia. A fault tolerant chinese check recognition system. *International Journal of Pattern Recognition and Artificial Intelligence*, pages 571–593, 1997.

[24] C.Y. Suen, L. Lam, D. Guillevic, N.W. Strathy, M. Cheriet, J.N. Said, and R. Fan. Bank check processing system. *International Journal of Imaging Systems and Technology*, 7:392–403, 1996.

[25] C.Y. Suen, Q.Z. Xu, and L. Lam. Automatic recognition of handwritten data on cheques – Fact or fiction? *Pattern Recognition Letters*, 20(13):1287–1295, 1999.

[26] Unisys Company. SoftCAR+ diamond edition capabilities overview. Technical Report, Unisys Company, 2005.

[27] Q.Z. Xu. Automatic segmentation and recognition system for handwritten dates on cheques. PhD Thesis, Computer Science, Concordia University, Montreal, November 2002.

[28] Q.Z. Xu, L. Lam, and C.Y. Suen. Automatic segmentation and recognition system for handwritten dates on canadian bank cheques. In *Proceedings of 7th International Con-*

ference on Document Analysis and Recognition, Edinburgh, Scotland, pages 704–708, 2003.

[29] X. Ye, M. Cheriet, C.Y. Suen, and K. Liu. Extraction of bankcheck items by mathematical morphology. *International Journal on Document Analysis and Recognition*, 2:53–66, 1999.

[30] L.Q. Zhang and C.Y. Suen. Recognition of courtesy amounts on bank checks based on a segmentation approach. In *Proceedings of International Workshop on Frontiers in Handwriting Recognition*, Niagara-on-the-Lake, pages 298–302, August 2002.

[31] J. Zhou. Segmentation of legal amount on bankchecks. Master Thesis, Computer Science, Concordia University, Montreal, September 2001.

20

Automatic Off-line Signature Verification by Computers

Bin Fang[1], Jing Wen[1], Y.Y. Tang[1] and Patrick S.P. Wang[2]

[1]*College of Computer Science, Chongqing University, Chongqing 400044, China; E-mail: {fb, wj, yyt}@cqu.edu.cn*
[2]*College of Computer and Information Science, Northeastern University, Boston, MA 02115, USA; E-mail: pwang@ccs.neu.edu*

Abstract

This chapter introduces latest development in automatic off-line signature verification. Techniques and algorithms used in different stages of the verification process are presented. The frame work consists of data acquisition and pre-processing feature extraction and pattern matching. Finally, some valuable research directions are suggested to help colleagues working in this field.

Keywords: rotation invariant, small size of samples, feature extraction, classification, off-line signature verification, personal identification, system security.

20.1 Introduction

As more and more organizations are looking forward to improving the security for user access, e-commerce, and other security application, more and more secure authentication methods are developed. The security field mainly uses three types of authentication: knowledge-based approaches, token-based approaches and biometric-based approaches, see Figure 20.1. Biometric is

Patrick Shen-Pei Wang (Ed.), Pattern Recognition and Machine Vision – In Honor and Memory of Professor King-Sun Fu, 319–333.

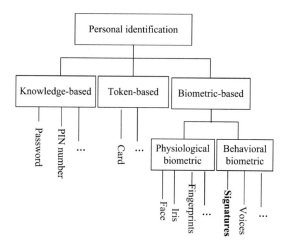

Figure 20.1 Personal identification

the most security and convenient authentication tool among them. In today society, biometric systems have been widely applied in various commercial and forensic applications as a means of establishing identity [42]. There are two types of biometric means: physiological biometrics and behavioral biometrics. Signatures belong to the latter. Comparing with the others, although signatures are by no means the most reliable means of personal identification, signatures are no invasive and inexpensive, and they are the most socially and legally accepted means of personal identification. In addition, signatures provide the direct link between the writer's identity and the transaction, therefore automatic signature verification system (ASV) can be incorporated transparently into the existing business processes requiring signatures such as banking transactions and credit card operations, etc. [28, 31].

In general, ASV can be categorized into two major areas according to the data acquisition method: on-line and off-line signature verification. In the former, data are obtained using digitizing tablet or other electronic devices that are able to record dynamic information including the pen's position pressure, handwriting order, etc., and it can be use in real-time security applications like credit card transaction, accessing to sensitive data, etc. In the latter, images of the signatures written on a paper are obtained using a scanner or a camera, and is useful in automatic verification of signatures found on bank checks, contract and documents. Unlike the on-line signature, where dynamic aspects of the signing action are captured directly in off-line signature is lost. Therefore, off-line signature verification has always been

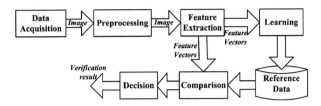

Figure 20.2 Process of automatic signature verification (ASV)

considered by researchers to be the more difficult approach [28]. Although the accuracy of the off-line systems reported are less than on-line systems, in fact, if the accuracy of the off-line systems promoted greatly, the off-line method offer more widespread application areas than that of the on-line one for off-line systems not requiring special devices to produce signatures.

The general process of ASV is shown in Figure 20.2, which sketches the five main stages: data acquisition, preprocessing, feature extraction, training and verification/comparison. After data acquisition, the preprocessing phase removes spurious noise from the input signature image.The discriminatory features are extracted in the feature extraction phase. During the training phase, the features extracted from the set of training signature samples are used to design the classifier.These features are matched against that of the test input signature in the verification/comparison phase. The result is used to validate the authenticity of the input signature. Every person has individuality of signature. However, no two signatures are identical from one writer, and variations of a typical signature also depend on time, age, the physical, mental state and practical conditions of a writer. Hence, signature verification is considered as a difficult verification problem. Although research into signature verification has been vigorously pursued for a number of years, it is still being explored due to his challenges and extensive application.

This chapter mainly presents a review of the state of the art in off-line signature verification from the process of ASV. It reports many recent advancements that have occurred in this field.

The organization of the paper as follows: the main aspects related to data acquisition and preprocessing are described in Section 20.2. Section 20.3 investigates the feature extraction of off-line signature. In Section 20.4, the different verification approaches are mentioned. The discussion and conclusion is given in Section 20.5.

20.2 Off-Line Signature Acquisition and Preprocessing

After the writing process, handwritten signatures are generally acquired from a bill, a check or a document and are converted to digital form by off-line acquisition devices such as an optical scanner or camera.

Typical preprocessing involves removal of spurious noise by filtering function [11] or morphologic operation [3, 4], signature location and extraction from background [30], size and rotation normalization [3, 27], binarization [24], image thinning [3, 12] and Segmentation [1, 17, 22, 29, 40], and so on. It is worth noting that these preprocessing operations are often performed as well depending on the requirements of the feature extraction phase.

In real-world applications, the extraction of signature from the background is still a difficult and complex task, because the signature images are inside the document, existence of texture and logotypes in the document background, especially in the case of bankcheck with a color pictorial background. Hence, signature images extracted from the bank check is a challenge and open problem, and is an area of continued research [30].

Segmentation is also an important step in the precessing stage. Via this step, signature image is divided into small parts. Segmentation approaches have a close relationship with some specific features, and strongly influences all the successive phases of signature verification. According to the requirements of the feature extraction, different segmentation approaches are developed.

Grid based segmentation scheme is a common segmentation method. In this scheme, the image is equally segmented into N equal-size cells with or without overlapping, and feature extraction is based on these cells [7].

Tree based segmentation is another approach, which describe a signature by a tree structure, e.g. Ammar et al. [1] obtained segments through the analysis of horizontal and vertical projection histograms, which identifies fundamental segments in the static image.

Component based segmentation obtain pertinent parts of the signature by structural analysis for making local measurements, which is the most difficult among all types of segmentation since no dynamic information is available, e.g. Fang et al. [17] used abrupt corners and cross points as break points based on the contour of the signature image, stroke segment is the sequence of border pixels between two break points. Figure 20.3 shows (a) the signature segmented into (b) stroke segments by Fang et al. [17]. An on-line segmentation model-guide technology has also been presented [44]. This approach

(a)   (b)

Figure 20.3 (a) The signature segmented into (b) stroke segments

used on-line reference data acquired as the basis for the segmentation process of the corresponding off-line data.

Other approaches segmented signature by radial segmentation. This approach is to divide the signature image into a number of concentric regions [29, 40]. Off-line signature segmentation by statistics of data has also been considered [23].

20.3 Feature Extraction

20.3.1 Categorization of Features

Feature extraction is one of the most crucial tasks in signature verification system, which is the process of obtaining some specific parameters from signatures that would enable the signatures of one person to be discriminated from the signatures of another person. Many researches have looked into feature extraction methodology. In general, three groups of features have been used in ASV: global, statistical, geometrical and topological features [27].

Global features are extracted from the whole signature, which describe an entire signature. Typical global feature include slant of the signature [3], the center of gravity (COG) [32], and coefficients obtained by mathematical transforms (e.g. the discrete wavelet transform [11], the Hough transform [32], Radon transform [6] and Gabor transform [18]). Although global features can be extracted easily and are robust to noise, they are dependent upon the position alignment and highly sensitive to distortion and style variations. Hence, they are used for random and simple forgeries detection.

Statistical features are to a signature image is representation by statistical distribution of points and strokes. This type of features is low complexity and have a minor tolerance to distortion and style variations. Some statistical features is based on region, e.g. [22, 23]. Projection is a popular statistical method, which is based on projection of signature image in different directions. This features include vertical and horizontal projection [3, 14], and central projection [37]. Some texture features based on statistics also have been extracted from signature images, which include the co-occurrence matrices of the signature image [3], and gray-level intensity features that provide useful pressure information [24]. In addition to the above features, counting the number of crossing or transition of a contour in a specified direction also belong to statistical features. Armand et al. [2] extracted the modified direction feature (MDF) which utilizes the location of transitions from background to foreground pixels in the vertical and horizontal directions of the boundary representation of an signature.

In many publications, various global and local properties of signature images can be represented by geometrical and topological features. They can tolerate high distortion and style variances, and they can also tolerate translation and rotation variations to some extent. The geometrical quantities is an approach of describing the geometrical and topological structure of signature images. Among these features, curvature information or direction of stroke is an important characteristic in signatures [22]. The measured geometrical quantities also can be approximated by a more compact geometrical set of features, which represent the envelope description and the interior strokes (e.g. [13]). Shape descriptor is also an important approach of describing the envelope of signature images, which represent two-dimensional information with one-dimensional coding. There are many versions of shape descriptor, e.g. in [37] the fractal code of the signature contour is applied. Zhang et al. [43] proposed a smoothness shape descriptor to encode the envelope of signature image. Mathematical morphology is a set theoretical approach for the analysis of geometric structures. Aguilar et al. [19] divided signature image into a number of blocks with overlapping, and then employed morphology method to extracted the local slant direction and envelope features of each signature block.

20.3.2 Non-Linear Rotation

Non-linear rotation of signature patterns inevitably exists and causes poor results. To avoid the effect of rotation, an orientation normalization is conventional employed during preprocessing process. However, normalization would be generated distortion of original signature image. Hence, many rotation-invariant approaches are developed in some works, see [6,37,40,43].

For instance, Figure 20.4 shows Ring-Peripheral Features (RPF) which are proposed by Wen et al. [40]. The important property of RPF is that they

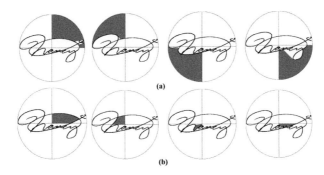

Figure 20.4 (a) Extraction of the REF feature, (b) extraction of the RIF feature

are able to detect interpersonal change by ring projection, and they are periodical to make the resulting feature vectors for the same writer identical with only phase shift in fact.

As can be observed from Figure 20.4, RPF consists of a REF and a RIF feature, in which REF is the ring projection area between the virtual circle frame and white-to-black pixel jump toward the center of circle. It is well suitable for describing the whole signature external shape. The RIF calculates the ring projection area between the first white pixel and the first white-to-black pixel jump beginning from the center of circle, consequently, it may represent more internal detail structure features such as stroke relative positions, etc. The extracted RPFs are not used directly to distinguish authentic signatures and forgeries because there exist phase shift of RPF vectors. In order to tackle rotation problems, two different approaches were applied to eliminate phase shift. One method employs Fast Fourier transform (FFT) making signature features consistent. The other approach is to construct a ring-HMM, directly evaluate match score for verification.

20.3.3 Combination of Different Types of Features

In researches, numerous types of features have been developed. Some system use only singular type feature, while others employ a combination of different types of features instead of singular type. A lot of experiments have proved that combination of different types of features allow better performance than singular type features because that combination can easily handle the intra-personal and inter-personal signature variations, such as a combination of global and local features [17, 19, 35], combination of static image

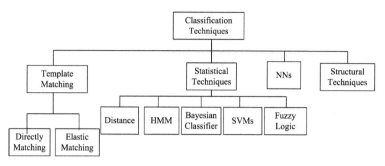

Figure 20.5 Classification techniques

pixel features and pseudo-dynamic features [24], and combination of global, statistical, geometrical and topological features [27, 35, 36].

20.4 Classification

In the previous section, feature extraction technologies have been described. Signature images can been represented by various feature sets. So far many kinds of pattern classification techniques have been developed based on various feature representations. These techniques can be investigated in four general approaches as shown in Figure 20.5. In ASV, these classification techniques are sometimes used individually or combined in many ways as part of the signature verification schemes.

20.4.1 Template Matching

Template matching is one of the simplest and earliest classified approaches [26], where the degree of similarity between two feature vectors is determined by matching operation. Basically, there are two groups of template matching: direct matching and elastic matching.

Direct matching is that a input image directly compared to a feature set of a reference image. This matching technique is usually a one-to-one comparison based on similarity measures [35, 38].

Elastic matching is defined as an optimization problem of two-dimensional warping specifying corresponding pixels between subjected images and is also known as deformable template or non-linear template matching. Its basic idea is to optimally match the unknown image against all possible compression or elastic stretching of each prototype. Elastic matching (EM) in ASV is used for solving the various distortions of signatures.

In [12], an elastic local alignment algorithm is applied to align corresponding feature points of signature images after a weak affine model is employed to globally register two signature skeletons. Finally, the similarity measurement is evaluated. Chen and Srihari [4] employed thin-plate splines warping to map every point on the reference image into one on the test image in each signature region. For every pair of images, the bending energy and dissimilarity score are measured.

Dynamic time warping (DTW) is also one of elastic matching methods. In [14], the one-dimensional projection profiles signature patterns are optimally matched using DTW, and then the position variations are measured by the resulting warping function. More recently, Güler and Meghdadi [21] presented an optimal dynamic time warping (ODTW) algorithm in comparison stage.

20.4.2 Statistical Techniques

Statistical techniques has been used successfully to design a number of commercial verification systems. In this approach, every pattern is represented by d-dimensional feature vector. The goal is to choose those features that allow pattern vectors belonging to different categories to occupy compact and disjoint regions in a d-dimensional feature space [26].

Distance classifiers are often used in ASV, including Euclidean distances [17], Mahalanobis distances [13, 15] and Kolmogorov–Smirnov statistics [36].

Hidden Markov Model (HMM) is another statistical approach used for signature verification. Due to the importance of the warping problem in signature verification as well as in handwriting recognition applications, the use of HMM is becoming more and more popular in both areas. In the field of off-line signature verification, HMM include two types structures: left-to-right and ring topology. Most approaches used the left-to-right HMM topology for signature verification [6]. Coetzer and du Preez [6] developed ring-HMM to deal with rotation problem for off-line signature verification. This model is similar to the left-to-right topology in which each state has a transition to itself and the next state, but a transition from the last state to the first state is also allowed in ring topology.

Comparatively speaking, other statistical techniques such as Bayesian classifiers [35], SVMs [35] and fuzzy logic [23], etc. are less used in ASV.

20.4.3 NNs

Another wide category of off-line signature verification systems is the systems based on neural networks (NNs) approach. Some literatures also classified NNs as statistical techniques. The main characteristic of an NN is that it has the ability to learn non-linear input-output relationships, use sequential training procedures, and adapt itself to the data.

In [28], Leclerc and Plamondon reported on the different projects dealing with neural network approaches in automatic signature verification from year 1989 to 1993. Afterwards, more and more NN models are wide used like Bayesian NNs [41], multilayer perceptrons (MLPs) [3, 24], self-organizing maps [29], backpropagation neural networks (BPNs) [10], radial basis functions (RBFs) [3] and so on.

Although NNs have good capabilities in generalization, they require large amounts of training data that are not always available like some statistical techniques.

20.4.4 Structural Techniques

Structural techniques adopt a hierarchical perspective. In this approach, a pattern is viewed as being composed of simple subpatterns which are themselves built from yet simpler subpatterns [1, 5, 6]. This approach is not particularly popular for the purpose of off-line signature verification. In [25], Huang and Yan constructed statistical models, which are based on both the pixel distribution and structural layout description. During verification, a structural feature verification algorithm is presented and compares the detailed structural correlation between the input and reference signatures.

20.4.5 Small Size of Samples

It is well known that when the number of training samples is less than the dimensionality of the feature space, the estimates of the statistical model parameters are unreliable. This problem is especially significant in off-line signature verification, because it is difficult to get sufficient signatures from each writer in a real problem (i.e. banking documents).

There are several ways to overcome the problem of small size of samples in signature verification, these techniques are categorized into three groups: (1) reducing the feature dimensionality through feature selection; (2) regularization of sample covariance matrix; (3) generating artificial additional training samples. Feature selection select a subset of the original feature set

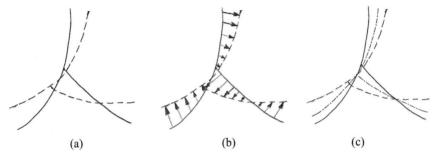

(a)   (b)   (c)

Figure 20.6 (a) Overlapped images of the original template (solid lines) and input (dashed lines) patterns. (b) The pair of signatures with the corresponding strokes identified and linked up by displacement vectors. (c) Additional signature is generated by making used the displacement vectors to 'interpolate' between the original pair of signatures.

to reduce the feature dimensionality to enhance the between class pattern variability and minimize the within class pattern variability and perform the best classification, e.g. principle component analysis (PCA) is used for reducing the feature dimension [9]. Neural network trainer is also employed for feature selection [3]. Although the above approaches could effectively reduce the feature dimensionality, they are not applicable in the study to estimate stable class statistics due to training samples come from genuine class for signature verification. Regularization techniques are also used in statistical-based methods [16, 40]. Artificially generating additional training samples is also an approach to avoid the problem of small size of samples. Oliveria et al. [8] used a convolution deformation model to generate additional samples. Traces of the signature are represented horizontally and vertically by two sequences, which are defined as polynomials. The resulting additional sample is produced by convolving these polynomials with deforming polynomials whose coefficients control the final layout the additional samples. In [24], linear transformations are used for automatically generated additional samples. The linear transformations include rotation, scale, slant distortion. However, these approaches do not mention that how much distortion should be applied. Fang et al. [13,16] used two-dimensional elastic structure matching algorithm to artificially generate additional training signature samples. Figure 20.6 illustrates a feasible guideline to measure possible distortion. Because the generated signatures are within bounds of variations of the original template and input samples, it is ensure that the generated signatures belong to the genuine samples.

20.5 Discussion and Conclusions

In recent years, along with the growth of economic and technology, the requirements of security continuous increase, and more and more organizations are devoted to the research of secure authorization approaches. Among these approaches, signature verification approach are considered since the signatures have well acceptance in our daily life [31]. Signature verification system consists of on-line and off-line signature verification system. Although on-line signature verification is more successful than off-line, off-line signature verification afford a further wide range of potential applications.

In this chapter, the main aspects related to process of off-line signature are investigated. In further research, research not only focuses on the further improvement of accuracy of system, but also focus on computation complex analysis and real-world implementation based on different scenarios.

Furthermore, in recent years, the extraction of the pseudo-dynamic features for off-line data under the guidance of the characteristic of on-line data still remains an interesting research area. On-line data provides a wealth of dynamic information, during the training stage, off-line signature verification system combines them, and a relative model is created [33,44]. Similar to the techniques of the off-line signature verification, off-line handwritten signatures are applied in other applications such as cryptography [20], which is a field is worthy of attention.

In addition, for combining the advantages of different classifiers and improving the accuracy of off-line signature verification system, research for the multi-expert system becomes a promising and important area. Different combination strategies have been proposed including parallel, serial and hybrid approach [34]. Some literatures have been addressed the validity of multi-expert systems [2,4,39].

Finally, most of researches mainly focus on the signatures from Western. However, many people from non-Western countries offer different signatures such as Chinese, Japanese, Arabian and so on, which contain special characters. Hence, some special approaches have been developed in the literatures (e.g. [9,29,38]). Thus, Non-Western signature verification is worthy of further research. Besides, in order to evaluate the performance of different off-line signature verification systems, in further research, it is necessary to develop the benchmark databases including signatures from Western and non-Western people.

In conclusion, nowadays, off-line signature verification systems are no longer limited in the research of the academics or research laboratories.

And with a further improvement of systems performance, automatic off-line signature verification systems would more widely apply to daily life.

Acknowledgements

This work was partly supported by the Program for New Century Excellent Talents of Educational Ministry of China (NCET-06-0762) and Natural Science Foundation Project of CQ CSTC under the grant No. CSTC2007BA2003.

References

[1] M. Ammar, Y. Yoshida, and T. Fukumura. Structural description and classificatio of signature images. *Pattern Recognition*, 23(7):697–710, 1990.

[2] S. Armand, M. Blumenstein, and V. Muthukkumarasamy. Off-line signature verification based on the enhanced modified direction feature with single and multi-classifier approaches. *IEEE Computational Intelligence Magazine*, 2(3):18–25, 2007.

[3] H. Baltzakis and N. Papamarkos. A new signature verification technique based on a two-stage neural network classifier. *IEngineering Applications of Artificial Interlligence*, 14(1):95–103, 2001.

[4] S. Chen and S. Srihri. A new off-line signature verification method based on graph matching. In *Proceedings of 18th International Conference Pattern Recognition*, Vol. 2, pages 869–872, 2006.

[5] Y. Chen, X.Q. Ding, and P.S.P. Wang. Dynamic structural statistical model based online signature verification. *International Journal of Digital Crime and Forensics*, 1(3):21–41, 2009.

[6] J. Coetzer and J.A. du Preez. Off-line signature verification. PhD Thesis, University of Stellenbosch, 2005.

[7] M.T. Das and L.C. Dülger. Off-line siganture verification with PSO-NN algorithm. In *Proceedings of the 22nd International Symposium on Computer and Information Sciences*, pages 1–6, November 2007.

[8] Claudio de Oliveira, C. Kaestner, F. Bortolozzi, and R. Sabourin. Generation of signatures by deformations. In *Advances in Document Image Analysis*, Vol. 1339, pages 283–298. Springer, Berlin, 1997.

[9] Y. Ding, Q. Chen, and J. Wang. Offline chinese signature verification system with information fusion. In *Proceedings of SPIE Third Internationa Symposium on Multispectral Image Processing and Pattern Recogntion*, Vol. 5286, pages 875–878, September 2003.

[10] J.P. Drouhard and R. Sabourin, and M. Godbout. Evaluation of a training method and of various rejection criteria for a neural network classifier used for off-line signature verification. In *Proceedings IEEE World Congress on Computational Intelligence, IEEE International Conference on Neural Networks*, Vol. 7, pages 4294–4298, 1994.

[11] E.A. Fadhel and P. Bhattacharyya. Application of a steerable wavelet transform using neural network for signature verfication. *Pattern Analysis & Applications*, 2:184–195, 1999.

[12] B. Fang, W.S. Chen, and X.G. You et al. Wavelet thinning algorithm based similarity evaluation for offline signature verification. In *Intelligent Computing in Signal Processing and Pattern Recognition*, KunMing, China, August 2006, Vol. 345, pages 547–555. Springer, Berlin, 2006.

[13] B. Fang, C.H. Leung, and Y.Y. Tang et al. Offline signature verification with generated training samples. In *IEE Proc. Vis. Image Signal Process*, Vol. 149, pages 85–90, 2002.

[14] B. Fang, C.H. Leung, and Y.Y. Tang et al. Off-line siganute verification by the tracking of feature and stroke positions. *Pattern Recognition*, 36:91–101, 2003.

[15] B. Fang and Y.Y. Tang. Reduction of feature statistics estimation error for small training sample size in offline signature verification. In *Proc. 1st Int. Conf. Biometric Authentication*, Berlin, Germany, Vol. 3072, pages 526–532. Springer, Berlin, 2004.

[16] B. Fang and Y.Y. Tang. Improved class statistics estimation for sparse data problems in offline signature verification. *IEEE Transactions on Systems, Man and Cybernetics – Part C:Applications and Reviews*, 35(3):276–286, 2005.

[17] B. Fang, Y.Y. Wang, C.H. Leung, and K.W. Tse. Offline siganture verification by the analysis of cursive strokes. *International Journal of Pattern Recognition and Artificial Intelligence*, 15(4):1659–673, 2001.

[18] J.B. Fasquel and M. Bruynooghe. A hybrid opto-electronic method for fast off-line handwritten signature verification. *International Journal on Document Analysis and Recognition*, 7:56–68, 2004.

[19] J. Fierrez-Aguilar, N. Alonso-Hermira, G. Moreno-Marquez, and J. Ortega-Garcia. An off-line signature verification system based on fusion of local and global information. In *Biometric Authentication*, Vol. 3087, pages 295–306. Springer, Berlin, 2004.

[20] M. Freire-Santos, J. Fierrez-Aguilar, and J. Ortega-Garcia. Cryptographic key generation using handwritten signature. In *Proc. of SPIE, the International Society for Optical Engineering*, Kissimmee, Florida, USA, Vol. 6202, pages 1–7, 2006.

[21] I. Güler and M. Meghdadi. A different approach to off-line handwritten signature verification using the optimal dynamic time warping algorithm. *Digital Signal Processing*, 18:940–950, 2008.

[22] J.K. Guo, D. Doermann, and A. Rosenfeld. Forgery detection by local correspondence. *International Journal of Pattern Recognition and Artificial Intelligence*, 15(4):579–641, 2001.

[23] M. Hanmandlu, M.H.M. Yusof, and V.K. Madsu. Offline signature verification and forgery detection using fuzzy modeling. *Pattern Recognition*, 38(3):341–356, 2005.

[24] K. Huang and H. Yan. Off-line signature verification based on geometric feature extraction and neural network classification. *Pattern Recognition*, 30:9–17, 1997.

[25] K. Huang and H. Yan. Off-line signature verification using structural feature correspondence. *Pattern Recognition*, 35:2467–2477, 2002.

[26] A.K. Jain, R.P.W. Duin, and J. Mao. Statistical pattern recognition: A review. *IEEE Transactions on Pattern Analysis and Machine Intelligence*, 22(1):4–37, 2000.

[27] M.K. Kalera, S. Srihari, and A. Xu. Offline signature verification and identification using distance statistics. *International Journal of Pattern Recognition and Artificial Intelligence*, 18(7):1339–1360, 2004.

[28] F. Leclerc and R. Plamondon. Automatic signature verification: The state of the art 1989–1993. *International Journal of Pattern Artifical Intelligence*, 3(1):643–660, 1994.

[29] H. Ma. Off-line Chinese-based signature verification using a threshold self-organizing map. *Journal of the Chinese Institute of Industrial Engineers*, 24(3):225–235, 2007.

[30] V.K. Madasu, M.H.M. Yusof, M. Hanmandlu, and K. Kubik. Automatic extraction of signatures from bank cheques and other documents. In *Proceedings VIIth Digital Image Computing: Techniques and Applications*, Sydney, December, pages 591–600, 2003.

[31] R. Plamondon and S.N. Srihari. On-line and off-line handwriting recognition: A comprehensive survey. *IEEE Transactions on Pattern Analysis and Machine Intelligence*, 22(1):63–84, 2000.

[32] P. Porwik and T. Para. Some handwritten signature parameters in biometric recognition process. In *Proceedings of the ITI 2007 29th International Conference on Information Technology Interfaces*, Cavtat, Croatia, pages 185–190, 2007.

[33] Y. Qiao, J. Liu, and X. Tang. Offline signature verification using online handwriting registration. In *Proceedings of IEEE conference on Computer Vision and Pattern Recognition*, pages 1–8, 2007.

[34] A.F.R. Rahman and M.C. Fairhurst. Multiple classifier decision combination strategies for character recognition: A review. *International Journal on Document Analysis and Recognition*, 5:166–194, 2003.

[35] S.N. Srihari, A. Xu, and M.K. Kalera. Learning strategies and classification methods for off-line signature verification. In *Ninth International Workshop on Frontiers in Handwriting Recognition*, pages 161 – 166, October 2004.

[36] H. Srinivasan, S.N.Srihari, and M.J. Beal. Signature verification using Kolmogorov–Smirnov statistic. In *Proceedings International Graphonomics Society Conference*, Salerno, Italy, pages 152–156, 2005.

[37] Y. Tao, T.R. Ioerger, and Y.Y. Tang. Extraction of rotation invariant signature based on fractal geometry. In *2001 IEEE International Conference on Image Processing*, Vol. 1, pages 875–878, 2001.

[38] K. Ueda. Investigation of off-line Japanese signature verification using a pattern matching. In *Proceedings of Seventh International Conference on Document Analysis and Recognition*, pages 951–955, August 2003.

[39] J. Wen, B. Fang, Y.Y. Tang, P.S.P. Wang, M. Cheng, and T. Zhang. Combining EODH and directional gradient density for offline signature verification. *International Journal of Pattern Recognition and Artificial Intelligence*, 23(6):1161–1177, 2009.

[40] J. Wen, B. Fang, Y.Y. Tang, and T.P. Zhang. Model-based signature verification with rotation invariant features. *Pattern Recognition*, 42:1458–1466, 2009.

[41] X.H. Xiao and G. Leedham. Signature verification using a modified Bayesian network. *Pattern Recognition*, 35(5):983–995, 2002.

[42] S.N. Yanushkevich, D. Hurley, and P.S.P. Wang. Pattern recognition and artificial intelligence in biometrics – Editorial. *International Journal of Pattern Recognition and Artificial Intelligence*, 22:367–369, 2008.

[43] T.P. Zhang, B. Fang, and B. Xu et al. Signature envelope curvature descriptor for offline signature verification. In *Proceedings of the 2007 International Conference on Wavelet Analysis and Pattern Recognition*, Beijing, China, pages 1262–1266, November 2007.

[44] A. Zimmer and L.L. Ling. Off-line signature verificaiton system based on the on-line data. *EURASIP Journal on Advances in Signal Processing*, 8(2):1–31, 2008.

21

Biometric Authentication

Chin-Hung Teng[1], Ho-Ling Hsu[2] and Wen-Hsing Hsu[3]

[1]Department of Information Communication, Yuan Ze University, 32003 Chung-Li, Taiwan, ROC; E-mail: chteng@saturn.yzu.edu.tw
[2]Department of Research and Development, STARTEK Engineering Inc., 30013 Hsinchu, Taiwan, ROC; E-mail: holing@mail.startek-eng.com
[3]Department of Electrical Engineering, National Tsing Hua University, 30013 Hsinchu, Taiwan, ROC; E-mail: whhsu@ee.nthu.edu.tw

Abstract

Biometrics is increasingly considered as a reliable mean of authentication. Its advantage and appeal lies in its convenience and easy to use, its level of apparent security, its high quality of performance and its nature of non-invasiveness. This chapter gives some fundamental concepts to biometric authentication, including how a biometric system is operated and how the performance of such a system is quantified. Because biometrics has gradually adopted by many large scale civilian projects, some issues for designing a large scale biometric system are also given. A number of standards concerning the specifications of biometrics are also addressed in this article. Finally, we briefly discuss the architecture and designing criteria of a large scale biometric project and introduce a potential mechanism for web-based biometric authentication.

Keywords: biometrics, authentication, verification, identification, biometric standard.

Patrick Shen-Pei Wang (Ed.), Pattern Recognition and Machine Vision – In Honor and Memory of Professor King-Sun Fu, 335–348.

21.1 Introduction

Biometrics is the technology of establishing the identity of an individual based on the biometric traits of an individual. Typically, biometric systems can be classified into two categories, biological- and behavioral- based, according to the characteristics employed. Examples of biological-based systems include face, fingerprint, iris, retinal scan, hand geometry, etc., while voice, signature, and keystroke recognition are behavioral-based biometric systems. A biometric system captures the biological or behavioral characteristics of a person via an appropriately designed sensor and then compares the biometric traits with the information stored in a database to determine the identity of this person. Compared to traditional authentication system, biometrics has been considered as a more secure and convenient approach for authentication and has been increasingly deployed in many situations including large scale civilian applications.

Identifying the identity of a person is necessary in many human activities. Traditional identity authentication is achieved by means of what an individual has or what an individual knows. The former is classified as token-based approach [11] where an individual presents a token (ID Card or key) to demonstrate that he has the claimed identity. The latter is recognized as knowledge-based approach where an individual verifies his identity by the information he knows, for example, the password. The security of these traditional systems is established on the assumption that only the "right" person has the "correct" token or knows "correct" information. However, tokens can be lost, stolen, or duplicated, while passwords can be either easily guessed or hard remembered. Thus, traditional authentication systems have their inherent limitations in the applications of authentication.

On the other hand, biometric systems use unique or near unique biometric traits to recognize an individual. Since these biometric traits are inherently associated with an individual, they can not be lost, be easily stolen or duplicated and do not need to be remembered. Thus, biometric systems advantage traditional token- or knowledge-based systems in terms of security and convenience. Unfortunately, in the past, many people are unwilling to provide their biometric traits for authentication because of privacy issue. This somewhat limits the development of biometric authentication. However, after 9/11 attack in 2001, more and more people has realized the importance of security and this pushes the public to treat the problem more serious. Today, more and more commercial biometric systems have appeared on the market. Ranging from personal use such as notebook and cellular phone login to public use

Figure 21.1 Three different types of fingerprint sensors. (a) Optical sensor. (b) Chip sensor. (c) Swipe sensor

such as national ID card, border crossing, and welfare disbursement, biometrics has gradually replaced traditional authentication systems. In fact, as the increase in the availability of inexpensive computing resources, advances in image processing and pattern recognition techniques, and cheaper sensing technologies, biometrics will definitely become the mainstream technologies in authentication applications.

In this chapter, we will not discuss deep mechanisms of various biometric technologies and further research issues in this field (one can refer to the introductory articles [2, 4] for more detailed information about these topics). Instead, after addressing some fundamental concepts to biometric authentication, we will focus on discussing the challenges of large scale biometric systems, the development of biometric standards, a large scale biometric project, and an emerging web-based biometric application.

21.2 Basic Operations of a Biometric System

In general, a biometric system has three distinctive modules: data acquisition (or sensor module), feature extraction, and matching. In sensor module, biometric raw data is acquired by an appropriately designed sensor. Some biometric data can be easily recorded by universal sensing devices. For example, face can be captured by a camera while voice can be recorded by a microphone. However, the acquisition of some biometric data such as fingerprint needs particularly designed sensor. According to different designing methodologies, even the same biometric trait can be acquired by very different sensors, generating biometric raw data with quite different qualities. For instance, Figure 21.1 illustrates three different types of sensors for fingerprint, the optical, chip, and swipe sensors. Traditionally, optical sensor can produce fingerprint image with better quality than the others. However, due to advances of sensing technologies, their gaps have been gradually bridged.

Normally, not all raw measurements from the sensor module could be used for biometric recognition. Some information is redundant and some will even interfere with the comparison of biometric data, leading to performance degradation. Which information should be preserved for authentication and which should be discarded is a critical issue for the design of a biometric system. Generally, a good biometric feature should be invariant for the input signals belonging to the same identity and differ maximally for those belonging to different identities [2]. The main task of feature extraction is to extract those useful biometric features for subsequent comparison and these extracted features are normally referred to as *biometric template*. Typically, one biometric measurement of a subject can produce one template, but some systems have an *enrollment* process which could collect more stable features from several measurements to organize a "good" template so as to improve recognition accuracy.

Generally, different biometric technologies use different features, but some standard features have emerged. For example, fingerprint minutiae (fingerprint ridge endings and bifurcations) are gradually become the standard features of fingerprint verification and dominate the fingerprint recognition market. Figure 21.2(a) shows an example of fingerprint minutiae where the left figure is the raw measurements of a fingerprint and the small circles on the right figure are the extracted fingerprint minutiae.

The matching module of a biometric system compares a pair of templates to determine a score indicating the similarity of the two templates. For example, the number of corresponding minutiae could be used to give a score of two fingerprints as illustrated in Figure 21.2(b). More corresponding minutiae indicate more similar the two fingerprints, and therefore higher comparing score. The comparing score can then be used to determine whether the templates belong to the same subject. According to physical applications, a biometric system can operate in two different matching modes: verification and identification.

Verification is a process of verifying an individual's biometric trait to examine whether it is consistent with the identity he claimed. The subject must first present his identity, normally a personal identity number (PIN), and then the system compares the two templates (one is from the subject and the other is from the database by the claimed identity) to verify his identity. This is a one-to-one matching. Identification is a process of identifying the "correct"

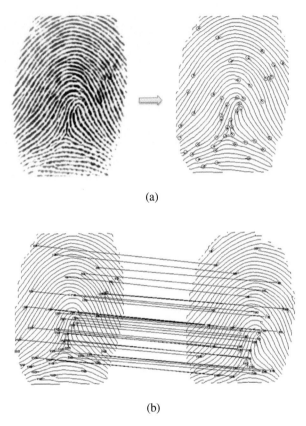

(a)

(b)

Figure 21.2 Fingerprint minutiae and a fingerprint matching based on minutiae

identity of an individual from a set of enrolled templates. The subject does not need to present his identity during the process. The system must compare all the templates in the database to determine the identity of the subject. This is a one-to-many matching. Two diagrams illustrating the operation flows of verification and identification are shown in Figure 21.3. Because identification involves the matching of a large number of templates, the processing time of identification is typically much longer than verification. In addition, because the security of a biometric system is directly in proportion to the number of comparisons, the overall security of identification is significantly reduced especially when the employed biometric traits are not sufficiently unique.

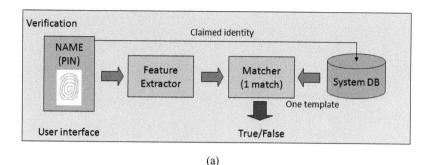

Figure 21.3 The block diagrams of verification and identification [4, 5]

21.3 Performance Evaluation

Because of distinct uniqueness of different biometric traits, some biometric systems have higher recognition accuracies than others. For instance, iris and retinal scan are typically known as more accurate than face, voice, and hand geometry recognition. In the field of biometrics, two indexes can be used to quantify the performance of a biometric system. False non-match rate (FNMR) or false reject rate (FRR) is the percentage of biometric templates from the same subject that are recognized as belonging to different subjects. This number can be roughly served as an index of the convenience of a biometric system. Lower FNMR indicates that the system will not reject a "correct" user easily. On the other hand, a system with higher FNMR means that a "correct" user may often need to try several times to enter the system. False match rate (FMR) or false accept rate (FAR) is the percentage of biometric templates from different subjects that are recognized as belonging to the same subject. FMR roughly represents the level of security of a biometric

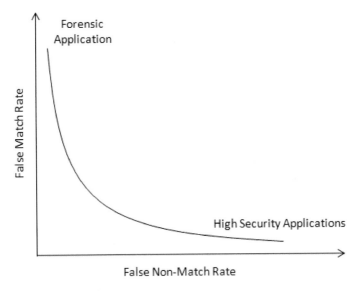

Figure 21.4 A Receiver Operating Characteristic (ROC) curve [2, 11]

system. The lower the FMR, the lower the probability the system falsely accepts an "incorrect" user.

Generally, different applications require different FNMR and FMR. For instance, in forensic applications we prefer a low FNMR to catch a criminal even at the expense of a high FMR. On the other hand, in ATM application we often require a low FMR to prevent the lost of illegal money transfer. Actually, the FNMR and FMR of a biometric system are not unique. We can modify the threshold of matching score to obtain different pairs of FNMR and FMR [2]. Typically, lowering the value of FNMR will lead to an increase of FMR, and vice versa. Therefore, to evaluate the performance of a biometric system, we must consider the two numbers simultaneously. In general, the two numbers can be plotted in a plane leading to a well-known curve, the Receiver Operating Characteristic (ROC) curve as illustrated in Figure 21.4. If the ROC curve of a biometric system is always below the ROC of another, then this system is considered to be more secure and convenient than the other.

21.4 Challenges of Large Scale Biometric Systems

In initial development of biometrics, system developers often do not consider the issue of compatibility between different biometric technologies and systems. In this stage, most biometric systems are designed for personal use such as laptop login. Because the sensor technology and recognition algorithm they used are identical, the variations of sensors and recognition technique are reduced to minimum, achieving a quite satisfactory recognition accuracy. However, as rapid development of biometric technologies and the availability of cheaper biometric sensors, some biometric technologies are gradually deployed in large scale applications, for example, the Personal Identity Verification (PIV) program of the United States Government. The goal of this program is to establish a secure authentication for Federal employees and contractors when they physically access to Federally controlled government facilities and electronically access to government information systems. Because of the huge scale of this project, the biometric devices they employed may come from different biometric vendors. How to make these intrinsically different biometric technologies to achieve a seamless interoperability is a challenge of this large scale project. In fact, even the same recognition algorithm could lead to performance degradation due to the variations and different measurement qualities of different sensors.

To prevent a large scale biometric system being locked-in a specific vendor, the employed biometric devices should be compliant with a predefined standard. Standard can regulate the biometric sensors to generate the data with the same format and produce the templates that are "sufficiently similar" for all compliant equipments and algorithms. However, besides the compatibility issue, there are other challenges for a large scale biometric system. Because of being an open system, the biometric sensors must produce the data with sufficient good quality for a variety of users under any possible environmental variations. Thus, for a large scale biometric system we must evaluate the effectiveness of biometric feature extraction from a more serious point to ensure that it could accommodate most users under any possible environments.

Moreover, implementing user identification in a large scale system should be very careful. Because of the huge number of users involved, performing identification is often very time-consuming. We often need an appropriately designed mechanism, such as template classification, to accelerate the process of identification. For example, fingerprint can be classified into arch, tented arch, left loop, right loop, and whorl according to its ridge patterns [5].

Another technique, the indexing (e.g., fingerprint indexing [1,5]), can also be applied to speed up identification. Indexing is a technique of comparing the templates using a rough but much fast matching algorithm. By indexing, we can obtain a ranking of the templates according to the results of rough matching so that the "correct" template can be identified by the normal matching algorithm as soon as possible.

Another issue of identification is the degradation of system security. A simple example can illustrate this problem. Suppose a biometric system has a FMR of 0.01%. If there are 10000 users enrolled in this system, then normally an identification will involve 10000 comparisons of templates, and by elementary probability theory the probability of positive response of an identification will approach 1. Actually, the more the comparisons of an identification, the more degraded the system security. To prevent system from breaking down, the number of comparisons must be reduced so that the resulting security is still in an acceptable range. Typically, a sophisticated classification algorithm can efficiently reduce the required number of comparisons in an identification. Unfortunately, not all biometric technologies have a companion classification algorithm with sufficient accurate classification rate. Another efficient approach that can be applied to overcome the problem of insufficient FMR is to use multimodal biometric systems [3, 4]. By integrating different biometric traits into a system, the FMR can be greatly reduced, leading to a highly secure biometric system. In fact, most large scale projects use multiple biometric traits to establish their biometric system. For example, the PIV program of the United States utilizes two fingers, normally the two indexes, to construct its biometric authentication system.

21.5 Biometric Standards

As mentioned above, the development of standard can improve the interoperability of biometric technologies from different manufacturers. Actually, biometric standards have also a great benefit to the creation of market as well as the reduction of technical risk the consumers face. With standards, biometric service providers need only focusing on developing the products that are compliant with the standards, and the purchasers will not be locked-in by a specific vendor due to technical limitations.

Several organizations have devoted themselves to the development of biometric standards, for instance, the International Organization for Standardization (ISO). In June, 2002, a joint technical committee SC37 by ISO and International Electrotechnical Commission (IEC) is established responsible for creating and maintaining standards in biometrics. A series of standards were then announced. For example, ISO/IEC 19794 defines a number of biometric data interchange formats, including finger image data, iris image data, signature/sign time series data, face image data, vascular image data, hand geometry silhouette data, etc. Because of the long history of fingerprint verification, finger data has received more attention so that the data formats of finger pattern spectral data, finger pattern skeletal data, and finger minutiae data are also defined in this standard.

Besides ISO/IEC, the InterNational Committee for Information Technology Standards (INCITS), which is accredited by American National Standard Institute (ANSI), has also released a series of similar biometric standards. For example, ANSI/INCITS 377, 378, and 381 define the data interchange formats for finger pattern, finger minutiae, and finger image. ANSI/INCITS 379, 385, 395, and 396 describe the interchange formats for iris image, face recognition, signature/sign, and hand geometry, respectively. The Common Biometric Exchange Formats Framework (CBEFF) is also defined in ANSI/INCITS 398 and ISO/IEC 19785.

In order to permit the cooperation of biometric sensors and recognition algorithms from different vendors, a common programming interface is also necessary. This leads to the standardization of biometric application programming interface (API), such as BioAPI, the famous biometric API developed by BioAPI Consortium. In fact, BioAPI has been approved by ISO and INCITS and has been standardized in ANSI/INCITS 358 and ISO/IEC 19784. The standardization of API allows system integrators to develop their biometric systems without deeply exploring complicated recognition algorithms. Moreover, integrating different biometric technologies into a system also becomes a simple task with standardized APIs.

In addition to the standardizations of data format and API, we also need a standard procedure to evaluate the performance of a biometric system. ISO/IEC 19795 and ANSI/INCITS 409 define how to evaluate a biometric system, including modality-specific testing, testing methodologies and interoperability performance testing. To ensure interoperability of biometric devices from different vendors, a cross-validation test is also required. For example, the National Institute of Standards and Technology (NIST) has organized a Minutiae Interoperability Exchange Test (MINEX) to measure the performance and interoperability of core fingerprint templates and matching capabilities. The goal of this test is to establish compliance for template encoders and matchers for PIV program. Only those vendors who pass this test can serve as the biometric technique providers of PIV program. Actually, in the future the commercial application market will be dominated by those biometric technologies that conform to standards.

21.6 Applications

Biometrics has been applied in many applications such as windows login, data encryption, password management, door access control, and safety box, etc. More recently, some large scale civilian applications have also been developed such as vein verification for ATM in Sumitomo Mitsui Bank of Japan, fingerprint verification for money transfer in Bhamashah program of India, PIV program of the United States Government, ID projects in Thailand, Hong Kong, Singapore, Malaysia and Philippine, border control in Japan, Singapore and Hong Kong, etc. Among these projects, PIV program is a typical model for large scale biometric applications. In addition, due to rapid development of Internet, web-based biometric authentication has become increasingly important. Thus, we briefly discuss these two biometric applications in the following.

21.6.1 Personal Identity Verification Program

In August 2004, President George W. Bush issued Homeland Security Presidential Directive 12 (HSPD-12) [10], aiming to call for a common standard for secure and reliable personal identification for all Federal employees and contractors. Based on this directive, NIST announced Federal Information Processing Standards (FIPS) 201 [9] as a guideline of the Personal Identity Verification (PIV) program of Federal employees and contractors. The United States Government intends to issue a personal identity, a smart card, to each of its 16 million federal employees and contractors. Each card records two types of biometric information, the photo and the templates of two fingerprints of the cardholder. Fingerprint is used as the major personal authentication mechanism while the photo, printed on the card, is used as a minor, visual-based authentication when the fingerprint verification is not available. The major goal of PIV program is to establish a biometric infrastructure allowing Federal employees and contractors for gaining physical access to Federally controlled facilities and logical access to Federally controlled information systems. The reader is referred to [9] for the detailed operation flow of PIV program.

FIPS 201 describes the overall architecture of PIV program. More detailed technical requirements are specified in a series of documents, the

Special Publication (SP) 800. For instance, SP800-73 [6] contains the interfaces and card architecture for storing and retrieving identity credentials from a smart card. SP800-76 [7] defines the interfaces and data formats of biometric information, and SP800-78 [8] states the cryptographic algorithms and key sizes for PIV. In fact, most technical requirements in SP800 are compliant with existing standards. For example, the requirements of live-scan fingerprint sensors such as resolution, geometric accuracy, modulation transfer function, gray-level uniformity, etc. should conform to FBI's Electronic Fingerprint Transmission Specification (EFTS). The data formats of biometric data of PIV program are derived from the standards of INCITS (e.g., ANSI/INCITS 378, 381, 385). The test procedures for biometric performance evaluation follow the standard of ISO/IEC 19795-4.

The reason PIV program receives more attention is that it creates a clear and feasible model for large scale biometric applications. Any standards it employed or developed have become a common reference of establishing any subsequent large scale biometric systems. For example, the MINEX program initiated by NIST has become a standard testing model and those vendors certificated by this test can directly serve as the biometric service providers of other biometric applications.

21.6.2 Web Service Authentication, WSA

Network authentication is a long-lasting problem for network security. Traditional network authentication is normally achieved by password, which is either insecure (simple password is easily guessed) or inconvenient (complex password is difficult to recall). Undoubtedly, biometrics is the best solution for network authentication in the future. A better approach to applying biometrics in network authentication is to embed biometric technologies into the World Wide Web. Equipping with a biometric sensor, a user can authenticate its identity via any networked devices such as notebook, PDA, or mobile phone without any difficulty. To achieve this purpose, two plugins must be devised. One plugin , which is responsible for biometric data acquisition, feature extraction, and the encryption of templates, is for client terminal. Another plugin, which deals with the communication with client and selects a matching server for biometric verification, is for web server. A system model for this web-based authentication is illustrated in Figure 21.5.

This web-based authentication has the following advantages: (1) Easy to install: Because biometric feature extraction and template encryption are embedded in the client plugin, the user do not need any additional procedures of

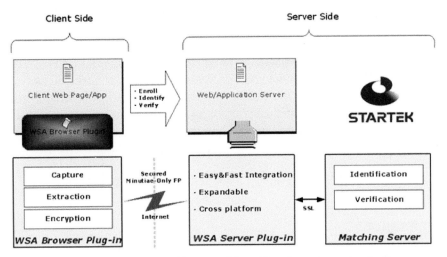

Figure 21.5 The system architecture of Startek's web service authentication

installation. When the user acquires the service first time, the required plugin is automatically downloaded from the server via the Internet. (2) Platform independence: Because of the attribute of platform independence of WWW, the resulting biometric service is also platform independent. (3) Easy to manage: Because each module in this web-based architecture has distinct functionality, management of this system is simple. Moreover, the number of matching servers could be increased to meet physical situations, thus improving the matching efficiency. Web server could balance the computational load by distributing the task to the most appropriate matching server. (4) Easy to develop any related applications: Because matching servers, client and server plugins are provided by biometric vendors, web developers do not need to understand any sophisticated biometric technologies. What web developers need to do is to embed the server-side plugin into their servers, and then their web page has the extended functionality of biometric authentication automatically.

In the future, the users need only selecting those biometric sensors that are compliant with industry standards and then he can use the function of biometric authentication anytime and anywhere on any platform. In fact, the latest windows system of Microsoft, Windows 7, has built in a windows biometric framework, aiming to provide biometric services in the windows. Thus, biometric authentication has gradually stepped into the stage of being a common service of any operating systems.

21.7 Conclusion

Biometrics is traditionally considered as a hi-tech, high cost technology and can only be afforded in forensics and high-security installations. However, as advances of biometric technologies, decreased price of biometric sensors and the development of biometric standards, biometric industry is entering a high growth stage. The more demands of biometric products from personal use and civilian applications declare that biometric authentication is becoming a necessary equipment for any IT systems. In visible future, biometrics will certainly dominate the security market of being a reliable and convenient authentication scheme and impact the way we conduct our daily business.

References

[1] J. Feng and A. Cai. Fingerprint indexing using ridge invariants. In *Proceedings of International Conference on Pattern Recognition*, Vol. 4, pages 433–436, 2006.

[2] A.K. Jain, R. Bolle, and S. PanKanti. *Introduction to Biometrics*. BIOMETRICS: Personal Identification in Networked Society, Kluwer Academic Publishers, 1999.

[3] A.K. Jain and A. Ross. Multibiometric systems. *Communications of the ACM* (Special Issue on Multimodal Interfaces), 47:34–40, 2004.

[4] A.K. Jain, A. Ross and S. Prabhakar. An introduction to biometric recognition. *IEEE Transactions on Circuit and Systems for Video Technology* (Special Issue on Image- and Video-Based Biometrics), 14:4–20, 2004.

[5] D. Maltoni, D. Maio, A.K. Jain, and S. Prabhakar. *Handbook of Fingerprint Recognition*. Springer-Verlag, New York, 2003.

[6] NIST Special Publication 800-73, Integrated Circuit Card for Personal Identity Verification, NIST, February 2005.

[7] NIST Special Publication 800-76, Biometric Data Specification for Personal Identity Verification, NIST, February 2006.

[8] NIST Special Publication 800-78, Cryptographic Algorithms and Key Sizes for Personal Identity Verification, NIST, March 2005.

[9] Personal Identity Verification of Federal Employees and Contractors, Federal Information Processing Standards Publication (FIPS PUB 201-1).

[10] Policy for a Common Identification Standard for Federal Employees and Contractors, August 27, 2004.

[11] A. Ross and A.K. Jain. Biometrics, Overview. In S.Z. Li (Ed.), *Encyclopedia of Biometrics*, pages 168–172. Springer, 2009.

22

Using Experimental Design Methods to Design Efficient Evolutionary Algorithms

Yuping Wang

School of Computer Science and Technology, Xidian University, Xi'an 710071, China; E-mail: ywang@xidian.edu.cn

Abstract

Experimental design methods are a branch of statistics which try to get as much information as possible by using as fewest samples as possible. In this chapter, the experimental design methods are briefly introduced first. Then the methods to design efficient evolutionary operators by using experimental design methods and quantization are described in details. Finally, some examples are given to demonstrate how to use experimental design methods to design crossover operators for different situations, to design the scheme of choosing the weight vectors in weighted-sum evolutionary algorithms for multi-objective optimization problems.

Keywords: genetic algorithm, experimental design, evolutionary computation, optimization algorithm.

22.1 Experimental Design Methods

There are many real world problems which need to generate a small, but representative number of samples in a specific domain. Experimental design methods were designed for this purpose and are a kind of sample methods which try to use a small number of samples to generate as much information as possible. They have been used to many real world problems and research

Patrick Shen-Pei Wang (Ed.), Pattern Recognition and Machine Vision – In Honor and Memory of Professor King-Sun Fu, 349–363.

fields [1–9], e.g., they have been used to design the effective crossover operator in GAs [4], the multimedia multicast routing scheme [9], the quality engineering [6], the global optimization method [1] and multi-objective evolutionary algorithm [5], etc.

We use an example to introduce the basic concept of experimental design methods. Suppose that the yield of rice depends on: (1) the plant density, (2) the amount of fertilizer, and (3) the kind of rice seeds (the kinds of rice seeds can be assigned positive integers). These three quantities are called the factors of the experiment. Each factor has five possible values, and we say that each factor has five levels. To find the best combination of levels for a maximum yield, we can do one experiment for each combination of levels, and select the best one. In the above example, there are $5 \times 5 \times 5 = 125$ combinations, and hence there are 125 experiments. In general, when there are N factors and each factor has Q levels, there will be Q^N combinations of levels. When N and Q are large, it may not be possible to do all Q^N experiments. Therefore, it is desirable to sample a small, but representative set of combinations for the experimentation. The experimental design methods are a kind of efficient methods for this purpose.

In the following, we briefly introduce the idea and concept of two experimental design methods: uniform design method and orthogonal design method. The main objective of uniform design and orthogonal design is to sample a small set of points from a given closed and bounded set $C \subset R^M$ such that the sampled points are uniformly scattered on C. The meaning of uniformly scattered points on an M-dimensional closed bounded set C is as follows. For more details, the reader is referred to [2, 4]. Suppose that for any subset D of set C, its volume is denoted as $V(D)$. For a set $P(q)$ of q points in C, let $NP(q)$ denote the number of points of set $P(q)$ falling in set D. The uniform design on C is then defined as determining a set $P(q)$ of q points in C such that the following discrepancy is minimized:

$$\sup_{D \subset C} \left| \frac{V(D)}{V(C)} - \frac{NP(q)}{q} \right|. \tag{22.1}$$

Finding a set of exactly uniformly scattered points on a complex set C, in general, is very difficult, but for some specific sets, there are some efficient methods, e.g., uniform design method and orthogonal design method, to look for a set of well approximately uniformly scattered points on some specific sets such as hypercube, hyper-sphere, surface of hyper-sphere and convex polyhedron, etc.

22.2 Uniform Design Methods and Applications

22.2.1 Uniform Design Methods in $[L, U]$

In this subsection, we introduce three widely used uniform design methods on Hyper-Cube [2]: Good-Lattice-Point method (GLP), Lattice-Square-Design method, and Power-Generated-Vector method. They can generate a set of q approximately uniformly scattered points, denoted by $O(q, n)$, on hyper-cube

$$[L, U] = \{x \in R^n \mid l_j \le x_j \le u_j, \ j = 1 \sim n\} \tag{22.2}$$

as follows, where x_j, l_j and u_j are the j-th component of x, L and U, respectively.

I. *Good-Lattice-Point method*

- Given a positive integer q, let

$$H_q = \{(h_1, h_2, \ldots, h_m) \mid h_i < q, (h_i, q) = 1, i = 1 \sim m\},$$

  where $(h_i, q) = 1$ represents the greatest common devisor of h_i and q being one. It is required that $m \ge n$.
- Let $G = (G_{ij})_{q \times m}$, where $G_{ij} = ih_j (\mathrm{mod}\, q)$, $i = 1 \sim q$, $j = 1 \sim m$.
- Define a $q \times m$ matrix $U_q(q^m)$ as follows: $U_q(q^m) = (\bar{v}_{ij})_{q \times m}$, where $\bar{v}_{ij} = (2G_{ij} + 1)/(2q)$, $j = 1 \sim m$, $i = 1 \sim q$.
- Construct a matrix $O(q, n) = (v_{ij})_{q \times n}$ by taking n columns from those of $U_q(q^m)$ such that the discrepancy of the set of the points consisting of the rows of $O(q, n)$ is minimized. Redefine the elements of $O(q, n)$ by $v_{ij} = l^j + v_{ij}(u^j - l^j)$, $j = 1 \sim n$, $i = 1 \sim q$. Then the set of points consisting of the rows of $O(q, n)$ is scattered uniformly in $[L, U]$.

Example 1: When $q = 5$, $m = 4$, $n = 3$ and $[L, U] = [0, 1]^3 = \{(x_1, x_2, x_3) \mid 0 \le x_i \le 1, \ i = 1 \sim 3\}$, $H_q = \{(h_1, h_2, h_3, h_4)\} = \{(1, 2, 3, 4)\}$,

$$G = \begin{bmatrix} 1 & 2 & 3 & 4 \\ 2 & 4 & 1 & 3 \\ 3 & 1 & 4 & 2 \\ 4 & 3 & 2 & 1 \\ 0 & 0 & 0 & 0 \end{bmatrix}, \quad U_q(q^m) = \begin{bmatrix} 0.3 & 0.5 & 0.7 & 0.9 \\ 0.5 & 0.9 & 0.3 & 0.7 \\ 0.7 & 0.3 & 0.9 & 0.5 \\ 0.9 & 0.7 & 0.5 & 0.3 \\ 0.1 & 0.1 & 0.1 & 0.1 \end{bmatrix}.$$

It can be confirmed that the first three columns of $U_q(q^m)$ are the best ones. Thus $O(q, n)$ can be given as

$$O(q, n) = \begin{bmatrix} 0.3 & 0.5 & 0.7 \\ 0.5 & 0.9 & 0.3 \\ 0.7 & 0.3 & 0.9 \\ 0.9 & 0.7 & 0.5 \\ 0.1 & 0.1 & 0.1 \end{bmatrix}.$$

Each row represents a point in $R^n = R^3$, and all five points (corresponding to five rows) uniformly distributed in $[0, 1]^3$.

II. *Lattice-Square-Design method*
We first introduce the concept of Latin square. For any $(x_1, x_2, \ldots, x_n) \in R^n$, a shift mapping $S : R^n \rightarrow R^n$ is defined as $S(x_1, x_2, \ldots, x_n) = (x_2, x_3, \ldots, x_n, x_1)$.

Definition 1. *An $n \times n$ matrix is called a Latin square of order n if the matrix satisfies the following conditions:*

(I) *Its first row $a = (x_1, x_2, \ldots, x_n)$, denoted as V_1, is a permutation of $(1, 2, \ldots, n)$, and*
(II) *Its i-th row, denoted as V_i, is given by $V_i = S(V_{i-1})$ for $i = 2 \sim n$.*

A Latin square of order n with a being its first row is denoted as $Ls(n, a)$, and its i-th row and j-th column element is denoted as G_{ij}.

Latin-square design is one of the most easily used uniform design methods. It can generate a set $O(n, n)$ of n points uniformly scattered in domain $[L, U]$ as follows. Choose a row vector $a \in R^n$ as the first row vector and generate a Latin square of order n, $Ls(n, a) = (G_{ij})_{n \times n}$. Then $O(n, n)$ can be defined as $O(n, n) = (v_{ij})_{n \times n}$, where $v_{ij} = l_j + (2G_{ij} - 1/2n)(u_j - l_j)$, $i, j = 1 \sim n$ and each row of $O(n, n)$ defines a point in $[L, U]$.
    Example 2: When $n = 5$, $a = (1, 2, 3, 4, 5)$ and $[L, U] = [0, 1]^5 = \{(x_1, x_2, \ldots, x_5) \mid 0 \leq x_i \leq 1, \ i = 1 \sim 5\}$, we can get $Ls(n, a)$ and $O(n, n)$ by

$$Ls(n, a) = \begin{bmatrix} 1 & 2 & 3 & 4 & 5 \\ 2 & 3 & 4 & 5 & 1 \\ 3 & 4 & 5 & 1 & 2 \\ 4 & 5 & 1 & 2 & 3 \\ 5 & 1 & 2 & 3 & 4 \end{bmatrix},$$

$$O(n, n) = \begin{bmatrix} 0.1 & 0.3 & 0.5 & 0.7 & 0.9 \\ 0.3 & 0.5 & 0.7 & 0.9 & 0.1 \\ 0.5 & 0.7 & 0.9 & 0.1 & 0.3 \\ 0.7 & 0.9 & 0.1 & 0.3 & 0.5 \\ 0.9 & 0.1 & 0.3 & 0.5 & 0.7 \end{bmatrix}.$$

Each row of $O(n, n)$ defines a point in $R^n = R^5$, and five rows of $O(n, n)$ define a set of points which are uniformly distributed in $[L, U] = [0, 1]^5$.

III. *Power-Generated-Vector method* [8]

- Given q and n, determine a number σ. This number can be gotten from the table given in [2].
- Generate a $q \times n$ integer matrix called uniform array denoted by $G(q, n) = [G_{ij}]_{q \times n}$, where $G_{ij} = (i\sigma^{j-1} \bmod q) + 1$, $i = 1 \sim q$, $j = 1 \sim n$.
- Define a $q \times n$ matrix $O(q, n)$ by $O(q, n) = \{(O_{i1}, O_{i2}, \ldots, O_{in}) \mid i = 1 \sim q\}$, where

$$O_{ij} = l^j + \frac{2G_{ij} - 1}{2q}(u^j - l^j), \quad j = 1 \sim n, i = 1 \sim q,$$

and $O(q, n)$ is a set of points in $[L, U]$. We also use $O(q, n)$ to denote the matrix whose rows are these points.

Example 3: When $q = 23$, $n = 2$ and $[L, U] = [0, 1]^n$, it can be found from the table in [2] that $\sigma = 7$. Thus,

$$G(q, n) = \begin{bmatrix} 2 & 3 & 4 & 5 & 6 & 7 & 8 & 9 & 10 & 11 & 12 & 13 \\ 8 & 15 & 22 & 6 & 13 & 20 & 4 & 11 & 18 & 2 & 9 & 16 \end{bmatrix}$$

$$\begin{bmatrix} 14 & 15 & 16 & 17 & 18 & 19 & 20 & 21 & 22 & 23 & 1 \\ 23 & 7 & 14 & 21 & 5 & 12 & 19 & 3 & 10 & 17 & 1 \end{bmatrix}^T, \quad (22.3)$$

and

$$O(q, n) = \{(0.0652, 0.3261) \quad (0.1087, 0.6303) \quad (0.1522, 0.9348)$$
$$(0.1957, 0.2391) \quad (0.2391, 0.5435) \quad (0.2868, 0.8478)$$
$$(0.3261, 0.1522) \quad (0.3696, 0.4565) \quad (0.4130, 0.7609)$$
$$(0.4565, 0.0652) \quad (0.5000, 0.3696) \quad (0.5435, 0.6739)$$
$$(0.5870, 0.9783) \quad (0.6304, 0.2826) \quad (0.6739, 0.5870)$$
$$(0.7174, 0.8913) \quad (0.7609, 0.1957) \quad (0.8043, 0.5000)$$
$$(0.8487, 0.8043) \quad (0.8913, 0.1087) \quad (0.9348, 0.4130)$$
$$(0.9789, 0.7174) \quad (0.0217, 0.0217)\}.$$

$$(22.4)$$

22.2.2 Application of Uniform Design Methods in Genetic Algorithm Design

Uniform design methods can be used to design efficient crossover operators and generate high quality initial population in genetic algorithms.

First, we use the uniform design methods in the above section to construct a crossover operator. This operator generates offspring of two parents and they are approximately uniformly scattered on a region containing their parents, and thus can effectively exploit the search space around their parents. Suppose that the problem dimension is M. For two parents $Y = (y_1, \ldots, y_M)$ and $X = (x_1, \ldots, x_M)$, we define two vectors

$$L = (l_1, \ldots, l_M) \quad \text{and} \quad U = (u_1, \ldots, u_M), \qquad (22.5)$$

where $l_i = \min\{x_i, \ y_i\}$ and $u_i = \max\{x_i, \ y_i\}$ for $i = 1 \sim M$. These two vectors define a hyper-cube,

$$[L, U] = \{(z_1, \ldots, z_M) \mid l_i \leq z_i \leq u_i, \ i = 1 \sim M\}. \qquad (22.6)$$

Choose a proper integer q_1. Any of three uniform design methods in the above subsection can be used to generate $q = q_1$ points approximately uniformly distributed on $[L, U]$, and these q_1 points can be regarded as offspring of two parents X and Y.

Note that the parameters n and q in Good-Lattice-Point method and Power-Generated-Vector method should satisfy $n < q$ [2]. Thus, these two uniform design methods cannot be directly used to the case $M \geq q$. To design an effective crossover operator by using uniform design in any case, we can adopt the following strategy:

Given n and $q = q_1$, the set of q_1 offspring, denoted by

$$O(q, M) = \{O^k \mid k = 1 \sim q_1\}$$
$$= \{(o_{k1}, o_{k2}, \ldots, o_{kM}) \mid k = 1 \sim q_1\} \qquad (22.7)$$

can be generated according to two cases: $M \leq q_1 - 1$ and $M > q_1 - 1$. The detail is given in the following algorithm:

Algorithm 1. (Crossover operator)

Step 1. For two parents X and Y, define two vectors L and U by formula (22.5) and a hyper-rectangle $[L, U]$ by formula (22.6). Choose a proper prime number q_1.

Step 2. If $M \leq q_1 - 1$, generate matrix $G(q, M)$. Then the k-th offspring $O^k = (o_{k1}, o_{k2}, \ldots, o_{kM})$ is generated by the k-th row of $O(q, M)$ in the three uniform design methods for $k = 1 \sim q_1$.

Step 3. If $M > q_1 - 1$, take $n = q_1 - 1$ and randomly divide L and U into n blocks of sub-vectors, respectively, in the following way:

$$L = (A^1, A^2, \ldots, A^{q_1-1}) \quad \text{and} \quad U = (B^1, B^2, \ldots, B^{q_1-1}), \qquad (22.8)$$

where A^j and B^j are sub-vectors of l and u in the same dimension. Generate matrix $G(q, n)$, then the k-th offspring

$$O^k = (O_{k1}, O_{k2}, \ldots, O_{kq_1-1}) \qquad (22.9)$$

is generated by

$$O_{kj} = A^j + \frac{2G_{kj} - 1}{2q_1}(B^j - A^j), \quad j = 1 \sim q_1 - 1$$

for $k = 1 \sim q_1$, where $G(q_1, M) = [G_{kj}]_{q_1 \times (q_1-1)}$ is defined by uniform design methods with $M = q_1 - 1$.

Example 1: Case 1. Let $M = 4$. If two parents are $X = (0, 4, 2, 0)$ and $Y = (6, 1, 5, -3)$, then $L = (0, 1, 2, -3)$ and $U = (6, 4, 5, 0)$. If we choose $q_1 = 5$, the set of q_1 offspring is generated by Step 2 since M and q_1 satisfy $M \leq q_1 - 1$. It can be found from the table in [2] that $\sigma = 2$. Thus, offspring are generated as follows:

$$O(4, 5) = \{O^k \mid k = 1 \sim q_1\}$$
$$= \big\{(1.8, 2.5, 4.7, -0.9), (3, 3.7, 4.1, -2.1), (4.2, 1.9, 3.5, -0.3),$$
$$(5.4, 3.1, 2.9, -1.5), (0.6, 1.3, 2.3, -2.7)\big\}.$$

Case 2. Let $M = 6$. If two parents are $X = (0.1, 1.2, 0.3, 1.4, 0.5, 1.6)$ and $Y = (1.1, 0.2, 1.3, 0.4, 1.5, 0.6)$, then $L = (0.1, 0.2, 0.3, 0.4, 0.5, 0.6)$ and $U = (1.1, 1.2, 1.3, 1.4, 1.5, 1.6)$.

If we also choose $q_1 = 5$, since M and q_1 satisfy $M > q_1 - 1$, the set of q_1 offspring is generated by Step 3. In this case, n should take the value of $n = q_1 - 1$ in $G(q_1, n)$ and $\sigma = 2$. Suppose without loss of generality that L and U are randomly divided into 4 blocks of sub-vectors as follows:

$$l = (A^1, A^2, A^3, A^4) \quad \text{and} \quad u = (B^1, B^2, B^3, B^4),$$

where

$$A^1 = (0.1, 0.2), \quad A^2 = (0.3, 0.4), \quad A^3 = (0.5), \quad A^4 = (0.6),$$

$$B^1 = (1.1, 1.2), \quad B^2 = (1.3, 1.4), \quad B^3 = (1.5), \quad B^4 = (1.6).$$

Then, offspring are calculated and given by

$$O(6, 5) = \{O^k \mid k = 1 \sim 5\}$$

$$= \{(0.4, 0.5, 0.8, 0.9, 1.4, 1.3), (0.6, 0.7, 1.2, 1.3, 1.2, 0.9),$$

$$(0.8, 0.9, 0.6, 0.7, 1.0, 1.5), (1.0, 1.1, 1.0, 1.1, 0.8, 1.1),$$

$$(0.2, 0.3, 0.4, 0.5, 0.6, 0.7)\}.$$

Uniform design methods also can be used to generate high quality initial population. Suppose that the population size in genetic algorithms is POP_s and the problem dimension is M. We first divide the search space $[L, U]$ into about $k = [POP_s/M]$ sub-hyper-cubes $[L_1, U_1], [L_2, U_2], \ldots, [L_k, U_k]$. In each sub-hyper-cube $[L_i, U_i]$, we can generate M points and totally generate POP_s points uniformly distributed in $[L, U]$ as follows.

Choose vector $a = (1, 2, \ldots, M)$, and generate matrix $Ls(M, M)$ by Lattice-Square-Design method. In each $[L_i, U_i]$, generate a set $O(M, M)$ of M points by Lattice-Square-Design method. All these points form the initial population which are uniformly distributed in $[L, U]$.

22.2.3 Using Uniform Design Methods in Evolutionary Algorithm Design for Multi-Objective Optimization

For the multi-objective optimization problem

$$\min_{x \in \Omega} \{f_1(X), f_2(X), \ldots, f_s(X)\} \tag{22.10}$$

One of the simplest and most efficient evolutionary algorithms is the weighted sum method which transforms the above multi-objective optimization problem into the following several single objective optimization problems:

$$\min_{x \in \Omega} \{v_{k1} f_1(X) + v_{k2} f_2(X) + \cdots + v_{ks} f_s(X)\} \tag{22.11}$$

where $V_k = (v_{k1}, v_{k2}, \ldots, v_{ks})$ is the k-th weight vector for $k = 1 \sim q$.

To solve the multi-objective optimization problem successfully, it is required to look for a uniformly scattered Pareto optimal solutions along the whole Pareto front. To achieve this purpose, one of the most important techniques is that the set of these weighted vectors should be scattered uniformly on the surface of a hyper-sphere in the objective space [5]

$$U(s) = \left\{(f_1, \ldots, f_s) \mid f_1^2 + \cdots + f_s^2 = 1, f_i \geq 0, i = 1, \ldots, s\right\} \tag{22.12}$$

The uniform design methods are often used to generate such a set of weight vectors. The main idea of uniform design to generate a set of q weight vectors, denoted by $D(s, q)$, on $U(s)$ is as follows according to the following two cases [2]:

Algorithm 2. (Case 1: When $s = 2m$ for some positive integer m, i.e., when s is an even number)

Step 1. Generate a set of q approximately uniformly distributed points $C(s - 1, q) = \{C_k = (c_{k1}, \ldots, c_{k,s-1}) : k = 1 \sim q\}$ on $C = [0, 1]^{s-1}$ or on a subset of C by one of the uniform design methods introduced in the previous section.
(The following steps are to map the points in $C(s - 1, q)$ to the points approximately uniformly distributed on $U(s)$.)

Step 2. Set $g_{km} = 1$ and $g_{k0} = 0, k = 1 \sim q$.

Step 3. For $k = 1, \ldots, q$, recursively compute $g_{kj} = g_{k,j+1} c_{kj}^{1/j}, j = m-1 \sim 1$.

Step 4. Compute

$$\begin{cases} d_{kl} &= \sqrt{g_{kl} - g_{k,l-1}}, \\ v_{k,2l-1} &= d_{kl} \cos(\tfrac{\pi}{2} c_{k,m+l-1}), \\ v_{k,2l} &= d_{kl} \sin(\tfrac{\pi}{2} c_{k,m+l-1}), \end{cases}$$

for $l = 1 \sim m$ and $k = 1 \sim q$. Then, $D(s, q) = \{V_k = (v_{k1}, \ldots, v_{ks}) : k = 1 \sim q\}$ is a set of weight vectors approximately uniformly distributed on $U(s)$.

Algorithm 3. (Case 2: When $s = 2m + 1$ for some positive integer m, i.e., when s is an odd number)

Step 1. Generate a set of q approximately uniformly distributed points $C(s - 1, q) = \{C_k = (c_{k1}, \ldots, c_{k,s-1}) : k = 1 \sim q\}$ on $C = [0, 1]^{s-1}$ or on a subset of C by one of the uniform design methods introduced in the previous section.

(The following steps are to map the points in $C(s - 1, q)$ to the points approximately uniformly distributed on $U(s)$.)

Step 2. Set $g_{km} = 1$ and $g_{k0} = 0$, $k = 1 \sim q$.

Step 3. For $k = 1, \ldots, q$, recursively compute $g_{kj} = g_{k,j+1} c_{kj}^{2/(2j+1)}$, $j = m - 1 \sim 1$.

Step 4. Compute $d_{kl} = \sqrt{g_{kl} - g_{k,l-1}}$, $l = 1 \sim m$, $k = 1 \sim q$.

Step 5. For $k = 1 : q$, do

$$
\begin{cases}
v_{k1} &= d_{k1}(1 - 2c_{km}), \\[2mm]
v_{k2} &= 2d_{k1}\sqrt{c_{km}(1 - c_{km})}\cos(\tfrac{\pi}{2}c_{k,m+1}), \\[2mm]
v_{k3} &= 2d_{k1}\sqrt{c_{km}(1 - c_{km})}\sin(\tfrac{\pi}{2}c_{k,m+1}).
\end{cases}
$$

$$For \quad l = 2 : m, \quad do$$

$$v_{k,2l} = d_{kl}\cos(\tfrac{\pi}{2}c_{k,m+1}),$$

$$v_{k,2l+1} = d_{kl}\sin(\tfrac{\pi}{2}c_{k,m+1})$$

$$End \quad do$$

End do

Then, $D(s, q) = \{V_k = (v_{k1}, \ldots, v_{ks}) : k = 1 \sim q\}$ is a set of points approximately uniformly distributed on $U(s)$.

22.3 Orthogonal Design Method

The *orthogonal design* provides a series of orthogonal arrays for different factors and levels. We let $L_M(Q^N)$ be an orthogonal array for N factors and Q levels, where M is the number of rows of $L_M(Q^N)$ and N is the number of columns. Every row represents a combination of levels. For convenience, we denote $L_M(Q^N) = [a_{i,j}]_{M \times N}$ where the j-th factor in the i-th combination has level $a_{i,j}$ and $a_{i,j} \in \{1, 2, \ldots, Q\}$.

In general, the orthogonal array $L_M(Q^N)$ has the following properties:

1. For the factor in any column, every level occurs M/Q times.
2. For the two factors in any two columns, every combination of two levels occurs M/Q^2 times.
3. For the two factors in any two columns, the M combinations contain the following combinations of levels: $(1, 1)$, $(1, 2)$, $\ldots$, $(1, Q)$, $(2, 1)$, $(2, 2)$, $\ldots$, $(2, Q)$, $\ldots$, $(Q, 1)$, $(Q, 2)$, $\ldots$, (Q, Q).
4. If any two columns of an orthogonal array are swapped, the resulting array is still an orthogonal array.
5. If some columns are taken away from an orthogonal array, the resulting array is still an orthogonal array with a smaller number of factors.

Consequently, the selected combinations are scattered uniformly over the space of all possible combinations and thus are the good representatives of all possible comninations.

22.3.1 Construction of Orthogonal Array

In the following, we introduce a simple method which can construct a special class of orthogonal arrays $L_M(Q^N)$ [4], where Q is odd and $M = Q^J$, where J is a positive integer fulfilling

$$N = \frac{Q^J - 1}{Q - 1}. \tag{22.13}$$

We denote the j-th column of the orthogonal array $[a_{i,j}]_{M \times N}$ by $\boldsymbol{a}_j$. Columns $\boldsymbol{a}_j$ for $j = 1, 2, (Q^2 - 1)/(Q - 1) + 1, (Q^3 - 1)/(Q - 1) + 1$, $\ldots, (Q^{J-1} - 1)/(Q - 1) + 1$ are called the *basic columns*, and the others are called the *nonbasic columns*. We first construct the basic columns, and then construct the nonbasic columns. The details are as follows:

Algorithm 4. *Construction of Orthogonal Array*

Step 1. Construct the basic columns as follows:
      FOR $k = 1$ TO J DO
      BEGIN

$$j = \frac{Q^{k-1} - 1}{Q - 1} + 1;$$

FOR $i = 1$ TO Q^J DO

$$a_{i,j} = \left\lfloor \frac{i-1}{Q^{J-k}} \right\rfloor \bmod Q;$$

END.

Step 2. Construct the nonbasic columns as follows:
FOR $k = 2$ TO J DO
BEGIN

$$j = \frac{Q^{k-1} - 1}{Q - 1} + 1;$$

FOR $s = 1$ TO $j - 1$ DO
FOR $t = 1$ TO $Q - 1$ DO

$$a_{j+(s-1)(Q-1)+t} = (a_s \times t + a_j) \bmod Q;$$

Step 3. Increment $a_{i,j}$ by one for all
$1 \le i \le M$ and $1 \le j \le N$.

22.3.2 Generation of Initial Population by Using Orthogonal Design

Note that an orthogonal array specifies a small number of combinations that are scattered uniformly over the space of all the possible combinations. Therefore, it is potential and possible for orthogonal design methods to generate a good initial population.

For a continuous optimization problem (either single objective or multiple objective problem), suppose that its dimension is n. We define x_j to be the j-th factor, so that each chromosome has n factors. These factors are continuous, but the orthogonal design is applicable to discrete factors only. To overcome this issue, we quantize the domain $[l_j, u_j]$ of x_j into Q levels $\alpha_{1,j}, \ldots, \alpha_{Q,j}$, where the design parameter Q is odd and $\alpha_{i,j}$ is given by

$$\alpha_{i,j} = l_j + (i - 1) \left(\frac{u_j - l_j}{Q} \right), \quad i = 1 \sim Q. \tag{22.14}$$

For convenience, we call $\alpha_{i,j}$ the i-th level of the j-th factor, and we denote $\alpha_j = (\alpha_{1,j}, \ldots, \alpha_{Q,j})$.

Note that we hope that the number of columns of $L_M(Q^N)$ is equal to the problem dimension n, i.e., $N = (Q^J - 1)/(Q - 1) = n$. However, there may not exist Q and J fulfilling this condition. We bypass this restriction as follows. We choose the smallest J such that

$$\frac{Q^J - 1}{Q - 1} \geq n. \qquad (22.15)$$

We execute Algorithm 4 to construct an orthogonal array with $N = (Q^J - 1)/(Q - 1)$ factors, and then delete the last $N - n$ columns to get an orthogonal array with n factors. The details are given in the following algorithm:

Algorithm 5. *Construction of $L_M(Q^n)$:*

Step 1. Select the smallest J fulfilling $(Q^J - 1)/(Q - 1) \geq n$.
Step 2. If $(Q^J - 1)/(Q - 1) = n$, then $N = n$ else $N = (Q^J - 1)/(Q - 1)$.
Step 3. Execute Algorithm 4 to construct the orthogonal array $L_{Q^J}(Q^N)$.
Step 4. Delete the last $N - n$ columns of $L_{Q^J}(Q^N)$ to get $L_M(Q^n)$ where $M = Q^J$.

Denote $L_M(Q^n) = [a_{i,j}]_{M \times n}$. To generate a high quality initial population, one good way is to take a relatively large population size. For example, we can set $Q = Q_1$, Q_2, Q_3, respectively. For each $Q = Q_k$ ($k = 1$, 2, 3), we get a sample of $M_k = Q_k^{J_k}$ combinations of levels of n factors. We apply these M_k combinations to generate the following M_k chromosomes as a part of initial population

$$\text{POP}_k = (v_{ij})_{M_k \times n}, \qquad (22.16)$$

where

$$v_{ij} = l_j + (i - 1)\frac{a_{i,j}}{Q_k}(u_j - l_j), \quad i = 1 \sim M_k, \; j = 1 \sim n.$$

Then we take the union of these POP_k as the initial population.

22.3.3 Design Crossover Operator Using Orthogonal Design Method

Each pair of parents should not produce too many offspring in order to avoid a large number of function evaluations in evolutionary algorithms. For this purpose, we can choose a small Q and use the same technique as that in Algorithm 1. The details are as follows:

Algorithm 6. (*Crossover Operator*)

Step 1. For two parents X and Y, define two vectors l and u by formula (22.5) and a hyper-rectangle $[L, U]$ by formula (22.6). Choose a small odd number $Q = 3$, $N = 4$ and $J = 2$.

Step 2. Generate matrix $L_9(Q^N) = [a_{i,j}]_{9 \times N}$.

Step 3. If $n \leq N$, Then the r-th offspring $O^r = (v_{r1}, v_{r2}, \ldots, v_{rn})$ is generated by formula (22.16) with $Q_k = Q$ and $M_k = 9$ for $r = 1 \sim 9$.

Step 4. If $n > N$, randomly divide L and U into N blocks of sub-vectors, respectively, in the following way:

$$L = (A^1, A^2, \ldots, A^N) \text{ and } U = (B^1, B^2, \ldots, B^N) \qquad (22.17)$$

where A^j and B^j are sub-vectors of l and u in the same dimension. Then the r-th offspring

$$O^r = (O_{r1}, O_{r2}, \ldots, O_{rN}) \qquad (22.18)$$

is generated by

$$O_{rj} = A^j + \frac{v_{rj}}{Q}(B^j - A^j), \quad j = 1 \sim N \qquad (22.19)$$

for $r = 1 \sim 9$.

22.4 Conclusions

In this chapter, the experimental design methods including three uniform design methods and one orthogonal design method are introduced first, then the methods and techniques to design efficient evolutionary algorithms by using them are described in details. These methods and techniques include designing crossover operators, generating high quality initial population, and constructing the weight vectors in weighted sum methods for multi-objective optimization problems. Furthermore, some examples are given to demonstrate how to use these methods and techniques.

Acknowledgement

This work is supported by the National Natural Science Foundation of China (No. 60873099).

References

[1] K.T. Fang and W. Li. A global optimum algorithm on two factor uniform design. Technical Report, Department of Mathematics, MATH-095, Hong Kong Baptist University, Hong Kong, 1995.

[2] K.T. Fang and Y. Wang, *Number-Theoretic Method in Statistics*. Chapman and Hall, London, 1994.

[3] C.R. Hicks. *Fundamental Concepts in the Design of Experiments*, 4th edn. Saunders College Publishing, Texas, 1993.

[4] Y.W. Leung and Y. Wang. An orthogonal genetic algorithm with quantization for numerical global optimization. *IEEE Trans. Evol. Comput.*, 5(1):41–53, February 2001.

[5] Y.W. Leung and Y. Wang. Multiobjective programming using uniform design and genetic algorithm. *IEEE Trans. Systems, Man, and Cybernetics, Part C: Applications and Review*, 30(3):293–304, 2000.

[6] Y.K. Lo. Application of the uniform design to quality engineering. Technical Report, MATH-129, Applied Mathematics Department, Hong Kong Polytechnic University, Hong Kong, 1996.

[7] D.C. Montgomery. *Design and Analysis of Experiments*, 3rd edn. Wiley, New York, 1991.

[8] Y. Wang and C. Dang. An evolutionary algorithm for global optimization based on level-set evolution and Latin squares. *IEEE Trans. Evol. Comput.*, 11(5):579–595, 2007.

[9] Q. Zhang and Y.W. Leung. An orthogonal genetic algorithm for multimedia multicast routing. *IEEE Trans. Evol. Comput.*, 3:53–62, April 1999.

PART V

MACHINE LEARNING AND PATTERN RECOGNITION APPLICATIONS

23

Application of Pattern Recognition Technology to Postal Automation in China

Yue Lu[1,2], Xiao Tu[1,2], Shujing Lu[1,2] and Patrick S.P. Wang[1,3]

[1]*Department of Computer Science and Technology, East China Normal University, 500 Dongchuan Road, Shanghai 200241, China; E-mail: ylu@cs.ecnu.edu.cn, tree_tx@yahoo.com.cn, shujing_l@hotmail.com*
[2]*ECNU-SRI Joint Lab of Pattern Analysis and Intelligence System, Shanghai Research Institute of Postal Science, China Post Group, Shanghai 200062, China*
[3]*College of Computer and Information Science, Northeastern University, Boston, MA 02115, USA; E-mail:pwang@ccs.neu.edu*

Abstract

Pattern recognition technology has achieved great progress in many applications, among which postal automation is one of its representative and successful practical areas. In this chapter, we present the applications of pattern recognition technology to postal automation in China, in particular, give the details of image acquisition, postcode and address segmentation and recognition which are key technologies in automatic letter sorting machines.

Keywords: postal automation, letter sorting machine, segmentation, character recognition, address recognition.

23.1 Introduction

With economic development, the volume of mail pieces received and delivered by post office increases rapidly. Manual sorting cannot meet the requirements of modern post, not only because of its time-consuming labor, but also because of its costly manpower.

Patrick Shen-Pei Wang (Ed.), Pattern Recognition and Machine Vision – In Honor and Memory of Professor King-Sun Fu, 367–381.

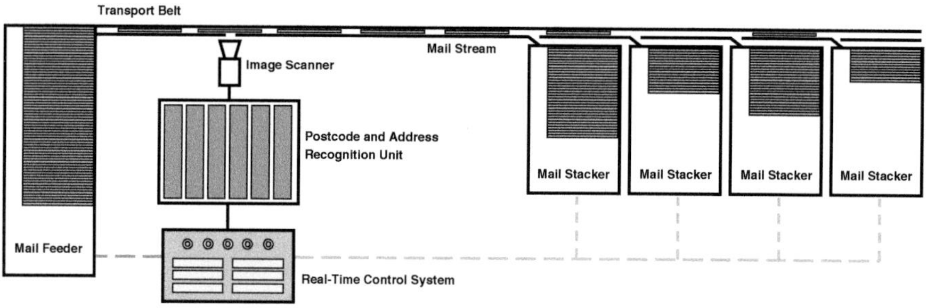

Figure 23.1 Diagram of automatic letter sorting machine.

To automatically sort mails, letter sorting machines employing pattern recognition technology are widely used in mail processing centers. For example, over 100 letter sorting machines have been deployed around China since 1990s. These machines were manufactured by Siemens (formerly AEG), NEC, Solystic (formerly Actel-Bell), and SRI (Shanghai Research Institute of Postal Science, China Post Group). SRI acts an important role for popularizing the automatic sorting technologies in China, because it has been involved with supplying almost 70% of the letter sorting machines around China.

For an automatic mail sorting machine, mail feeder, image scanner, mail stacker, real-time control system, postcode and address recognition are basic modules, as illustrated in Figure 23.1. The function of mail feeder module is to feed the letters to machine one by one, which generates a mail stream in the transport belt. As a mail passes by the CCD camera in the image scanner module, its image is captured and sent to the postcode and address recognition unit. The real-time control system traces each one in the mail stream, and controls the mail to its corresponding stacker according to the result of the postcode and address recognition unit.

Character and document recognition has been a very successful area of pattern recognition [4, 7, 13], which is one of the key technologies in developing automatic letter sorting machines. In the postcode and address recognition unit, pattern recognition technology is applied to automatically read and understand the information written or printed on the envelopes. The information involved in the envelopes includes: (a) postcode in postcode frames, (b) postcode in destination address block, (c) address in destination address block.

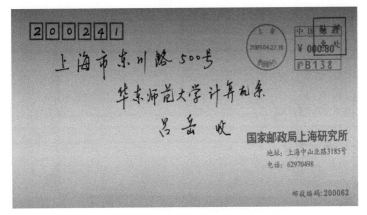

Figure 23.2 Gray image of envelope.

23.2 Image Acquisition

While an envelope passes by a camera, the envelope's image will be captured by the camera for downstream processing. The camera is to create an image from light focused on the image plane by its lens. The most important component of the camera is its digital sensor, i.e. CCD (charge-coupled device).

To achieve high throughput, for example 40,000 pieces per hour, the working speed of transport belt is more than 4 meters per second. A line sensor, therefore, is usually employed, which creates an image with one pixel width. With the relative movement of the envelope in front of the camera, many lines are assembled into a two-dimensional image.

Due to the fast movement of the envelopes, the exposure time of each pixel is obviously limited. Consequently, the usage of line CCD sensor requires very bright illumination. Since LEDs (light-emitting diode) have many practical advantages, such as their longevity, using comparatively little power and producing little heat, they are currently the primary illumination technology and are commonly used in letter sorting machines. Traditional incandescent lamps have been fallen into disuse in the recent years.

To ensure the characters on the envelopes are clear enough for recognition system, the resolution of captured images is normally 200 DPI (dots per inch) or more. In general, gray image is common, but color image has been becoming popular. Figure 23.2 shows an envelope's gray image with 300 DPI captured by line CCD sensor camera in a letter sorting machine.

23.3 Image Segmentation and Recognition

The images captured by camera are delivered to computer for understanding information written or printed on the envelopes. Suppose the camera acts as human's eyes, the image processing unit is our brain for understanding the captured images. The procedure of image processing plays an essential role for automatic letter sorting machines. In general, there are three steps to extract information from the image, i.e. image segmentation, character recognition, and interpretation.

23.3.1 Image Segmentation

The purpose of image segmentation is to extract regions in the image that correspond to the objects we are interested in. The information which is useful for letter sorting machine generally includes postcode and destination address.

Postcode frame detection and postcode character extraction   In many countries, in particular in the eastern countries such as China, Japan, India, Bangladesh, etc. postcode frames are pre-printed on the envelopes for re-minding people to write the postcode of the mail address. On the other hand, postcode is included in the address in the western countries [11], such as USA, UK, Canada, etc.

In the first case, postcode frames detection is essential to extract postcode characters. Firstly, edge detection methods are used to extract vertical and horizontal edges of postcode frames. We use the Sobel operator to calculate the gradient and its direction of the image intensity at each point. According to the gradient direction, regular vertical edges and horizontal edges, postcode frames are detected as shown in Figure 23.3.

At the first sight, it is easy to extract postcode numbers if postcode frames have been detected exactly. Figure 23.2 is showed a nice writer, who wrote the numbers just in the center part of each frame without any adherence. In this case, after removing the frame, we can get the intact image of each number. However, in other cases, postcode characters may overlap the frames. Strokes of postcode characters are detected for separating the character pixels from the frames. However, color image is the best solution for this issue because the colors of characters and frames are not the same in general.

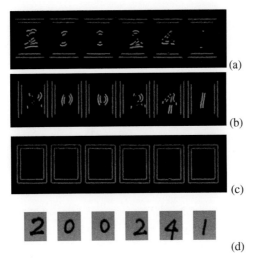

Figure 23.3 Postcode frame detection and postcode character extraction: (a) horizontal edges, (b) vertical edges, (c) frames, (d) extracted postcode numerals

Destination address block localization   Destination address block localization is to extract the area which includes addressee's address information [9]. Based on connected components analysis, this step is to separate envelope image to several text lines and select the text lines of destination address block.

Binarization is commonly used for extracting objects in the envelope image, which converts an image of up to 256 gray levels to a black and white image. The simplest way is to choose a threshold, and classify all pixels with values above this threshold as white, while all other ones as black. The problem then is how to select the correct threshold. In many cases, finding one threshold compatible to the entire image is very difficult, and even impossible sometimes. Therefore, adaptive image binarization is needed where an optimal threshold is chosen for each sub-area.

Otsu's method is one of most effective and efficient binarization algorithms. The algorithm assumes that the image is divided by a threshold into two categories of pixels (e.g. foreground and background) then calculates the optimum threshold separating those two categories so that their intra-class variance is minimal. Figure 23.4 is the binary image of Figure 23.2.

For the purpose of extracting the destination address block, bottom-up strategy is employed, in which connected components (CCs) are extracted and labeled for destination address block analysis. The aim of connected

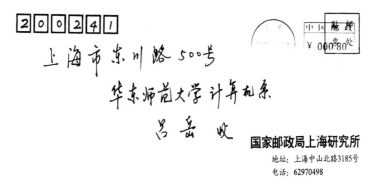

Figure 23.4 Binary image of envelope

component labeling is to transform a binary image into a symbolic image in order that each connected component can be distinguished by assigning a unique label. The issue of labeling connected components has been extensively studied, with numerous algorithms reported so far [5, 12]. We use a run-based algorithm for labeling connected components. Runs are extracted from the image row by row. The positional relations among the runs of current rows and the runs of their preceding rows are represented utilizing trees, where each tree corresponds to a connected component. Only one-pass scan is required for obtaining the characteristics of the connected components, such as bounding rectangle, area, number of pixels. Figure 23.5 shows all connected components of the above image. By analyzing geometric features of the connected components, they were categorized to three types: noise, graph and character. Both noise and graph connected components will be detected and removed.

A CC with small width and small height is defined as a noise which can be either real noise or tiny strokes of characters. Graph CCs contain straight lines, curve lines or tables, which match one of the following conditions:

1. $\frac{\text{Width(CC)}}{\text{Height(CC)}} > T_{\text{wd}}$,

2. $\frac{\text{Height(CC)}}{\text{Width(CC)}} > T_{\text{hi}}$,

3. $\text{Density(CC)} = \frac{\sum_{\text{pixels}}}{\text{Width(CC)} * \text{Height(CC)}} > T_D$ or

4. $\sum_{\text{CCs}} > T_N$

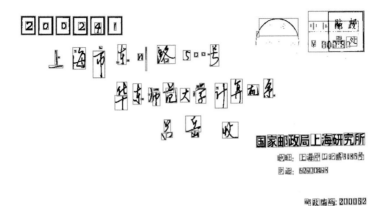

Figure 23.5 Connected component in the binary image

where Width(CC) is the width of the current CC, Height(CC) is the height of the current CC, $\sum_{\text{pixels}}$ is the number of foreground pixels in the current CC, and $\sum_{\text{CCs}}$ is the number of CCs which are covered by the current CC. T_{wd}, T_{hi}, T_D and T_N are predefined thresholds. Figure 23.6 gives the result of the above two steps.

A Chinese character consists of several character CCs which either overlap or do not. Based on the structure of Chinese characters, the overlapped or nearly-adhered CCs are merged first. The mean and standard deviation values of the merged CCs' area, height-to-width aspect ratio and the nearest-distance with others of the CCs are calculated then. Based on these features, a clustering algorithm is carried out to finally obtain the destination address block, as shown in Figure 23.7.

23.3.2 Postcode Recognition

In the Chinese postal system, a postcode is composed of six numerals. The normalized image of an individual numeral character is fed to different independent classifiers. The outputs of them are combined to obtain final recognition results.

23.3.2.1 Character Image Normalization

The performance of character recognition is largely dependent on character shape normalization, which aims to regulate the size, position, and shape of character images so as to reduce the shape variation between the images of the

Figure 23.6 Character Connected Components.

Figure 23.7 Localization of destination address blocks

same class. Generally, mapping the input character image onto a standard image plane can make all the normalized images have the same dimensionality. Also, it is desirable to restore the deformation of characters. Linear normalization (LN) can scale the image despite its inner structure, and the nonlinear normalization (NLN) method based on line density equalization is popularly used now. The practical process of shape normalization may vary for different systems. In general, LN is used for numeral character normalization while NLN is used for Chinese character normalization.

23.3.2.2 Classifiers for Isolated Numeral Characters

Multiple-layer Perceptron (MLP) One feed forward neural network (multiple-layer perceptron) is trained as a numeral classifier and the back-propagation algorithm is applied during the training procedure. The directional codes histogram features and gray-scale transformation features [1] are employed as the input to the neural network. A rectangular frame enclosing the contours of the numeral is divided into 4×6 rectangular grids. In each grid, a local histogram of the contour chain codes is calculated. These local histograms compose the directional codes histogram features. The contour direction is quantized into one of four possible values (H,V,L,R), thus the feature has 96 components for the image of one numeral character. The gray-scale transformation features are generated by applying an averaging filter to the binary image of numerals three times on a 3×3 window. The entire image is divided into 8×12 grids, and the normalized average value in each grid is chosen to be its feature. The network is structured as 192 nodes in the input layer corresponding to the feature space (96 nodes for the directional codes histogram features and 96 nodes for the gray-scale transformation features), 30 nodes in the hidden layer, and 10 nodes in the output layer corresponding to the 10 numerals.

Tree Classifier Based on Topological Features (TCTF)   In this method, a tree classifier based on topological features and contour information is employed for recognizing numerals, which is similar to the one in [2]. A contour-following algorithm is carried out on the images of numeral characters, and then the numerals are divided into three groups according to their topological properties.

In the first group, the characters have only one contour, which include the numerals '1', '2', '3', '4', '5', '6', '7', etc. The numerals in this group are recognized by their Fourier descriptors of their outer contour. In calculating the Fourier descriptors, the contour pixel with the furthest distance from the contour centroid is chosen as the starting point, and then 36 sampled points are obtained from the contour by an equal sampling interval. The discrete Fourier transform is done based on the coordinates of the sample pixels, in which the coordinates are treated as a complex numbers. The complex coefficients including 72 parameters constitute the normalized Fourier descriptors of the contour.

In the second group, the characters have two contours, which include the numerals '0', '2', '4', '5', '6', '9', etc. The numeral '5' is recognized first according to the topological relationship between the two contours as the first level of recognizing the numeral of this group. Because the two contours are disjunct only for the numeral '5' in this group, while for the other numerals in this group, one contour is always inside the other contour. In the second level, the features including the mean and variance of the normalized distance function are utilized to discriminate the numeral '0' from the others. In the third level, the relative position of the two contours' centroid is used for discrimination of the numeral '4' and '9' from the numeral '2' and '6', because the interior contour centroid of '4' and '9' are always in the upper part of their images while that of '2' and '6' are always in the lower part of their images. In the fourth level, the Fourier descriptor of the outer contour is applied to discriminate numeral '4' from numeral '9'. The numeral '2' and '6' are discriminated by the difference of the relative position of the outer and interior contours, because the interior contour centroid of numeral '2' is always at the left side of its outer contour centroid, while the interior contour centroid of numeral '6' is always at the right side of its outer contour centroid. In the third group, basically there is only the numeral '8' that has three contours.

Havnet   The third classifier is based on HAVNET [10, 15], which is a neural network employing Hausdorff distance as a similarity metric. Two-dimensional binary images are used directly for training and recognition after preprocessing of noise elimination and scale normalization. HAVNET consists of three layers: the plastic layer, the Voronoi layer, and the Hausdorff layer. The input of HAVNET is a two-dimensional binary image. The numeral image is scaled to an image with the size of height = 12 and width = 10. Plastic layer weights are trained on samples first to represent all sub-class patterns. In the recognition procedure, the Voronoi layer and Hausdorff layer are used to compute the overall dissimilarity between the input image and the learned patterns in the format of Hausdorff distance. The output layer of the network consists 10 nodes, one representing each numeral category. The HAVNET neural network outputs the directed Hausdorff distance between the input image and the patterns stored in the channels of plastic layer which have been pre-trained prior to the recognition procedure.

Support Vector Machine (SVM) classifier   Support vector machine is the fourth classifier that is used in numeral recognition. One against all approach is used to extend SVMs for multi-class classification where N classifiers are performed to separate one of N mutually exclusive classes from all other classes. RBF kernel is utilized to map the input data to a higher dimensional feature space so that the problem becomes linearly separable. Sixty-four local directional features and 16 local density features are the input data of SVMs for both training and recognition. Firstly, feature vectors with their sub-class patterns are input to Support Vector Machine to train the SVMs model parameters. Then a numeral image can be recognized with the computing of its feature vector and the trained SVMs Model parameters.

Threshold-Modified Bayesian Classifier (TMBC) The fifth method for numeral recognition is based on Threshold-modified Bayesian classifier. Binary images are used directly for training and recognition after noise removing and scale normalization. The covariance matrix of each class is calculated to obtain the corresponding eigenvalues and eigenvectors. Supposing the data distribution is Gaussian distribution, the classification capability is optimal when the total eigenvectors are calculated. The smaller the eigenvalue, the better the classification performance. Because the smaller eigenvalue reflects the convergence of within-class, the corresponding eigenvector is more important. But, in practice, there exist some small eigenvalues that are close to zero, which causes a problem to calculate the inverse matrix of the covariance matrix and then Bayesian classifier becomes invalid. To deal with this problem, a threshold is used to replace all of eigenvalues which are less than it. Then the downstreaming process can be continued and a numeral image can be classified in valid.

23.3.2.3 Multiple Classifier Combination
High recognition rate is certainly required for a practical system. On the other hand, high recognition reliability is crucial too [8]. Combination of multiple classifiers is used in many applications [3]. So multiple classifier combination is applied to improve the performance of handwriting numeral recognition. Figure 23.8 describes the process of combining the results of several classifiers. Supposing K classifiers are employed to recognize numeral character x_n, the outputs of the K classifiers, $S_n^1, S_n^2, \ldots, S_n^K$, are combined by a voting algorithm to get the final recognition result c_n.

23.3.3 Address Recognition

Apart from the postcode, the destination address provides additional delivery information. Address recognition is implemented by employing an address library.

23.3.3.1 Classifier for Isolated Chinese Characters
Compared with numerals and alphabets, the number of Chinese characters is extremely large. Therefore, the recognition processing is separated into two stages: rough classification and fine classification.

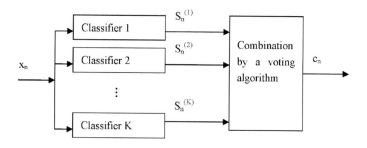

Figure 23.8 Combination of multiple classifiers

Rough classification The purpose of rough classification is to select a few candidates from the large number of categories as rapidly as possible. Peripheral features of a normalized binary character image are the input information of rough classification. City block distance (CBD), as a discriminant function, is used to select 50 candidates for an unknown pattern from the total 3755 categories. In fine classification, one candidate will be determined from these 50 candidates.

Fine classification The candidates selected by rough classification are usually characters with similar structure. In order to distinguish similar characters, it is necessary to express the original distributions of similar characters as distinctly as possible. Thus, the directional element feature (DEF), which is considered suitable for Chinese character recognition, is extracted to implement fine classification. By considering the property of distributions of Chinese, the asymmetric Mahalanobis distance (AMD) is proposed for fine classification [6]. The AMD is a function that can express distributions of images with a small number of parameters. For fine classification, a function that designates the distribution of samples is necessary. The Mahalanobis distance is derived by a probability density function of multivariate normal distribution, so it is considered as an appropriate function if the distribution of samples is multivariate normal. However, a lot of distributions of Chinese characters are asymmetric rather than normal. Therefore, a modified Mahalanobis distance, named asymmetric Mahalanobis distance, that adds a bias to eigenvalues is proposed for fine classification.

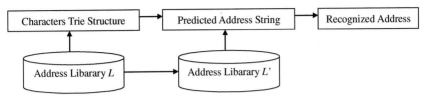

Figure 23.9 Schematic diagram for address interpretation

23.3.3.2 Address Interpretation

An address library L is constructed by all available addresses provided by China Post. An address-driven method is employed for Chinese postal address interpretation. Figure 23.9 shows a schematic diagram to illustrate the processes.

Character recognition results are put into a trie structure. Recognition for the address image becomes to identify an optimal combination path in the trie structure. The recognition is divided into two stages, i.e. coarse selection and fine selection. In coarse selection stage, the number of characters occurring in each available address is accounted. Only if the ratio of occurred characters to the total number of an address' characters is more than a predefined threshold, the address is selected to a new address library L' as address candidate. On the one hand most of addresses with lower probability in library L are filtered; on the other hand the reasonable nodes in trie structure are to be identified. Unnecessary nodes and their characters can be positively filtered. In fine selection stage, the inexact string matching technology is used to evaluate the probability of address candidates in L'. Finally, the address with highest probability are the target address.

23.4 Conclusions and Future Work

Postal automation is one of representative applications of pattern recognition. On the other hand, such successful application in turn has supported and promoted the rapid development of pattern recognition technology.

With economic development, automatic letter sorting machines have been widely deployed in mail processing centers. Shanghai Research Institute of Postal Science of China Post Group acts an important role for popularizing the automatic sorting technologies in China. It cooperated with East China Normal University to establish the Joint Lab of Pattern Analysis and Intelligence System focusing on applying pattern recognition technology to postal automation. Its targeted research vigorously pushed forward the application and promotion of letter sorting machine in China. Moreover, its research on Bangla recongition has been successfully applied to the letter sorting machine in Bangladesh Post [14].

Nowadays, telephone, cellphone and the Internet have become popular modern communication fashion. However, it is well known that the volume of mailpieces has kept increasing, either in the developed countries or in the developing countries. This is, to a certain extent, due to the extensive use of business mails including bill-statement letters and advertising letters. There is still a strong demand for the use of automatic letter sorting machines in post office. Furthermore, postal automation requires more advanced pattern recognition technology to promote its further development. For example, higher recognition performance will help to increase economic efficiency of postal enterprises. The technique of image-based virtual ID-tag will enable post office to sort and trace mailpieces at low cost.

References

[1] J. Cao, M. Ahmadi, and M. Shridhar. Recognition of handwritten numerals with multiple feature and multistage classifier. *Pattern Recognition*, 28(2):153–160, 1995.

[2] D. Cheng and H. Yan. Recognition of handwritten digits based on contour information. *Pattern Recognition*, 31(3):235–255, 1998.

[3] X. Feng, X. Ding, Y. Wu, and P.S.P. Wang. Classifier combination and its application in iris recognition. *International Journal of Pattern Recognition and Artificial Intelligence*, 22(3):617–638, 2008.

[4] H. Fujisawa. Forty years research in character and document recognition – An industrial perspective. *Pattern Recognition*, 41(8):2435–2446, 2008.

[5] R.H. Haralick. Some neighborhood operations. In M. Onoe, K. Preston, and A. Rosenfeld (Eds.), *Real Time/Parallel Computing Image Analysis*. Plenum Press, New York, 1981.

[6] N. Kato, S. Omachi, H. Aso, and Y. Nemoto. A handwritten character recognition system using directional element feature and asymmetric Mahalanobis distance. *IEEE Trans. on Pattern Analysis and Machine Intelligence*, 21(3):258–262, 1999.

[7] S.W. Lee, Y.Y. Tang, and P.S.P. Wang (Eds.). *Advanced Aspects of Oriental Language Recognition and Processing*. World Scientific, 1998.

[8] Y. Lu and C.L. Tan. Combination of multiple classifiers using probabilistic dictionary and its application to postcode recognition. *Pattern Recognition*, 35(12):2823–2832, 2002.

[9] Y. Lu, C.L. Tan, P. Shi, and K. Zhang. Segmentation of handwritten Chinese characters from destination addresses of mail pieces. *International Journal of Pattern Recognition and Artificial Intelligence*, 16(1):85–96, 2002.

[10] R.G. Rosandich. HAVNET: A new neural network architecture for pattern recognition. *Neural Networks*, 10(1):139–151, 1997.

[11] S.N. Srihari. Handwritten address interpretation: A task of many pattern recognition problems. *International Journal of Pattern Recognition and Artificial Intelligence*, 14(5):663–674, 2000.

[12] K. Suzuki, I. Horiba, and N. Sugie. Linear-time connected-component labeling based on sequential local operations. *Computer Vision and Image Understanding*, 89(1):1–23, 2003.

[13] P.S.P. Wang (Ed.). *Character and Handwriting Recognition – Expanding Frontiers.* World Scientific, 1991.

[14] Y. Wen, Y. Lu, and P. Shi. Handwritten Bangla numeral recognition system and its application to postal automation. *Pattern Recognition*, 40(1):99–107, 2007.

[15] X. Wu and P. Shi. Unconstrained handwritten numeral recognition using Hausdorff distance and multi-layer neural network classifier. In *Proceedings of the Fifth International Conference on Document Analysis and Recognition*, Bangalore, India, pages 249–252, 1999.

24

A Lossless Data Hiding Method by Histogram Shifting Based on an Adaptive Block Division Scheme

Che-Wei Lee[1] and Wen-Hsiang Tsai[1,2]

[1]*Department of Computer Science, National Chiao Tung University, Hsinchu 30010, Taiwan, ROC; E-mail: lcw.cs94g@nctu.edu.tw*
[2]*Department of Information Communication, Asia University, Taichung 41354, Taiwan, ROC; E-mail: whtsai@cs.nctu.edu.tw*

Abstract

A lossless data hiding method based on histogram shifting is proposed, which employs a scheme of adaptive division of cover images into blocks to yield large data hiding capacities as well as high stego-image qualities. The method is shown to break a bottleneck of data-hiding-rate increasing at the image block size of 8×8, which is found in existing histogram-shifting methods. Four ways of block divisions are designed, and the one which provides the largest data hiding capacity is selected adaptively. A series of experiments have been conducted, and superiority of the proposed method is shown by comparing the experimental results with those of other histogram-shifting based methods. A good property of the proposed method observed in the experiment is that data hiding rates are increased without degrading the stego-image quality expressed by the peak signal-to-noise ratio measure.

Keywords: lossless data hiding, data hiding, histogram shifting, block division.

Patrick Shen-Pei Wang (Ed.), Pattern Recognition and Machine Vision – In Honor and Memory of Professor King-Sun Fu, 383–396.

24.1 Introduction

The development of information hiding techniques provides a way to protect digital media. Such techniques may be employed to embed private or secret information into *cover media* in such a way that the existence of the hidden information is imperceptible but known to the pre-concerted recipient. The cover media can be of various types of digital data, such as image, text, video, etc. Information like private annotations, business logos, and critical intelligence can be embedded into a cover image in an invisible form so that many applications, like ownership claim of digital contents, copyright protection of media, covert communication between parties, etc., can be fulfilled. Information techniques used for covert communication are often called *steganography*, and those for ownership or copyright protection are often called *watermarking*.

In the early phase, conventional steganography [1, 9, 17–19] emphasizes exploring higher hiding payload and pursuing lower quality degradation in the *watermarked image* (also referred to as the *stego-image*, in contrast with the *cover image*). Bender et al. [1] proposed the technique of least-significant-bits (LSBs), in which a secret message is embedded in the least significant bits of the pixel values of cover images. The method yields high data hiding capacities and low computational complexities. Mielikainen [9] proposed a modified method of LSB replacement which embeds as many bits but with a smaller number of changed pixel values. Wu and Tsai [17] proposed a steganographic method for images by pixel-value differencing (PVD). The difference values between pixel value pairs are classified into a number of ranges, and the range into which the difference value between a pair of pixels falls decides how many bits can be embedded into the pixel pair. A method improving that by Wu and Tsai [17] was later proposed by Wu et al. [18], which utilizes the PVD technique to embed data in the smooth areas of the cover image and applies LSB replacement in the edged areas. This hybrid way of data hiding improves the hiding capacity and yields nearly equal stego-image quality, compared with the original PVD method [17]. Afterward, Yang et al. [19] proposed an adaptive k-LSB substitution method in which larger values of k are adopted in edge areas and smaller values of k are used for smooth areas. Also, the range of the PVD values of two consecutive pixels is divided into different levels. And the value of k is adaptively decided according to the level into which the PVD value falls. In this way, larger hiding capacities with higher stego-image quality can be obtained.

Besides the previously mentioned techniques of data hiding in the spatial domain, Wang et al. [15] transformed image block contents into coefficients in the frequency domain by the discrete cosine transform (DCT). The new AC values for the central block of every nine 8×8 image blocks are recomputed. Secret bits are embedded by modifying the magnitude relation between the new AC values and the original ones. In addition to data embedding using the DCT technique, the discrete wavelet transform (DWT) [8, 14, 16] and the discrete Fourier transform (DFT) [11, 12] have also been used.

The methods mentioned above yield permanent distortion in the stego-image. In general, a small amount of content distortion is usually imperceptible to human vision. However, such distortion is not preferred in some applications, such as legal documentation, medical imaging, military reconnaissance, high-precision scientific investigation, etc., because it may lead to risks of incorrect decision making. In view of this, a kind of novel data hiding technique, which is referred to as *reversible, invertible, lossless*, or *distortion-free*, has been developed in recent years. In this study, a reversible data hiding method which produces stego-images with good qualities and high data hiding capacities is proposed.

Reversible data hiding techniques can be employed to restore stego-images to their pristine states after the hidden data are extracted. Such techniques can be classified into three groups: (1) based on data compression [2, 3]; (2) based on pixel-value difference expansion [6, 13]; and (3) based on histogram shifting [4, 5, 7, 10]. The strategy used in the techniques of the first group is to compress the data to be embedded as well as the related information for data recovery, and then to embed these compressed data directly into the cover image. Celik et al. [3] proposed a high-capacity lossless hiding method in which each image pixel is quantified by a so-called L-level scalar quantization technique, and a lossless compression algorithm called CALIC is applied to compress the residues yielded by the quantified image pixels. Then, the compressed residues integrated with the secret bits are embedded into the quantified image by the LSB replacement technique. The second group of reversible data hiding methods aims to explore the redundancy of pixel values in images. Tian [13] proposed a technique of PVD expansion by performing fundamental arithmetic operations on pairs of pixels to discover hidable space for data embedding. However, not all pairs can be expanded for data hiding. A location map is used to indicate whether and where pairs are expanded or not. An enhanced PVD expansion method proposed by Kim et al. [6] used a refined location map and a new concept of expandability to achieve higher hiding capacities while keeping the resulting image distortion as low as the Tian method.

The last group of reversible data hiding methods, which the proposed method belongs to, is based on histogram shifting. Ni et al. [10] proposed a reversible data hiding method which shifts slightly the part of the histogram between the maximum point (also called the *peak point*) and the minimum one to the right side by one pixel value to create an empty *bin* besides the peak point for hiding the input message. The knowledge of the maximum point and the minimum point of the histogram is necessary for retrieving the hidden data and restoring the stego-image losslessly to the original state. In addition, the coordinates of the pixels whose gray values are equal to the gray value of the minimum point *b* need be memorized as overhead information when the value of *b* is not zero. Fallahpour [4] later proposed the idea of decomposing the entire cover image into blocks and using the peak point of the histogram of each block to hide data. Also, Hwang et al. [5] proposed the concept of slightly adjusting the pixel values located at both sides of a histogram peak to embed data. An advantage of this method is that it is unnecessary to record the knowledge of the location of the peak point because the peak location will not be changed after data hiding. But a location map is still required to store the information for restoring the cover image. The hiding capacity, when compared with that of Ni et al. [10], is smaller, and the peak signal-to-noise ratio (PSNR) of the stego-image gets worse in some cases [7]. Later, Kuo et al. [7], similarly to Fallahpour [4], used the block division technique, which is also employed in this study, to increase the hiding capacity of Hwang et al. [5].

An important characteristic of reversible data hiding methods based on histogram shifting is that more peaks imply higher hiding capacities. Therefore, in this study we try to explore the possibility of using a larger number of peaks, instead of just one, in a block to increase the data hiding capability and decrease the distortion in the stego-image. We propose a new reversible histogram-shifting data hiding method which uses an adaptive block division scheme for improving the data hiding capacity and stego-image quality. In the proposed method, each non-overlapping square block in a cover image is divided by four ways of sub-block decompositions. And the way providing the highest data hiding capacity is chosen adaptively. Compared with the existing histogram-shifting based methods, much larger hiding capacities with lower stego-image quality degradations can be achieved.

The remainder of this paper is organized as follows. In Section 24.2, more details of the proposed method are described. Experimental results and some discussions are included in Section 24.3. Conclusions are made finally in Section 24.4.

24.2 Adaptive Block Division for Histogram-shifting Based Data Hiding

As mentioned above, the use of more peaks yields larger hiding capacities. In Ni et al. [10], the number of pixels constituting the peak in the histogram of a cover image is equal to the hiding capacity because only a single peak in a cover image is used. As an improvement, Fallahpour [4] divided the cover image into blocks so as to generate a respective peak for each block. This technique of block division successfully enhances the hiding capacity because the total volume of data that can be hidden in the multiple blocks is generally larger than that which can be hidden in a single cover image, as mentioned previously. Furthermore, the location of the peak in the histogram indicates generally that a great number of pixels are 'centralized' in the neighboring area around the peak point. For this reason, Hwang et al. [5] used the two neighboring points beside the peak point to embed data. On the other hand, the block division technique is also used by Kuo et al. [7] for improving the performance of Hwang et al. [5]. A cover image of size 512×512 was divided into four blocks by Kuo et al. [5].

In this study, to see the trend of data-hiding-rate increasing by block divisions, we divide the cover image into equal-sized sub-blocks from size 256×256 to 2×2 and implemented the method of Kuo et al. [7] to test them. A surprising result was observed in the experimental results, that is, the hiding rate increases from the size of 256×256 through 8×8 and then turns to *decrease* from 4×4 to 2×2. Figure 24.1 shows an example of this trend by bar charts for the image "Lena". On the other hand, the trend of the PSNR values of the stego-image of Lena is also shown as red curve over the bar charts in Figure 24.1. This trend shows that the PSNR value keep *increasing* from the size of 256×256 all way down to 2×2, contrary to the intuition that hiding more data will result in worse stego-image quality. Similar trends of the data-hiding rates and the PSNR values found for some other test images and that of Lena together are summarized in Table 24.1.

From the above observation, the block size of 8×8 is seen to be the best choice for maximizing the data hiding capacity while minimizing the image distortion. However, this size may be contrarily regarded as a *bottleneck* in the data-hiding-rate increasing trend. It is desired to break this bottleneck in this

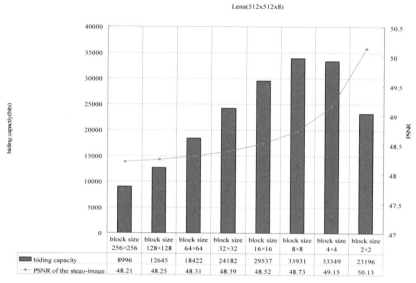

Figure 24.1 Trend of hiding capacities and PSNR values versus block sizes for image 'Lena'

Table 24.1 Statistics of experimental results of implementation of Kuo et al.'s method [7] which show a bottleneck of data-hiding-rate increasing at block size 8×8

512×512	256×256	128×128	64×64	32×32	16×16	8×8	4×4	2×2
(No. of blocks)	(4)	(16)	(64)	(256)	(1024)	(4096)	(16384)	(65536)
Lena	8996	12645	18422	24182	29537	33931	33349	23196
(PSNR)	(48.21)	(48.25)	(48.31)	(48.39)	(48.52)	(48.73)	(49.15)	(50.13)
Airplane	25555	30296	35109	41780	46945	50920	49690	38718
(PSNR)	(48.38)	(48.43)	(48.54)	(48.68)	(48.86)	(49.15)	(49.71)	(50.94)
Baboon	6044	7731	9474	11454	12426	12909	12042	8309
(PSNR)	(48.18)	(48.20)	(48.23)	(48.27)	(48.33)	(48.46)	(48.78)	(49.65)
Tiffany	16409	21617	28875	35272	40449	44490	43022	30901
(PSNR)	(48.28)	(48.35)	(48.45)	(48.55)	(48.69)	(48.92)	(49.38)	(50.39)
Boat	12931	18897	26875	33562	38382	40379	38234	25955
(PSNR)	(48.25)	(48.32)	(48.41)	(48.50)	(48.62)	(48.81)	(49.19)	(50.13)

study, and the investigation result gives the answer of "*yes*" – by the proposed method the data hiding rate will be increased again beyond the size of 8 × 8! It is also found in the experimental results of the proposed method that the PSNR value still keeps going up with the decreasing block size, as will be shown later.

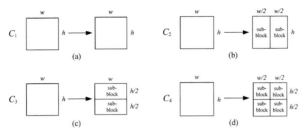

Figure 24.2 Four ways of block divisions

134	126	129	126	133	130	129	131
134	126	129	126	133	130	129	131
134	126	129	126	133	130	129	131
134	126	129	126	133	130	129	131
134	126	129	126	133	130	129	131
132	125	128	130	130	133	131	130
132	126	133	133	130	130	131	127
132	134	133	134	131	134	131	131

Figure 24.3 Pixel values of a block of size 8 × 8 in image 'Lena'

The basic concept behind the proposed method is to divide each 8×8 block by four ways into 4 × 8, 8 × 4, 4 × 4, and the originally-adopted size 8 × 8. The four ways of block divisions are illustrated in Figure 24.2. The best way among the four, which provides the largest volume of space for data hiding, is then chosen adaptively to yield the best data hiding capacity.

Figure 24.3 shows a block in the image 'Lena' which we use to illustrate a case that a block of 8 × 8 can provide a larger data hiding capacity after it is divided into two 4×8 sub-blocks. Originally, the peak point of the entire block is found at the gray value of 126 with a hiding capacity of only 2 bits. This can be seen from the fact that there are only two pixels with gray values 125 and 127 next to 126 at its two sides. But after dividing the block horizontally by the way of block division of C_3 shown in Figure 24.2, totally a hiding capacity of 13 bits can be generated from the upper and the lower sub-blocks with peak points 129 and 131, respectively, as can be seen from the gray values in the two sub-blocks where there are 13 pixels with gray values 128, 130, and 132, which are located next to 129 and 131.

In the following, the proposed method is described in details as two algorithms, the first for data embedding and the second for data extraction.

A *key* is generated by the data embedding algorithm (Algorithm 1). The key is used in the data extraction algorithm (Algorithm 2) for security control; without the key, the embedded message data cannot be extracted successfully.

Algorithm 1 – Data Embedding.

Input: a cover image I divided into blocks with size $n \times n$, and a message bit string M to be embedded.

Output: a stego-image I' with M embedded, a key K in the form of an integer number sequence, and a location map S_M.

Steps:

Step 1 (*Block decomposition*) divide each block B in I by the four ways C_1, C_2, C_3, and C_4, respectively, as shown in Figure 24.4.

Step 2 (*Estimation of data hiding capacities*) compute the data hiding capacity for each case C_i of block divisions, $i = 1, 2, 3$, and 4, by the following way.

    2.1 For each sub-block I_j in C_i, perform the following operations.

        (a) Generate the histogram h of I_j.

        (b) Find the peak in h and its location x_0.

        (c) Sum up $h(x_0 - 1)$ and $h(x_0 + 1)$ as d_j.

    2.2 Sum up all d_j to get the data hiding capacity D_i for C_i.

Step 3 (*Optimal decision*) select the C_i with the maximum capacity D_i among D_1 through D_4, denote it as C_m, and record integer m as a *component k* in the key K for the currently-processed block B.

Step 4 (*Data embedding*) for each sub-blocks I_j in C_m, perform the following operations.

    4.1 Generate the histogram h of I_j.

    4.2 Find the peak in h and its location x_0.

    4.3 Collect all pixels in I_j with gray values smaller than x_0 as a set S_L, and those with gray values larger than x_0 as a set S_R.

    4.4 Collect all pixels in S_L whose gray values are zero as a set S_L^0, and all those whose gray values are 255 in S_R as a set S_R^{255}.

    4.5 (*Histogram shifting*) decrement the gray value of each pixel in S_L by one except those in S_L^0, and increment the gray value of each pixel in S_R by one except those in S_R^{255}.

    4.6 (*Location map generation*) put S_L^0 and S_R^{255} into a set S_M, and call it a *location map*.

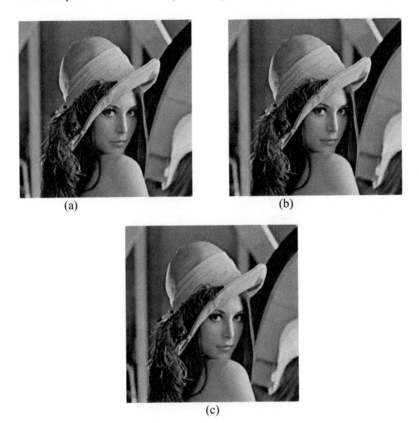

Figure 24.4 An example of experimental results. (a) Cover image of Lena. (b) Stego-image of Lena generated from Algorithm 1. (c) Losslessly-recovered image of Lena generated from Algorithm 2.

4.7 (*Bit embedding*) take a data bit m_ℓ sequentially from M, scan the pixels in I_j in a raster-scan order, take an unprocessed pixel with gray value v, and perform the following task, until all bits in M are exhausted:

(a) if $v = x_0 - 2$, then
increment v by one if $m_\ell = 1$ or keep v unchanged if $m_\ell = 0$;

(b) if $v = x_0 + 2$, then
decrement v by one if $m_\ell = 1$ or keep v unchanged if $m_\ell = 0$.

Step 5 (*Key generation*) concatenate the values k's of all the blocks in I in the block-processing order into an integer number sequence $k_1 k_2 k_3 \ldots$ as the desired key K.

The proposed data extraction process is described in the following as an algorithm, which is basically a reverse process of Algorithm 1. The input includes the output data yielded by Algorithm 1 and the number of message data bits embedded in the stego-image.

Algorithm 2 – Data Extraction.
Input: a stego-image I' with blocks of size $n \times n$, the number N of message data bits, the location map S_M, and the key K.
Output: the message data M embedded in I' and the original cover image I losslessly recovered.
Steps:

Step 1 (*Initialization*) set $M = \varepsilon$ (an empty string).
Step 2 Take out the blocks in I' in order, and divide each of them, denoted by B', in the way of the block division specified by the corresponding component k in the key K.
Step 3 (*Data extraction*) for each sub-blocks I'_j in B', perform the following operations.

   3.1 Generate the histogram h' of I'_j.
   3.2 Find the peak in h' and its location x'_0.
   3.3 (*Bit extraction*) scan the pixels in I'_j in a raster-scan order, take an unprocessed pixel with gray value v, and perform the following task, until all the N message bits are extracted:

   (a) if $v = x'_0 - 2$ or $x'_0 + 2$, then extract a bit "0" and append it to the end of M;
   (b) if $v = x'_0 - 1$, then extract a bit "1," append it to the end of M, and decrement v by one;
   (c) if $v = x'_0 + 1$, then extract a bit "1," append it to the end of M, and increment v by one.

   3.4 Collect all pixels in I'_j with gray values smaller than x'_0 as a set S'_L, and those with gray values larger than x'_0 as a set S'_R.
   3.5 (*Reverse histogram shifting*) perform the following steps to recover the cover image I losslessly.

   (a) Increment the gray value of each pixel in S'_L by one, and decrement the gray value of each pixel in S'_R by one.

(b) Use the content (S_L^0, S_R^{255}) of the location map S_M to restore the original gray values of the pixels by setting the gray value of each pixel in S_L^0 to be 0 and that of each pixel in S_R^{255} to be 255.

By Algorithm 2, the hidden data can be extracted successfully, and the original cover image recovered *losslessly*.

24.3 Experimental Results

Each test image used in our experiment is a grayscale one of the size 512×512 with gray values ranging from 0 through 255. For the purpose of comparing our results with those of other methods, we implemented additionally the algorithms of Ni et al. [10], Hwang et al. [5], and Kuo et al. [7].

As mentioned previously, as the size of blocks in each test image becomes smaller, the PSNR of the resulting stego-image keeps increasing but the data hiding capacity increases until a bottleneck at the block size of 8×8, as shown in Table 24.1 and as illustrated by Figure 24.1. Therefore, we processed the tested images, starting mainly from the block size of 8×8; and chose adaptively for each block one of the four sizes 8×8, 4×8, 8×4, and 4×4 to optimize the data hiding capacity, using Algorithms 1 and 2. Some experimental results are shown in Tables 24.2 and 24.3.

Table 24.2 Comparison of results of proposed method and Kuo et al.'s method [7]

Method		Kuo et al.'s method		Proposed method (initial block size of 8×8)	
Block size		Blocks of size 8×8		Combination of blocks with sizes 8×8, 8×4, 4×8 and 4×4	
Hiding capacity	Quality of stego-image	Bits	PSNR	Bits	PSNR
Lena (512×512)		33931	48.73	41257	48.90
Airplane (512×512)		50920	49.15	59397	49.37
Baboon (512×512)		12909	48.46	17455	48.57
Tiffany (512×512)		44490	48.92	52297	49.10
Boat (512×512)		40379	48.81	46833	48.95

Table 24.3 Comparison of results of proposed method and related methods

Method		Ni's		Hwang's		Kuo's		Proposed method	
Image or block size		Image of size 512×512		Image of size 512×512		Blocks of size 8×8		Combination of blocks with sizes 8×8, 8×4, 4×8 and 4×4	
Hiding Capacity	Quality of stego-image	bits	PSNR	bits	PSNR	bits	PSNR	bits	PSNR
Lena (512×512)		5409	48.22	5304	48.18	33931	48.73	41257	48.90
Airplane (512×512)		18700	48.45	17502	48.28	50920	49.15	59397	49.37
Baboon (512×512)		5932	48.28	5793	48.18	12909	48.46	17455	48.57
Tiffany (512×512)		10153	48.30	9942	48.21	44490	48.92	52297	49.10
Boat (512×512)		10546	49.25	9709	48.21	40379	48.81	46833	48.95

As can be seen from Table 24.2, the resulting data hiding capacity and the PSNR value for each of the five test images are both raised, compared with those yielded by Kuo et al. [7]. For example, for the image of Lena, the bottleneck of the data hiding capacity is 33931 bits with the PSNR 48.73 dB, and the resulting capacity of the proposed method now is 41257 bits with the PSNR value being 48.90 dB. In Figure 24.4, the image processing results of the test image Lena are shown. Both the imperceptibility of the data hidden in the stego-image and the reversibility of the proposed method (i.e., the capability of *lossless* data recovery) can be verified from the images.

A more complete comparison of our results with those the related methods, including Ni et al. [10] and Hwang et al. [5], in addition to Kuo et al. [7], is shown in Table 24.3. As can be seen, the proposed method produces better results both in hiding capacities and in image qualities. For a clearer visualization of the data included in Table 24.3, we use a bar-chart diagram to show the results of image Lena in Figure 24.5 as an example.

It also can be observed from the above experimental data that the data hiding rates are increased by the proposed method without degrading the

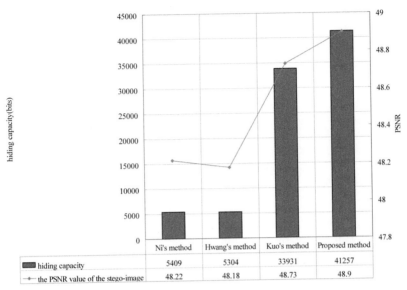

Figure 24.5 Illustration of comparison of hiding capacity versus image distortion in mentioned methods using image 'Lena'

stego-image quality expressed by the peak signal-to-noise ratio measure, which is an unusual phenomenon in data hiding research results.

24.4 Conclusions

A lossless data hiding method based on histogram shifting using an adaptive block division scheme has been proposed, which not only embeds large-volume data into cover images but also produces stego-images with high qualities. The bottleneck for data-hiding-rate increasing tendency at the block size of 8×8 found in the existing methods is broken by the proposed method. Experimental results show the effectiveness of the proposed method. Future researches may be directed to investigating more block division types and recursive division schemes for further improvement on the data hiding capacity, applying the histogram shifting technique to other information hiding methods and applications, etc.

References

[1] W. Bender, D. Gruhl, N. Morimoto, and A. Lu. Techniques for data hiding, *IBM Systems Journal*, 35(3–4):313–336, 1996.

[2] M.U. Celik, G. Sharma, A.M. Tekalp, and E. Saber. Reversible data hiding. In *Proceedings International Conference on Image Processing*, Vol. 1, pages 71–76, 2002.

[3] M.U. Celik, G. Sharma, A.M. Tekalp, and E. Saber. Lossless generalized-LSB data embedding, *IEEE Trans. Image Process.*, 14(2):253–266, 2005.

[4] M. Fallahpour and M.H. Sedaaghi. High capacity lossless data hiding based on histogram modification. *IEICE Electron. Express*, 4(7):205–210, 2007.

[5] J.H. Hwang, J.W. Kim, and J.K. Choi. A reversible watermarking based on histogram shifting. *Proceedings IWDW 2006*, Lecture Notes in Computer Science, Vol. 4283, pages 348–361. Springer, 2006.

[6] H.J. Kim, V. Sachnev, Y.Q. Shi, J. Nam, and H.G. Choo. A novel difference expansion transform for reversible data embedding. *IEEE Trans. Information Forensics and Security*, 3:456–465, 2008.

[7] W.C. Kuo, D.J. Jiang, and Y.C. Huang. A reversible data hiding scheme based on block division. In *Proceedings CISP 2008*, Vol. 1, pages 365–369, 2008.

[8] A. Lumini and D. Maio. A wavelet-based image watermarking scheme. In *Proceedings International Conference on Information Technology*, 2000.

[9] J. Mielikainen. LSB matching revisited. *IEEE Signal Processing Letters*, 13(5):285–287, 2006.

[10] Z. Ni, Y.Q. Shi, N. Ansari, and W. Su. Reversible data hiding. *IEEE Trans. Circuits Syst. Video Technol.*, 16(3):354–362, 2006.

[11] S. Pereira and T. Pun. Robust template matching for affine resistant image watermarks. *IEEE Trans. Image Process.*, 9:1123–1129, 2000.

[12] C.-M. Pun. A novel DFT-based digital watermarking system for images. *Proc. Signal Processing*, 2, 2006.

[13] J. Tian. Reversible data embedding using a difference expansion. *IEEE Trans. Circuits Syst. Video Technol.*, 13(8):890–896, 2003.

[14] S.-H. Wang and Y.-P. Lin. Wavelet tree quantization for copyright protection watermarking. *IEEE Trans. Image Process.*, 13:154–165, 2004.

[15] Y.N. Wang and A. Pearmain. Blind image data hiding based on self reference. *Pattern Recognition Lett.*, 25:1681–1689, 2004.

[16] Y.P. Wang, M.-J. Chen, and P.-Y. Cheng, Robust image watermark with wavelet transform and spread spectrum techniques. In *Proceedings Thirty-Fourth Asilomar Conference on Signals, Systems and Computers*, Vol. 2, pages 1846–1850, 2000.

[17] D.C. Wu and W.H. Tsai. A steganographic method for images by pixel-value differencing. *Pattern Recognition Lett.*, 24(9–10):1613–1626, 2003.

[18] H.C. Wu, N.I. Wu, C.S. Tsai, and M.S. Hwang. Image steganographic scheme based on pixel-value differencing and LSB replacement methods. *Proc. Inst. Elect. Eng., Vis. Images Signal Process.*, 152(5):611–615, 2005.

[19] C.-H. Yang, C.-Y. Weng, S.-J. Wang, and H.-M. Sun. Adaptive data hiding in edge areas of images with spatial LSB domain systems. *IEEE Trans. Information Forensics and Security*, 3(3), 2008.

25

IS – A Novel Handwriting Input Device Embedded in the TFT Array of Display Panel

Sen-Shyong Fann[1], Nae-Jye Hwang[2], Sheng-Tai Liaw[2] and Huey-Liang Hwang[2]

[1]*Department of Electrical Engineering, Lee-Ming Institute of Technology, 200 Taipei, Taiwan, ROC; currently also at Integrated Digital Technologies Inc.*
[2]*Integrated Digital Technologies Inc., Hsinchu Science Park, 300 Hsinchu, Taiwan, ROC; E-mail: hlh@idti.com.tw*

Abstract

This chapter presents an innovative handwriting input device embedded in the thin film transistor (TFT) array of a display panel. The structure of this innovative input device is patented. This device, Interactive Screen (IS), is described beginning with a survey of the state-of-the-art of the hand touch panel. IS consists of some readout electronics and a two-dimensional photo detector array which is made of hydrogenated amorphous silicon TFT (a-Si:H TFT). The detector array is developed while the TFT array of a Liquid Crystal Display (LCD) panel is being fabricated. As a consequence, cost is very attractive as compared with resistive and capacitive touch panels. The intrinsic properties of the a-Si:H material and the characteristics of a-Si:H TFT are briefed. After description of a-Si:H TFT, the structure and operating principles of IS are described. Both the TFT array costs and the readout electronics costs are very low-cost due to the simplicity and efficient structure of the innovative input device. Finally, some of the advantages stemming from in-cell touch technolgies are addressed.

Patrick Shen-Pei Wang (Ed.), Pattern Recognition and Machine Vision – In Honor and Memory of Professor King-Sun Fu, 397–404.

Keywords: touch panel, hydrogenated amorphous silicon (a-Si:H), TFT array, Interactive Screen (IS).

25.1 Introduction

Display devices traditionally serve the role of displaying the information or the output from a system, while other input device or devices provide inputs to the system. Interactive devices, such as a touch panel that can receive a user's input via touching the display panel, combine both input and output functions and allow a user to interact with the display, or to interact with the system which is coupled to the display. As an example, devices such as personal digital assistants (PDAs), mobile phones, personal computers (PCs), tablet PCs, etc., have incorporated touch panels for users with more choices in providing inputs to operate a system.

Conventional touch panels or touch screens have a number of different designs, such as resistive type, surface-wave type, capacitive type, electro-magnetic inductance type, and infrared-ray type designs. However, all these designs typically require combining a display device with a separate touch panel screen or structure, which may affect the quality of the display, increase the weight and size of an existing display device, and, usually unavoidably, significantly increase the manufacturing cost of the combined device. To avoid drawbacks of the design of a touch panel which has an extra panel, the optoelectronic type panel which employ photo detectors embedded in the display panel is considered [1–5]. However, the various photo detector designs also possess some drawbacks. For example, certain designs require the panel first to store received input information in a capacitor before such information is subsequently read. Such requirement may occupy additional area to provide the input-maintaining capacitors. The design is not acceptable, since it may reduce the aperture ratio of a display device as its area is limited while proper display efficiency or brightness is desired.

Therefore, there is a need for the display device, such as Interactive Screen (IS), that is interactive while preserving some advantages but avoiding drawbacks of the traditional touch panels.

Figure 25.1 illustrates the typical TFT array of an active matrix liquid crystal display (AMLCD) [8–10]. The matrix consists of pixels which are accessed by gate and data lines. A TFT with an accompanying capacitor comprise a pixel. TFT plays the role of a switching component. One gate line is selected to turn on the TFTs connected to this gate line and activate all the pixels on this gate line at each time. Upon activation of the pixel via

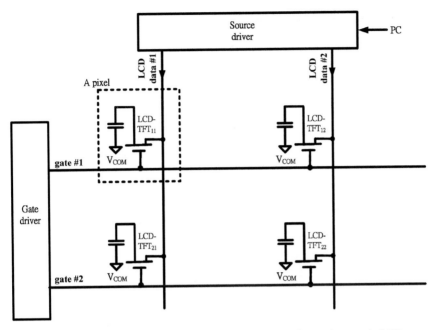

Figure 25.1 Schematic diagram of a typical TFT array of an active matrix LCD

the gate line, the data line transfers charge to the capacitor of the pixel to set the voltage on the liquid crystal. When all gate lines of the matrix are scanned, all the pixel voltages are set and the displayed image of the panel is defined.

In addition to the typical TFT array mentioned above, an IS panel also consists of a photosensitive detector array including a plurality of photosensitive transistors, photo-detector TFTs, and a plurality of switching transistors, switch-TFTs, formed in rows and columns as the configuration of TFT array of LCD. In the presence of an input optical signal, the photosensitive transistor generates a current if the gate line is selected. The current goes into the readout electronics and the spatial location of signal is then determined. The TFT array fabrication processes for the LCD which has an embedded IS will be exactly the same as those for an ordinary LCD since both switch-TFT and photo-detector TFT of IS are made at the same time for the TFT of LCD. As a result, IS cost is very low compared with resistive and capacitive touch screens due to it simple manufacturing process and no cost for an extra panel, and the elimination of the subsequent integration, calibration costs, etc.

25.2 Photo-I-V Characteristics of a-Si:H TFT

Hydrogenated amorphous silicon (a-Si:H) has a remarkably low density of localized state [6, 7]; hence the conductivity can be altered by several orders of magnitude by changing the gate voltage. A-Si:H thin film transistor operation has been demonstrated successfully in hydrogenated devices prepared from the rf glow discharge of SiH4, or rf discharge of SiH4 and SiF4 gas mixtures, etc. Active matrix arrays made of a-Si:H TFT are used for imaging and display. Examples are liquid crystal displays, optical scanners, and radiation imaging arrays. These devices contain many individual elements (pixels) which must be addressed or read out. The problem of addressing or reading out data with large numbers of pixels is solved by having a grid structure of interconnecting lines such as gate lines and data lines. Each pixel is at the intersection of a gate and data line, which are connected to a switch transistor at the pixel. Application of a voltage to a gate line activates a row of switch transistors, and allows the associated pixels to be activated. The complete array is accessed by sequentially addressing all the gate lines.

Since amorphous silicon so far offers the most important technical advantage of being deposited inexpensively and uniformly over a very large area, and it is more photosensitive than thin film poly-Si for visible light, both the photo detectors and switch TFTs are all made of amorphous silicon can get a lot of benefits on cost and photo-electric performance.

The photo characteristics of the photo-detector TFT illuminated by different light intensities are shown in Figure 25.2. The TFT is employed in IS as photo-detector TFT or switch TFT. The illumination varies from 0 lux (dark) to 6270 lux. The photo-current of this photo-detector TFT increases as the intensity of input incident light increases. Layer structure of photo-sensor-embedded LCD module is also shown in Figure 25.3.

Compared with crystalline Si which has indirect band gap, a-Si:H behaves optically like a direct band gap. Consequently, the thin film device made of a-Si:H material is more photo sensitive than the one made of crystalline Si. For some applications, such as LCD, we need the operation of TFT far away from the influence of the ambient light as well as the back light, all TFTs must be shielded by a black matrix. However, since a-Si:H TFT is sensitive to light, a-Si:H TFT can also be a good photo detector, and the detector TFT is exposed to the light of ambient or the light from something in the ambient.

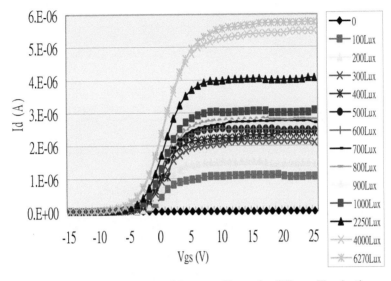

Figure 25.2 Experimental data of I_d versus Vgs under different illuminations

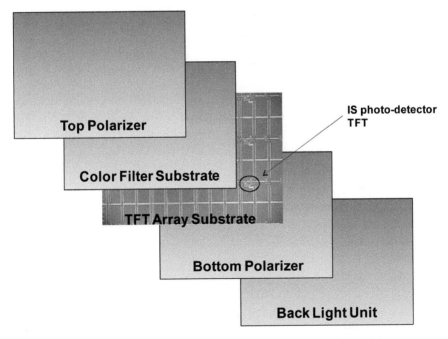

Figure 25.3 Layer structure of photo-sensor-embedded LCD module

Figure 25.4 The prototype of a 5-inch QVGA IS. We used a light pen to write directly on the display panel

25.3 Advantages of Photo-Sensor Embedded Touch Screen

As we mentioned at the beginning of this chapter, due to the simplicity and efficient structure of the photo-sensor-embedded touch screen, both the TFT array cost and the readout electronics costs are very low. In addition, other advantages of in-cell touch technologies including:

1. No additional film/glass thickness.
2. No additional weight.
3. Standard housing/tooling
4. No size limitation
5. Higher optical transmission
6. Better display performance (no haze, etc.)
7. Longer lifetime
8. No assembly/calibration required.
9. Ideal solution to Palm Rejection
10. Genuine Multi-Touch Solution.
11. Unique remote input capability.

25.4 Conclusions

This chapter presents a device, IS, constructed from TFT transistors and light sensors. For IS, the light sensor indeed is made of a-Si:H TFT to simplify the fabrication process. IS comprises a photosensitive detector array, normally it is arranged that one display pixel with one detector, and a readout electronics which scans the detector array and then determines the spatial location of the signal. The photosensitive detector array includes a plurality of photosensitive transistors and a plurality of switching transistors formed in rows and columns. Alternatively, the photosensitive transistor can work simultaneously as a switching component to further simplify the detector array. A photo detector array includes photo detectors disposed on a substrate and arranged to receive and sense a photo signal from light pen or a shadow from something on a specified area. Each photo detector represents a pixel or a sub-pixel for this area. The detectors each have a photo TFT that generates a photocurrent in response to the light intensity irradiates on it. The photo TFT may be manufactured using the common processes for TFTs in forming the flat panel. Operation of the switch TFT is enabled by a select line which is coupled to the gate electrode of the readout TFT.

IS was realized by idti (Integrated Digital Technologies, Inc., Hsinchu Science Park, 300 Hsinchu, Taiwan), it works very successfully, and its volume production prevails in the 4th quarter of 2009.

References

[1] A. Abileah et al. Integrated optical touch panel in a 14.1" AMLCD. In *Proceedings SID'04*, pages 1544–1547, 2004.

[2] W. den Boer. Image sensor with photosensitive thin film transistors. United States Patent Application Publication, US 2003/0218116 A1, 27 November 2003.

[3] W. den Boer et al. Active matrix LCD with integrated optical touch screen. In *Proceedings SID'03*, pages 1494–1497, 2003.

[4] S.-S. Fann. Photodetector array with background current compensation means. United States Patent, US 7,525,078 B2, 28 April 2009.

[5] S.-S. Fann. A novel thin film photo detector array for applications of handwrite input device. *Journal of Lee-Ming Institute of Technology*, 19(1):65–69, April 2007.

[6] P.G. LeComber, W.E. Spear, and A. Ghaith. Amorphous-silicon field-effect device and possible applications. *Electron. Lett.*, 15(6):179–181, March 1979.

[7] A. Madan and M.P. Shaw. *The Physics and Applications of Amorphous Semiconductors*. Academic Press, 1988.

[8] A. Nathan, P. Servati, K.S. Karim, D. Striakhilev, and A. Sazonov. Device physics, compact modeling, and circuit applications of a-Si:H TFTs. In *Thin Film Transistors, Material and Processes*, Y. Kuo et al. (Eds.). Kluwer Academic Publishers, 2004.

[9] T. Nose, N. Ikeda, H. Kanoh, H. Ikeno, H. Hayama, and S. Kaneko. A 28-cm-diagonal UXGA TFT-LCD for legible A4-size document viewers. In *Proceedings 1999 SID Symposium Digest*, pages 180–183, 1999.

[10] S. Vrablick. Low-temperature poly-silicon TFT-LCDs for emerging portable applications. In *Proceedings FID Conference*, 2000.

26

Speech Recognition Based on Attribute Detection and Conditional Random Fields

Chi-Yueh Lin and Hsiao-Chuan Wang

Department of Electrical Engineering, National Tsing Hua University, Hsinchu 30013, Taiwan, ROC; E-mail: bancolin@gmail.com, hcwang@ee.nthu.edu.tw

Abstract

The integration of phonetic knowledge in a speech recognizer which is based on data-driven technique can further improve its performance. This article presents a cascaded structure which consists of attribute detection and conditional random field approach to make use of phonetic knowledge within the phone decoding process. The attribute detection can be implemented by using any effective feature extraction methods. In this study, an HMM-based method is applied for attribute tagging of Mandarin speech. Then a conditional random field method which uses attribute labels as input vectors is applied to perform the speech recognition. The preliminary experiment shows that the proposed method is very promising and worthy for further investigation.

Keywords: attribute detector, conditional random field, speech recognition.

26.1 Introduction

The speech recognition based on hidden Markov models (HMM) has been the most popular technique for decades. What makes the HMM so attractive is that not only its structure resembles the speech production of human beings, but also it possesses rigorous mathematical background. Over the

Patrick Shen-Pei Wang (Ed.), Pattern Recognition and Machine Vision – In Honor and Memory of Professor King-Sun Fu, 405–415.

years speech recognition systems based on the HMM have demonstrated its practicability in real world applications. Even under various noisy conditions, many novel methods have been proposed to improve the robustness of HMM-based systems. Bilmes [1] mentioned that, given enough hidden states and a sufficiently rich class of observations, an HMM can accurately model any real world probability distributions. However, the accuracy of an HMM heavily relies on the assumption of the observation independence. To be specific, observation $X(t)$ is independent of all previous observations and all previous hidden states for a given state variable $Q(t)$. The statement is too strong to meet a real world speech signal. The contradiction between reality and theory results in the imperfect recognition rate of an HMM-based system. In fact the improvement of recognition performance has slowed down for many years, and it seems that there is a bottleneck to be broken through. To overcome this difficulty, two streams of maneuvers stand on their own principles. The first maneuver keeps investigating more sophisticated methods to polish HMM-based system. The second maneuver, on the contrary, tries to search for an alternative modeling structure. It is arguable which one has the opportunity to overcome the difficulty, but one would agree that both of them are challenging and worthy of investigation.

This chapter addresses the issue of alternative modeling structure, which is due to the recent significant improvement in machine learning algorithms and to the possibility of incorporating acoustic-phonetic knowledge into a speech recognition system [7, 11]. Recent researches have focused on schemes that can take phonetic knowledge into consideration. The purpose of incorporating phonetic knowledge into a speech recognition system is that once contrastive speech events are highlighted, one could hypothesize the speech sounds being produced. Recognizing speech in this fashion is different from that in a conventional HMM-based system, where all speech units are modeled in the same manner, i.e. static and dynamic cepstral coefficients. Instead, one is encouraged to design several phonetic-dependent detectors to locate various speech landmarks. Those detectors can be based on different modeling techniques, and run at different detection rates. Apparently the system structure is composed of two major parts in tandem: a front-end that has many detectors running in parallel or sequentially, and a back-end that gathers phonetic information coming from the detectors to decode the speech. This kind of manipulation has raised many open issues. Taking front-end detectors for example, it is reasonable to apply different modeling techniques to different phonetic units. The fact is that it is highly impossible to have a single modeling technique that can describe all phonetic units equally well. One of the well-known examples is the stop consonant. The closure-burst transition of a stop consonant is one of the most brief acoustic events in speech. The conventional setting, due to the time-frequency resolution tradeoff, is not a good choice for detecting this event. As to back-end processing, how to merge or fuse detection results coming from the front-end detectors is definitely an important issue. The results can be provided in probabilistic, non-probabilistic, or the integration of both fashions.

In this study two modeling techniques are cascaded to achieve a complete decoding process for converting the speech signal to phone labels. The front-end system consists of a set of attribute detectors running in parallel to transform speech signal into intermediate attribute/event labels. The back-end system employs the conditional random fields (CRF) [4] technique to incorporate those attribute labels for the hypothesis of phone identities. The Mandarin speech recognition is implemented based on the proposed method to show the efficiency of this cascaded structure. The HMM-based attribute detectors are implemented for experiments. In fact, more sophisticated techniques can be employed for the front-end processing. The other focus is to introduce the CRF technique into speech processing.

26.2 Proposed Speech Recognition System

As mentioned before, the proposed Mandarin speech recognizer is composed of two cascaded systems. The front-end is a group of attribute detectors. In this study, the HMMs are used for attribute detection. They do not recognize phone identities directly, but transform feature vectors into attribute labels. The following back-end CRF-based system makes use of these attribute labels to produce the final recognition results. Figure 26.1 illustrates the block diagram of the proposed speech recongition system. In this cascaded structure, each attribute detector extracts a specific speech attribute and outputs its corresponding attribute label. These N attribute labels are input to the back-end CRF-based system to serve as evidence of which phone has been pronounced.

26.2.1 Front-End Attribute Detectors

Due to the syllabic nature of Mandarin, a conventional Mandarin speech recognition system usually adopts basic units of Initials and Finals, not phones as in English. A typical Mandarin syllable consists of a leading Initial and a following Final. A system that uses phones as basic units should

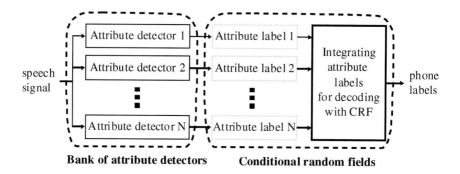

Figure 26.1 The cascaded system

consider the context dependency in modeling those phones. For Mandarin, a similar technique is applied to model the context dependency. However, only right-context dependency (RCD) is considered in Mandarin. The left-context dependency only bears little or no information on contrastive phonetic knowledge. To be specific, every Mandarin Final is modeled in a context in-dependent fashion, while each Mandarin Initial has various context dependent variations.

The speech attribute detectors are initialized by existing RCD-HMMs. For a faster training, the RCD-HMMs that belong to the same attribute are tied together and re-estimated five times using the same training data. In our RCD-HMMs each Initial is modeled in a left-to-right 3-state topology, and each Final is modeled in a left-to-right 5-state topology. Under some circumstances, Initials and Finals are pooled together for model tying, such as the model of N/A attribute (Not Available). Due to the fact that the numbers of states used in these two units are different, it is not possible to tie them directly. In practical implementation, an N/A attribute model is generated for Finals and another N/A attribute model is generated for Initials. During the recognition phase, the output labels will be the same, however.

There are various ways to formulate speech attributes. In accordance with phonetic knowledge of Mandarin, six speech attributes are set up for each Mandarin Initial and Final. These six speech attributes are the manner of articulation, the place of articulation, the Final onset type, the Final ending type, the aspiration, and the voiced. It should be noted that some attributes are only meaningful to Initials, and some are to Finals. Thus it is not necessary to use all attributes at the same time to determine the phone identity. Based on

Table 26.1 HMMs for each attribute detector

Attribute	HMMs
manner (Final)	Finals ended with static vocal tract, Finals ended with a vowel, Finals ended with a nasal
manner (Initial)	affricate, fricative, stop, nasal, lateral
Final onset type	/a/, /e/, /o/, /yi/, /yu/, /wu/, /eh/, N/A.
Final ending type	/ai/, /ei/, /yi/, /ao/, /ou/, /wu/, /an/, /en/, /ang/, /eng/, /yu/, /a/, /o/, /e/, /eh/, N/A
place	bilabial, labial-dental, front coronal, middle coronal, back coronal, dorsum, back (velar), N/A.
aspiration	aspiration, non-aspiration, N/A
voiced	voiced, unvoiced

the previous attributes, each Final can be determined uniquely by three speech attributes: the manner of articulation, the Final onset type, and the Final ending type. On the other hand, each Initial can be determined by the manner of articulation, the place of articulation, the aspiration, and the voiced. Table 26.1 lists the attributes to be used and the attribute HMMs within each attribute. The following are descriptions of these speech attributes:

Manner of articulation In this category there are three HMMs for Finals and five HMMs for Initials. The borderline between three HMMs for Finals depends on how a Final ends. A Mandarin Final can be ended in either the following three ways: with a static vocal tract configuration, with a vowel, or with a nasal. It is arguable the necessity to further divide a single "Final" category into three sub-categories. As we will see later, Mandarin Finals possess complex structures so that it is preferable to use several refined categories instead of only a single category. As to those five manners of articulation for Mandarin Initials, we follow the classification rule that is commonly applied to consonants. They are affricate, fricative, stop, nasal, and lateral.

Final onset types This attribute classifies Finals into several groups according to its onset pronunciation. The onset portion of Mandarin Finals should be pronounced with /a/, /e/, /o/, /yi/, /yu/, /wu/, or /eh/. Sounds not belonging to Finals are classified into a special group called N/A (Not Available).

Final ending types Similar to the Final onset types, the ending portion of a Mandarin Final should be pronounced with /ai/, /ei/, /yi/, /ao/, /ou/, /wu/, /an/, /en/, /ang/,/eng/, /yu/, /a/, /o/, /e/, or /eh/. Comparing to the Final onset types, the ending types are more complicated. Articulators are fixed during the pronunciation in the former case, however, in pronouncing the compound Finals the articulators are changing with time. Similar to the previous case, there is still a N/A group for those sounds not belonging to Finals.

Place of articulation Each Mandarin Initial can be classified into one of the following groups: bilabial, labial-dental, front coronal, middle coronal, back coronal, dorsum, or back (velar). The classification of Mandarin Initials depends on where the constriction is formed inside the oral cavity. Sounds belonging to finals are grouped as N/A.

Aspiration Aspiration and non-aspiration attributes are mainly for stops and affricates, and all other sounds have no such attribute. Thus all sounds but stop and affricate are grouped into N/A.

Voiced This is the most simple and straightforward attribute. All finals are voiced, whereas all initials but /m/, /n/, /l/, and /r/ are unvoiced.

To decode speech into attribute labels with the previous attribute HMMs, each input utterance is framed with 25 ms frame size and in 10 ms frame rate. Then each frame is parameterized to a 42-dimensional feature vector that includes 13-order MFCC, normalized log-energy, and their corresponding delta and delta-delta coefficients. Techniques such as the pre-emphasis and cepstral mean subtraction are also applied in the pre-processing stage. Six sets of attribute HMMs described in Table 26.1 execute in parallel to decode the utterance into six attribute representations. In other words, after this decoding phase the output of each frame will be a vector of six dimensions, and these six dimensions correspond to six attribute labels respectively. These attribute labels are regarded as intermediate representations and are then processed by the subsequent CRF-based system to transform the attribute labels into their final representation form, Mandarin Initials or Finals.

26.2.2 Back-End CRF-Based System

Conditional Random Fields (CRFs) [4] belong to undirected graphical models. The main advantage of CRFs is their flexibility to include a wide variety of arbitrary and dependent features. It has been widely applied to natural language processing, especially in the task of POS tagging and NP-chunking. According to the Hammersley–Clifford theorem, CRFs define the conditional probability of a set of output values given a set of input values to be proportional to the product of potential functions on cliques of a graph,

$$P_\Lambda(\mathbf{s}|\mathbf{o}) = \frac{1}{Z_o} \prod_{c \in C(\mathbf{s},\mathbf{o})} \Phi_c(\mathbf{s}_c, \mathbf{o}_c) \tag{26.1}$$

where $\Phi_c(\mathbf{s}_c, \mathbf{o}_c)$ is clique potential on clique c, and Z_o is a normalization factor for output values. Z_o is also known as the partition function.

Among the family of CRFs, the most popular one is linear-chain CRF, which can be expressed as follows:

$$p(\mathbf{s}|\mathbf{o}) = \frac{1}{Z(\mathbf{o})} \exp\left(\sum_{t=1}^{T} \sum_{k=1}^{K} \lambda_k f_k(s_{t-1}, s_t, \mathbf{o}, t)\right) \tag{26.2}$$

where $\mathbf{o} = \{o_1, o_2, \ldots, o_T\}$ is the observation vector spanning T frames, $\mathbf{s} = \{s_1, s_2, \ldots, s_T\}$ is its corresponding state sequence, $f_k(s_{t-1}, s_t, \mathbf{o}, t)$ represents the k-th feature function that measures the degree to which state s_t is compatible with a transition from state s_{t-1} and with observation $\mathbf{o}$, λ_k is the weighting factor for the feature function f_k, and $Z(\mathbf{o})$ is a normalization constant. In practice, feature function f_k can be designed manually or generated automatically [5] for domain-specific applications. In order to tell the difference between HMM and CRF, it should be noted that in a CRF the transition between adjacent states depends on the whole observations and there is no assumption about the distribution over observations.

While training CRF, weighting factor λ_ks which reflects the relative importance of feature function f_k are the only parameters to be considered. Given a set of training samples of the form $(\mathbf{o}_i, \mathbf{s}_i)$, where $\mathbf{o}_i = (o_{i,1}, o_{i,2}, \ldots, o_{i,T})$ and $\mathbf{s}_i = (s_{i,1}, s_{i,2}, \ldots, s_{i,T})$, the CRF can be trained to maximize the log-likelihood of the training data, with a regularized term to prevent the over-fitting. Let $\Theta = \{\lambda_1, \ldots, \lambda_K\}$ denote all the tunable parameters in the model. Then the objective function to be maximized is

$$J(\Theta) = \log \prod_{i=1}^{N} p(\mathbf{s}_i|\mathbf{o}_i) - \sum_{k=1}^{K} \frac{\lambda_k^2}{2\sigma^2} \tag{26.3}$$

$$= \sum_{i=1}^{N} \sum_{t=1}^{T} \sum_{k=1}^{K} f_k(s_{i,t-1}, s_{i,t}, \mathbf{o}_i, t) - \sum_{i=1}^{N} Z(\mathbf{o}_i) - \sum_{k=1}^{K} \frac{\lambda_k^2}{2\sigma^2}$$

The last term in the right-hand side of Equation (26.3) is the regularized term, and it can be viewed as performing a MAP estimation of Θ, where Θ is assigned a Gaussian prior with mean zero and covariance $\sigma^2\mathbf{I}$. The partial derivative of objective function $J(\Theta)$ is

$$\frac{\partial J(\Theta)}{\partial \lambda_k} = \sum_{i=1}^{N}\sum_{t=1}^{T} f_k(s_{i,t-1}, s_{i,t}, \mathbf{o}_i, t) \tag{26.4}$$

$$-\sum_{i=1}^{N}\sum_{t=1}^{T}\sum_{s,s'} f_k(s, s', \mathbf{o}_i, t)p(s, s'|\mathbf{o}_i) - \frac{\lambda_k}{\sigma^2}$$

Some gradient ascent methods, such as the iterative scaling and the limited memory Broyden–Fletcher–Goldfarb–Shanno (L-BFGS) algorithm, can be applied to find the optimal value of each λ_k. Here we choose the L-BFGS as our training algorithm because it runs faster than the iterative scaling and the conjugate gradient [9].

To incorporate CRF in our task, observation vectors $\mathbf{o}$ are attribute labels generated from the attribute detectors and state variables $\mathbf{s}$ are recognized Initial/Final labels. At time t, we determine the output Initial/Final label s_t according to its nearby observations at time t. How these nearby observations are interacted to determine the current output label is described by the feature functions and their corresponding weighting factors. The flexibility of including observations over a longer time span is one of advantages of a CRF over an HMM. The feature functions designed here are described as follows:

- The manner of articulation, the Final onset type, and the Final ending type at current frame, previous frame, and next frame.
- The manner of articulation, the place of articulation, the aspiration, and the voiced at current frame, previous frame, and next frame.
- Six attributes individually at current frame, previous frame, and next frame.

In other words, we apply a sliding window of three frames centered at each time step over observation $\mathbf{o}$. Transition from previous state s_{t-1} to current state s_t actually depends on observations o_{t-1}, o_t, o_{t+1}, not the entire observation. Apply a wider sliding window would include more observation vectors into consideration, though complexity will increase during training and decoding phases. For simplicity, feature functions used here are chosen to be binary-valued. The training and decoding processes of conditional random fields are with help of CRF++ package [3].

Table 26.2 Recognition results of six sets of attribute detectors

Attributes	Correction(%)	Accuracy(%)
manner	81.39	76.34
Final onset type	86.12	82.59
Final ending type	87.14	82.89
place	83.82	80.77
aspiration	88.61	84.72
voiced	86.71	83.93

26.3 Experiments

Six sets of attribute HMMs are trained on TCC300 Mandarin speech corpus. Each utterance in the corpus was recorded in a 16-bit PCM format with 16 kHz sampling rate in office environments. The collected utterances were divided into the training set and the evaluation set. There are 24,742 utterances in the training set and the accumulated time is about 24 hours. On the other hand, there are 2,595 utterances in the evaluation set and roughly 2.5 hours in total. Training attribute HMMs directly from the raw speech data is a time-consuming task, so a feasible way to obtain sets of attribute HMMs is via the existing RCD-HMMs. Following the tying policy mentioned in Section 26.2.1, the training time is greatly reduced.

Table 26.2 lists the recognition results of each set of attribute detectors. A further analysis of each performance in attribute detectors shows that there are some unsatisfactory results. In the manner of articulation, the most often mis-recognized sound is fricative (about 71% in correction rate). It is often confused with affricate or stop. In the place of articulation, three coronal relative attributes are often mis-classified one another. It makes some Initials hard to be classified. For the Final ending type, those ending with a nasal sound are easily confused with others. The indiscrimination of these sounds is due to pronunciation errors in Mandarin. Mis-classification of these front-end attributes does affect the performance of the back-end CRF in a certain degree because the feature functions are chosen to be binary-valued. Any missed detection of an attribute label that should occur will force the feature function to be zero. No matter how large its weighting factor to be, the influence of that feature function becomes negligible. Thus the accuracy of the back-end CRF will unavoidably degrade.

Table 26.3 gives a comparison between two conventional HMM recognition systems and the proposed cascaded system. Both correction rate and accuracy rate are obtained based on the grammar-free process. In the context-independent HMM case (CI-HMM), there are 22 context-independent Initial

Table 26.3 Results of Mandarin Initials/Finals recognition task by three different systems. No linguistic constraints are applied during the decoding phase

System	Correction (%)	Accuracy (%)	Num. Models
CI-HMM	69.61	54.91	61
RCD-HMM	71.84	44.12	138
Proposed	61.60	58.25	45

models and 39 context-independent Final models. In the RCD-HMMs case, however, Initial models are extended to 99 according to its right context. The number of Final models in the RCD-HMMs is held fixed. Note that the results given in the table are based on a free grammar decoding. No phonotactic constraints are made in the decoding process. The results will be better if some linguistic constraints are considered in the decoding phase. The most promising result comes from the accuracy rates given by the cascaded system, though the correction rate is the lowest of the three systems. The small difference between the correction rate and accuracy rate implies that the cascaded system does not over-estimate the phone candidates. The low accuracy rate in the RCD-HMM case is due to a higher insertion rate. Usually a system with a large number of models, while decoding without language models, suffers from such high insertion error problem.

Since CRF provides a flexible way to formulate the interaction between different feature functions, it is worth investigating several interaction formations and examining their influences on the overall performance. In various experiments it is found that the decoding results are sensitive to the selection and the formation of feature functions. A small change of the formation may result in a big different result. However, it seems that there is no standard procedure to follow and no guarantee of how performance will be. From some experimental trials, it is found that the more distinct features are included, the more reliable results one could obtain.

Another possibility for incorporating attribute labels is the use of other CRF structures. The linear-chain CRF adopted here is the simplest topology among the CRF family. There are other CRF structures, like the semi-Markov CRF [8], the hidden-CRF, and the Dynamic-CRF [10]. Although some of them have not been tested on speech recognition tasks, it is possible that they have potential for bringing benefit to speech processing. The most interesting one is that, from a graphical model point of view, the graphical topology of the hidden-CRF is almost identical to the HMM except that the hidden-CRF is an undirected model. The hidden-CRF reserves all the characteristics that an HMM has, such as being capable of including hidden variables and modeling in a segmental fashion. The tasks of phone classification [2] and gesture recognition [6] have confirmed that the hidden-CRF has such a great potential.

26.4 Conclusion

In this study, we have implemented a prototype of attribute-based speech recognition system. There still exist many open issues to be addressed under this framework. The front-end detectors should be further investigated to find out which kind of detector is more suitable for some specific speech attribute, and how many attributes should be included to identify a target phone. For the back-end system, the linear-chain CRF can be replaced by other types of CRF for a better modeling of speech characteristics. In conclusion, the attribute-based speech recognition system is a promising approach and has potential to overcome the bottleneck faced nowadays.

References

[1] A. Bilmes. What HMMs can do. *IEICE Trans. Information and Systems*, E89-D(3):869–891, March 2006.

[2] A. Gunawardana, M. Mahajan, A. Acero, and J.C. Platt. Hidden conditional random fields for phone classification. In *Proceeding of INTERSPEECH*, pages 1117–1120, 2005.

[3] T. Kudo. CRF++: Yet another CRF toolkit. `http://crfpp.sourceforge.net`.

[4] J. Lafferty, A. McCallum, and F. Pereira. Conditional random fields: Probabilistic models for segmenting and labeling sequence data. In *Proceedings of International Conference of Machine Learning*, pages 282–289, 2001.

[5] A. McCallum. Efficiently inducing features of conditional random fields. In *Proceedings of Uncertainty in Artificial Intelligence*, pages 403–410, 2003.

[6] L. Morency. Context-based visual feedback recognition. Technical Report, CSAIL Lab., MIT, 2006.

[7] J. Morris and E. Fosler-Lussier. Combining phonetic attributes using conditional random fields. In *Proceedings of INTERSPEECH 2006*, pages 597–600, 2006.

[8] S. Sarawagi and W.W. Cohen. Semi-markov conditional random fields for information extraction. In *Proceedings of Advanced in Neural Information Processing Systems*, 2004.

[9] F. Sha and F. Pereira. Shallow parsing with conditional random fields. In *Proceedings of Human Language Technology*, pages 134–141, 2003.

[10] C. Sutton, K. Rohanimanesh, and A. McCallum. Dynamic conditional random fields: Factorized probabilistic models for labeling and segmenting sequence data. In *Proceedings of International Conference on Machine Learning*, page 99, 2004.

[11] H. Yu and A. Waibel. Integrating thumbnail features for speech recognition using conditional exponential models. In *Proceedings of ICASSP 2004*, pages 893–896, 2004.

27

Image Point Set Registration: Modeling and Algorithms

Shaoyi Du and Nanning Zheng

Institute of Artificial Intelligence and Robotics, Xi'an Jiaotong University, Xi'an, Shaanxi Province 710049, China; E-mail: {sydu, nnzheng}@aiar.xjtu.edu.cn

Abstract

As one of the key technologies in image registration, point set registration has become a hot topic in computer vision, pattern recognition and image analysis. This chapter presents the modeling and algorithms of point set registration. First of all, the general model is formulated for point set registration, and then its theory research is discussed. With a thorough study of point set registration, this chapter discusses rigid and non-rigid point set registration. We introduce the problem formulation of rigid registration firstly and present a typical method, the iterative closest point (ICP) algorithm, the fundamental idea of which is summarized. Secondly, the development history of non-rigid point set registration is introduced from two aspects, namely, whether they are based on the ICP algorithm or not. Following that, modeling and algorithms of non-rigid registration of point sets are presented.

Keywords: image registration, point set registration, rigid registration, non-rigid registration, iterative closest point.

27.1 Introduction

With the development of digital image, image registration especially feature-based image registration has become a key topic in computer vision, pattern

Patrick Shen-Pei Wang (Ed.), Pattern Recognition and Machine Vision – In Honor and Memory of Professor King-Sun Fu, 417–433.

recognition and image analysis. When salient features of an image are represented as geometric entities, such as point, line and surface, the aim of image registration is to register the images with the optimal or suboptimal geometric transformation. Due to development of last several decades, geometric features have got a great success, e.g. corner points, intersection points, local maximum points, contour, lane. Of all features, point feature, the simplest pattern, is to locate the feature position. As point feature is a component of other complex features like line and surface, it is regarded as the basic feature of image. Therefore, point set registration is a fundamental problem in image registration.

In recent years, 3D measuring devices have been applied widely such as range scanner, stereoscopic plotter and medical equipment. With the development of 3D measuring devices, it is of great importance to register range images and build 3D geometric model for recognition.

In a word, whichever 2D images or 3D range images are, image point set registration has been a hot topic in computer vision, pattern recognition and image analysis due to its wide applications such as simultaneous localization and mapping (SLAM), 3D reconstruction, biometrics recognition, medical image analysis, remote sensing image analysis and digital inspection.

This chapter discusses the models and algorithms of image point set registration. The rest of this chapter is organized as follows. Firstly, the general model and theory of point set registration is introduced in Section 27.2. Following that is Section 27.3 in which a rigid registration problem of point sets is presented and a typical algorithm, the iterative closest point (ICP) algorithm is given. In Section 27.4, non-rigid point set registration is reviewed briefly, and then modeling and algorithms are presented. Finally, a summary is drawn.

27.2 Point Set Registration

27.2.1 Problem Formulation

For two images, the goal of image point set registration is to define a measure function and search a geometric transformation, with which two image point sets are in the best alignment. Figure 27.1 presents a simple example of point set registration. To solve this problem, a general statement is described first as follows.

Given two m dimensional (m-D) point sets in $\mathbb{R}^m$, a data point set $P \triangleq \{\vec{p}_i\}_{i=1}^{N_p}(N_p \in \mathbb{N})$ and a model point set $M \triangleq \{\vec{m}_i\}_{i=1}^{N_m}(N_m \in \mathbb{N})$. To register

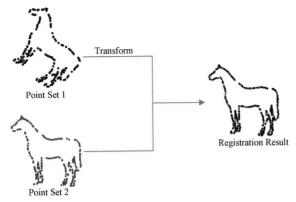

Figure 27.1 An example of point set registration

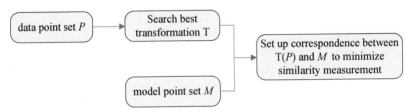

Figure 27.2 The framework of point set registration

between two point sets is to find a transformation T, with which P and M are matched and the similarity measurement is minimized. The framework of point set registration can be seen in Figure 27.2.

To formulize this problem, the mathematic model is built as follows. Assume the correspondence between two point sets P and M is a map $C : P \rightarrow M$. Hence, the formulation is based on the following criterion:

$$\min_{T,C} J(T(P), C(P)) \qquad (27.1)$$

where $J(\cdot, \cdot)$ is the similarity measurement between two point sets.

From this function, we see that it is of great difficulty in three aspects: (1) point correspondence is unknown, (2) transformation is unknown, and (3) there is not enough information about point sets.

27.2.2 Theory Research

Point set registration is a fundamental problem and its theory research is focused on the similarity measurement J and geometric transformation T. This section will discuss these two aspects.

27.2.2.1 Similarity Measurement J

The purpose of point set registration is to set up the best correspondence between two point sets, which is an optimization process. It focuses on the formulation of objective function which will effect final results, that is the correspondence of point sets C. Hence, it is important to choose similarity measurement J in Equation (27.1) which is a distance function. Currently, the common distance measurements are Euclidean distance, Mahalanobis distance, Hausdorff distance, probabilistic distance, etc. As Euclidean distance is simplest, we will introduce it as follows.

In Euclidean space, many metrics can be used for similarity measurement, such as point-to-point, point-to-line and point-to-face distance. Point-to-point distance is used more frequently and is introduced as follows: point-to-point distance represents the real distance of $\vec{x} = (x_1, x_2, \ldots, x_m)^T$ and $\vec{y} = (y_1, y_2, \ldots, y_m)$ in $\mathbb{R}^m$. It is expressed as

$$d_E = \|\vec{x} - \vec{y}\|_2 = \sqrt{\sum_{i=1}^{m}(x_i - y_i)^2},$$

where $\|\cdot\|_2$ denotes a 2-norm.

This distance denotes the dissimilarity between two points. As it is simple to be computed, we always use it for similarity measurement. With this distance, the objective function (27.1) is usually expressed as least square (LS) criterion [2, 31]. For two point sets $P \triangleq \{\vec{p}_i\}_{i=1}^{N_p}$ and $M \triangleq \{\vec{m}_i\}_{i=1}^{N_m}$, the criterion function of their registration is

$$J(T(P), C(P)) = \sum_{i=1}^{N_p} \|T(\vec{p}_i) - \vec{m}_{j(i)}\|_2^2 \tag{27.2}$$

where $j(i)$ represents that the ith point of P corresponds to the jth point of M.

27.2.2.2 Geometric Transformation T

In problem (27.1), the choice of geometric transformation T is important, which is determined by the deformation between two point sets. According to the deformation, geometric transformations are always classified into rigid and non-rigid transformation. Non-rigid transformation includes similar transformation, scaling transformation, affine transformation, projective transformation and elastic transformation. In the following, these transformations will be introduced.

1. Rigid transformation

Rigid transformation, the simple and common transformation, consists of rotation and translation transformations. For two points $\vec{x} \in \mathbb{R}^m$ and $\vec{y} \in \mathbb{R}^m$, the rigid transformation is expressed as $\vec{y} = \mathbf{R}\vec{x} + \vec{t}$.

$\mathbf{R}$ is a rotation matrix which has many forms, such as Euler angular, unit quaternion, orthonormal matrice and dual quaternions. Orthonormal matrice is a well-known representation method, which is expressed as $\mathbf{R}^T \mathbf{R} = \mathbf{I}_m$, $\det(\mathbf{R}) = 1$, where $\mathbf{I}_m$ is an m-dimensional identity matrix.

The dimension of rotation matrix is $m(m-1)/2$ and the dimension of translation matrix is m. In the research of image point set registration, rigid registration is often used. A typical method named the iterative closest point (ICP) algorithm [5, 7, 35] is known for solving this problem for its good accuracy and fast speed. It has been widely used in a variety of fields such as medical images, document images, fingerprint images and face images, etc.

2. Non-rigid transformation

(a) Similar transformation

It is also called isotropic scale transformation, which is composed of scale scalar and rigid transformation. For any points $\vec{x} \in \mathbb{R}^m$ and $\vec{y} \in \mathbb{R}^m$, it can be represented as $\vec{y} = s\mathbf{R}\vec{x} + \vec{t}$, where s is a scalar.

From this equation, we find that similar transformation has just one more dimension than rigid transformation. Since the parameter of similar transformation is relatively less, it is used quite often in similar registration [9, 34, 36].

(b) Scaling transformation

It is also called anisotropic scale transformation, which is an extension of similar transformation. It is composed of scale matrix and rigid transformation: $\vec{y} = \mathbf{R}\mathbf{S}\vec{x} + \vec{t}$, where $\mathbf{S}$ is a diagonal matrix.

In this equation, scaling transformation has m dimensions more than rigid transformation. Because it is relatively simple, it is always used in scaling registration [10].

(c) Affine transformation

It is relatively common used in non-rigid transformations. It includes isotropy scale transformation, anisotropy scale transformation, shear transformation, and so on. Compared with transformation mentioned above, it has more degrees of freedom, which is expressed as $\vec{y} = \mathbf{A}\vec{x} + \vec{t}$, where $\mathbf{A}$ is a linear matrix and its dimension is m^2. As $\mathbf{A}$ is non-singular, its determinant cannot be 0, the constraint is $\det(\mathbf{A}) \neq 0$.

Many researchers have studied affine registration of point sets, such as Feldmar and Ayache [12], Ho [15], and Du [11].

(d) Projective transformation

It is more general than affine transformation. It transforms a straight line into a straight line, but there is no need to preserve the parallel of lines. According to the distance of projection center and projection plane, projection is divided into orthogonal projection and perspective projection. In orthogonal projection, the distance of projection center and projection plane is infinite while it is finite in perspective projection. Projective transformation sets up the correspondence of $m + 1$ dimensional points and m dimensional points. Projecting 3D scenes into 2D plane is always applied in projective registration, which is a relatively independent research field.

(e) Elastic transformation

It is named non-linear transformation, too. Unlike the above non-rigid transformation, it transforms a straight line to a curve, which reflects the distortion of image points. The simplest formulation of elastic transformation is algebraic polynomial function. According to Weierstrass theorem, we can see that the continuous function which is at closed interval can be approximated by the polynomial function. Hence, the polynomial function is often used for the local deformation. For example, the thin plate spline (TPS) [32] is always applied to elastic registration.

Recently, elastic registration is a hot issue especially in medical image registration. Since elastic transformation has more degrees of freedom, it is one of the most complicated problems of point set registration [1, 8, 21–23, 25].

27.3 Rigid Point Set Registration

27.3.1 Rigid Registration Problem

The goal of rigid registration is to find a rigid transformation, with which data point set $P \triangleq \{\vec{p}_i\}_{i=1}^{N_p} (N_p \in \mathbb{N})$ is registered to be in the best alignment with model point set $M \triangleq \{\vec{m}_i\}_{i=1}^{N_m} (N_m \in \mathbb{N})$ in Euclidean space, that is, let T of Equation (27.1) be rotation and translation transformations and J be a LS criterion respectively. Hence, the rigid registration between two point sets is

$$\min_{\mathbf{R},\vec{t}, j \in \{1,2,\dots,N_m\}} \left(\sum_{i=1}^{N_p} \left\| (\mathbf{R}\vec{p}_i + \vec{t}) - \vec{m}_j \right\|_2^2 \right) \tag{27.3}$$

$$s.t. \qquad \mathbf{R}^{\mathsf{T}}\mathbf{R} = \mathbf{I}_m, \ \det(\mathbf{R}) = 1$$

where $\mathbf{R} \in \mathbb{R}^{m \times m}$ is a rotation matrix, $\vec{t} \in \mathbb{R}^m$ is a translation vector.

27.3.2 The ICP Algorithm

The ICP algorithm proposed by Besl and McKay [5, 7, 35] is an efficient method to tackle rigid registration between two point sets. The ICP algorithm achieves registration with good accuracy and fast speed, and it has two steps.

Firstly, set up correspondence $\{\vec{p}_i, \vec{m}_{c_k(i)}\}$ between two point sets according to the $(k-1)$th rigid transformation $(\mathbf{R}_{k-1}, \vec{t}_{k-1})$:

$$c_k(i) = \underset{j \in \{1,2,\dots,N_m\}}{\arg\min} (\left\| (\mathbf{R}_{k-1}\vec{p}_i + \vec{t}_{k-1}) - \vec{m}_j \right\|_2^2), \quad \text{for } i = 1, 2, \dots, N_p \tag{27.4}$$

Secondly, compute the new transformation between two point sets $\{\mathbf{R}_{k-1}\vec{p}_i + \vec{t}_{k-1}\}_{i=1}^{N_p}$ and $\{\vec{m}_{c_k(i)}\}_{i=1}^{N_p}$ by minimizing the squared distance:

$$(\mathbf{R}^*, \vec{t}^*) = \underset{\mathbf{R}^{\mathsf{T}}\mathbf{R}=\mathbf{I}_m, \det(\mathbf{R})=1, \vec{t}}{\arg\min} \left(\sum_{i=1}^{N_p} \left\| \mathbf{R}(\mathbf{R}_{k-1}\vec{p}_i + \vec{t}_{k-1}) + \vec{t} - \vec{m}_{c_k(i)} \right\|_2^2 \right) \tag{27.5}$$

Update the kth transformation $\mathbf{R}_k$ and $\vec{t}_k$ as follows:

$$\mathbf{R}_k = \mathbf{R}^*\mathbf{R}_{k-1}, \quad \vec{t}_k = \mathbf{R}^*\vec{t}_{k-1} + \vec{t}^* \tag{27.6}$$

The simplest way for the problem (27.4) is to search all points of model point set for any point of data point set. There are many methods for improving the search speed, such as k-d tree [4, 24], the nearest point search based on Delaunay tessellation [3, 6]. Therefore, the focus of the ICP algorithm is to solve the problem (27.5). Many closed-form methods are known to be used to compute the rigid transformation: singular value decomposition (SVD) [2, 31], unit quaternions [16], orthonormal matrices [17] and dual quaternions [33]. An overview of these methods and a comparative analysis is given in [20]. In the following, the method based on SVD is introduced briefly.

Theorem 1 *Given two m-D point sets $\{\vec{q}_i\}_{i=1}^N$ and $\{\vec{n}_i\}_{i=1}^N$, then the function*

$$F(\vec{t}) = \sum_{i=1}^N \left\| \vec{q}_i + \vec{t} - \vec{n}_i \right\|_2^2$$

has the minimum when

$$\vec{t} = \frac{1}{N} \sum_{i=1}^N \vec{n}_i - \frac{1}{N} \sum_{i=1}^N \vec{q}_i.$$

Suppose

$$\vec{p}_i' \triangleq \mathbf{R}_{k-1} \vec{p}_i + \vec{t}_{k-1}$$

according to Theorem 1, we get

$$\vec{t} = \frac{1}{N_p} \sum_{i=1}^{N_p} \vec{m}_{c_k(i)} - \frac{1}{N_p} \sum_{i=1}^{N_p} \mathbf{R} \vec{p}_i'$$

Let

$$\vec{q}_i \triangleq \vec{p}_i' - \frac{1}{N_p} \sum_{i=1}^{N_p} \vec{p}_i' \quad \text{and} \quad \vec{n}_i \triangleq \vec{m}_{c_k(i)} - \frac{1}{N_p} \sum_{i=1}^{N_p} \vec{m}_{c_k(i)}$$

Therefore,

$$F(\mathbf{R}, \vec{t}) = \sum_{i=1}^{N_p} \vec{q}_i^T \vec{q}_i - 2 \sum_{i=1}^{N_p} \vec{n}_i^T \mathbf{R} \vec{q}_i + \sum_{i=1}^{N_p} \vec{n}_i^T \vec{n}_i \qquad (27.7)$$

To minimize the objective function 27.7 is to minimize $- \sum_{i=1}^{N_p} \vec{n}_i^T \mathbf{R} \vec{q}_i$, which is solved by a method based on SVD [2, 31].

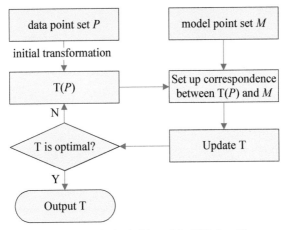

Figure 27.3 The basic idea of the ICP algorithm

Theorem 2 *Suppose* $\vec{n}_i \in \mathbb{R}^m$, $\vec{q}_i \in \mathbb{R}^m$ *and* $\mathbf{R}$ *is a rotation matrix. Let* $\sum_{i=1}^{N_p} \vec{q}_i \vec{n}_i^T = \mathbf{U}\Lambda\mathbf{V}^T$. *Minimizing the following objective function:*

$$\min_{\mathbf{R}}\left(-\sum_{i=1}^{N_p} \vec{n}_i^T \mathbf{R}\vec{q}_i\right)$$

$$s.t. \ \mathbf{R}^T\mathbf{R} = \mathbf{I}_m, \ \det(\mathbf{R}) = 1$$

we get

$$\mathbf{R}^* = \mathbf{V}\mathbf{D}\mathbf{U}^T.$$

where

$$\mathbf{D} = \begin{cases} \mathbf{I}_m & \det(\mathbf{U})\det(\mathbf{V}) = +1 \\ \text{diag}(1, 1, \dots 1, -1) & \det(\mathbf{U})\det(\mathbf{V}) = -1 \end{cases}$$

According to Theorems 1 and 2, we get the solution of the problem (27.5). Hence, the ICP algorithm is completed.

Reviewing the ICP algorithm, we find it is completed by iterating two main steps which are to solve the correspondence of point sets and rigid transformation respectively. We summarize the basic idea of the ICP algorithm in Figure 27.3.

To understand the algorithm much thoroughly, the demo of the algorithm process is displayed in Figure 27.4.

From Figures 27.3 and 27.4 the ICP algorithm is found to reduce the mean square error (MSE) effectively. The ICP algorithm is a local convergent algorithm, which is stated in the following theorem [5].

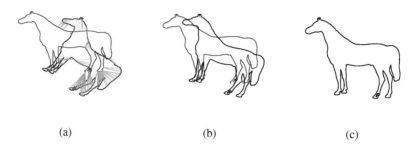

(a)   (b)   (c)

Figure 27.4 The procedure of the ICP algorithm. (a) Build up correspondence between two point sets. (b) Compute the new rigid transformation by minimizing squared distance, and then apply this new transformation. (c) Registration result of the ICP algorithm

Theorem 3 *The ICP algorithm converges monotonically to a local minimum with respect to MSE criterion from any given initial parameters.*

According to Theorem 3, to obtain the desired global minimum, the usual way is to find the minimum of all local minima. However, it is difficult to characterize precisely and generally that the registration state space is partitioned into local minima regions, for the encountered shapes are quite different. Therefore, it is hard to estimate initial parameters well. Besl and McKay [5] and Lee et al. [19] have detailed the initial rotation matrix.

In the past decade, a number of scholars have studied the traditional ICP algorithm on how to speed it up. Fitzgibbon [13] employed the Levenberg–Marquardt algorithm to accelerate ICP, and Jost et al. [18] combined a coarse to fine multi-resolution technique with the neighbor search algorithm into ICP to improve the registration. Moreover, many scholars have introduced other methods into ICP for its more robustness. Lee et al. [19] proposed a measure for estimating the reliability of ICP. Invariant features were described by Sharp et al. [28] to decrease the probability of being trapped in a local minimum. Granger et al. [14] added Expectation-Maximization principles to ICP and used a coarse-to-fine approach based on an annealing scheme to improve the robustness. Silva et al. [29] introduced genetic algorithms and evaluation metric into ICP for more precise registration. As for the overview of data sets registration using ICP method, the reader is referred to [26,27,30].

27.4 Non-Rigid Point Set Registration

27.4.1 Brief Review

The ICP algorithm is known to solve rigid registration problem for its good accuracy and fast speed. Meanwhile, non-rigid registration algorithms have been developed rapidly. Currently, some algorithms extend the ICP algorithm, the others do not. We will introduce non-rigid point set registration from these two aspects.

At the end of the last century, Feldmar and Ayache [12] studied the non-rigid registration problem. Firstly, they completed rigid registration by introducing the normal and the principal curvatures to the ICP algorithm. Then, they searched the best global affine transformation for affine registration. Finally, assuming the transformation of each point is affine transformation, they finished non-rigid registration by smoothing the local affine transformations. Zha et al. [34] used extended signature images to estimate the global scale and applied it to traditional ICP for registration, while Zinßer et al. [36] directly estimated the scale in the ICP algorithm for similar registration. Amberg et al. [1] employed an additional stiffness term and a landmark term for locally affine regularization, which was based on the ICP algorithm. Moreover, Du et al. [9, 10] proposed an extension of the ICP Algorithm which integrated the scale factor with boundaries into the original ICP algorithm for non-rigid registration.

In addition, some scholars managed to solve the non-rigid registration problems without using the ICP algorithm. Paragios et al. [25] built an isotropic scale registration model, and then changed the translation part to finish the non-rigid registration. Ho et al. [15] tried to reduce the general affine registration problem to that of the orthogonal case with covariance matrices, and then they only needed to solve the rigid registration problem. Moreover, some scholars have tried to use probabilistic point set registration methods which soft-assigned correspondence for registration. The typical work was thin plate spline – robust point match (TPS-RPM) proposed by Chui and Rangarjan [8]. The algorithm worked iteratively in a similar way as the ICP algorithm, and it defined the transformation in two parts: affine transformation and the kernel function of TPS. With the soft assigning of weights, it completed the non-rigid registration robustly. Furthermore, some similar works like probabilistic point matching (PPM) [21] and coherent point drift (CPD) [22, 23] were proposed respectively.

27.4.2 Modeling and Algorithms

In this section, to illustrate the non-rigid registration problems and algorithms, we will give one simple example to solve the similar registration, which can be easily extended for other non-rigid registration problems.

The original ICP algorithm does not take scale factor into account in the LS problem, while the scale factor always exists in point set registration. The similar registration problem is a much complicated problem. A matching between two point sets without noise is minimizing (27.1) which yields an optimal result of 0, but the scale transformation is not unique. One case is that two point sets are best registered with certain scale factor while the other case is that the scale factor is close to 0, which means that points of a set converge to a point of the other set. To solve this ill-posed problem, we consider adding the constraint condition that the scale factor is bounded. Hence, we always need to consider the following LS problem:

$$\min_{\substack{s,\mathbf{R},\vec{t} \\ j \in \{1,2,\dots,N_m\}}} \left(\sum_{i=1}^{N_p} \left\| (s\mathbf{R}\vec{p}_i + \vec{t}) - \vec{m}_j \right\|_2^2 \right) \tag{27.8}$$

$$s.t. \quad \mathbf{R}^\mathsf{T}\mathbf{R} = \mathbf{I}_m, \ \det(\mathbf{R}) = 1, \ s \in [a,b]$$

Actually, we can solve this problem in the similar way to the ICP algorithm by iteration. In each iterative step, two steps are mainly included:

Step 1, build up the set of correspondences by the current similarity transformation $(s_{k-1}, \mathbf{R}_{k-1}, \vec{t}_{k-1})$:

$$c_k(i) = \underset{j \in \{1,2,\dots,N_m\}}{\arg\min} \left(\left\| (s_{k-1}\mathbf{R}_{k-1}\vec{p}_i + \vec{t}_{k-1}) - \vec{m}_j \right\|_2^2 \right) \tag{27.9}$$

Step 2, compute the new similarity transformation $(s^*, \mathbf{R}^*, \vec{t}^*)$:

$$(s^*, \mathbf{R}^*, \vec{t}^*) = \underset{\substack{\mathbf{R}^\mathsf{T}\mathbf{R}=\mathbf{I}_m, \det(\mathbf{R})=1, \\ s \in [a/s_{k-1}, b/s_{k-1}], \vec{t}}}{\arg\min} \left(\sum_{i=1}^{N_p} \left\| s\mathbf{R}(s_{k-1}\mathbf{R}_{k-1}\vec{p}_i + \vec{t}_{k-1}) + \vec{t} - \vec{m}_{c_k(i)} \right\|_2^2 \right)$$

$$\tag{27.10}$$

Update s_k, $\mathbf{R}_k$ and $\vec{t}_k$:

$$s_k = s^* s_{k-1}, \ \mathbf{R}_k = \mathbf{R}^* \mathbf{R}_{k-1}, \ \vec{t}_k = s^* \mathbf{R}^* \vec{t}_{k-1} + \vec{t}^* \tag{27.11}$$

To compute the scale, rotation and translation transformation in step 2, we simplify the problem (27.10) as the ICP algorithm and obtain:

$$F(s, \mathbf{R}) = s^2 \sum_{i=1}^{N_p} \vec{q}_i^T \vec{q}_i - 2s \sum_{i=1}^{N_p} \vec{n}_i^T \mathbf{R} \vec{q}_i + \sum_{i=1}^{N_p} \vec{n}_i^T \vec{n}_i \qquad (27.12)$$

1. For any given s, minimizing $F(s, \mathbf{R})$ is equivalent to maximizing $\sum_{i=1}^{N_p} \vec{n}_i^T \mathbf{R} \vec{q}_i$, which has been solved by Theorem 2.

2. For any given $\mathbf{R}$, the function $F(s, \mathbf{R})$ is known to be a parabola with respect to s. If $s \in [a/s_{k-1}, b/s_{k-1}]$, the minimum can be achieved at the point which is nearest to the vertex of this parabola, we get the scale s^* of the similar ICP algorithm

$$s^* = \underset{s \in [a/s_{k-1}, b/s_{k-1}]}{\arg\min} \left| s - \sum_{i=1}^{N_p} \vec{n}_i^T \mathbf{R} \vec{q}_i \Big/ \sum_{i=1}^{N_p} \vec{q}_i^T \vec{q}_i \right|.$$

The similar ICP algorithm shares the same procedure as the ICP algorithm. Although the points in the data point set cannot find the correct corresponding points in the model point set within just one step, they move to the correct points much closer by iteration. Similar to the ICP algorithm, a convergence theorem for the similar ICP algorithm above is stated as follows.

Theorem 4 *The similar ICP algorithm converges monotonically to a local minimum with respect to MSE criterion from any given initial parameters.*

Theorem 4 can be easily proved in the way that the ICP algorithm does. As this algorithm converges locally, it is important to estimate initial parameters well and give the constraints. In the following, we will discuss the initial scale factor with corresponding constraints.

Variability is able to be described by covariance matrix from which eigenvectors and eigenvalues are calculated. The eigenvector corresponding to the largest eigenvalue denotes the direction of largest variance of the data set and the eigenvectors are orthonormal. Assume that points are sampled enough from two shapes and their covariance matrices are $\mathbf{C}_P$ and $\mathbf{C}_M$. The square roots of $\mathbf{C}_P$'s and $\mathbf{C}_M$'s eigenvalues are $\lambda_1, \ldots, \lambda_m$ and $\mu_1, \ldots, \mu_m$ respectively. If the data point set P registers the model point set M, the eigenvalues of data point set will scale to the corresponding eigenvalues of model point set. Hence, the initial scale matix s_0 and its constraint are estimated as

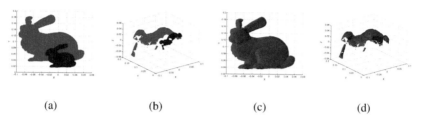

| (a) | (b) | (c) | (d) |

Figure 27.5 Registration result of the similar ICP algorithm. (a) Point sets with 2D view point. (b) Point sets with 3D view point. (c) Registration result with 2D view point. (d) Registration result with 3D view point

follows:

$$s_0 = \frac{1}{m} \sum_{i=1}^{m} \frac{\mu_i}{\lambda_i} \tag{27.13}$$

$$\frac{1}{m} \sum_{i=1}^{m} \frac{\lambda_i}{\mu_i} - \delta \leqslant s \leqslant \frac{1}{m} \sum_{i=1}^{m} \frac{\lambda_i}{\mu_i} + \delta \tag{27.14}$$

where δ is a given value which means the tolerance of the deformation. For example, we may set it $0.1s_0$.

To understand the algorithm much comprehensively, experimental result of the algorithm is presented in Figure 27.5.

Although the similar ICP algorithm discusses isotropic scale, it can be easily extended to anisotropic scale case, which can solve scaling registration problem [10]. Furthermore, the similar ICP algorithm can be extended to the affine ICP algorithm [11] for affine registration of point sets.

27.5 Summary

We present an overview of registration techniques for image point sets in this chapter. Firstly, we build the fundamental formulation for point set registration, and then its theory research is studied. Following that, rigid registration problem is modeling, and as an important algorithm of rigid registration, the ICP algorithm is introduced. We also review the development of non-rigid registration of point sets. In particular, the model of similar registration is discussed and the corresponding algorithm, the similar ICP algorithm is introduced.

Though point set registration has gained great success heretofore, there is still plenty of room for further numerous research issues. Future work will center on theory analysis of these point set registration methods, such as stability and speed. Apart from that, the existing methods also need to be done in conjunction with other algorithms for registration in practice.

References

[1] B. Amberg, S. Romdhani, and T. Vetter. Optimal step nonrigid ICP algorithms for surface registration. In *Proceedings of IEEE Conference on Computer Vision and Pattern Recognition (CVPR)*, pages 1–8, 2007.

[2] K.S. Arun, T.S. Huang, and S.D. Blostein. Least-squares fitting of two 3-D point sets. *IEEE Transactions on Pattern Analysis and Machine Intelligence*, 9(5):698–700, 1987.

[3] C.B. Barber, D.P. Dobkin, and H. Huhdanpaa. The quickhull algorithm for convex hulls. *ACM Transactions on Mathematical Software*, 22(4):469–483, 1996.

[4] J.L. Bentley. K-D trees for semidynamic point sets. In *Proceedings of Sixth ACM Annual Symposium Computational Geometry*, pages 187–197, 1990.

[5] P.J. Besl and H.D. McKay. A method for registration of 3-D shapes. *IEEE Transactions on Pattern Analysis and Machine Intelligence*, 14(2):239–256, 1992.

[6] H.M. Chen and T.H. Lin. An algorithm to build convex hulls for 3-D objects. *Journal of the Chinese Institute of Engineers*, 29(6):945–952, 2006.

[7] Y. Chen and M. Gerard. Object modelling by registration of multiple range images. *Image and Vision Computing*, 10(3):145–155, 1992.

[8] H. Chui and A. Rangarajan. A new point matching algorithm for non-rigid registration. *Computer Vision and Image Understanding*, 89(2-3):114–141, 2003.

[9] S. Du, N. Zheng, S. Ying, and J. Wei. ICP with bounded scale for registration of M-D point sets. In *Proceedings of IEEE International Conference on Multimedia and Expo (ICME)*, pages 1291–1294, 2007.

[10] S. Du, N. Zheng, S. Ying, Q. You, and Y. Wu. An extension of the ICP algorithm considering scale factor. In *Proceedings of 14th IEEE International Conference on Image Processing (ICIP)*, Vol. 5, pages 193–196, 2007.

[11] S. Du, N. Zheng, G. Meng, and Z. Yuan. Affine registration of point sets using ICP and ICA. *IEEE Signal Processing Letters*, 15:689–692, 2008.

[12] J. Feldmar and N. Ayache. Rigid, affine and locally affine registration of free-form surfaces. *International Journal of Computer Vision*, 18(2):99–119, 1996.

[13] A.W. Fitzgibbon. Robust registration of 2D and 3D point sets. *Image and Vision Computing*, 21(13–14):1145–1153, 2003.

[14] S. Granger and X. Pennec. Multi-scale EM-ICP: A fast and robust approach for surface registration. In *Proceedings of European Conference on Computer Vision (ECCV)*, Vol. 1253, pages 418–432, 2002.

[15] J. Ho, Y. Ming-Hsuan, A. Rangarajan, and B. Vemuri. A new affine registration algorithm for matching 2D point sets. In *Proceedings of IEEE Workshop on Applications of Computer Vision (WACV)*, pages 25–28, 2007.

[16] B.K.P. Horn. Closed-form solution of absolute orientation using unit quaternions. *Journal of the Optical Society of America, Series A*, 4(4):629–642, 1987.

[17] B.K.P. Horn, H.M. Hilden, and S. Negahdaripour. Closed-form solution of absolute orientation using orthonormal quaternions. *Journal of the Optical Society of America, Series A*, 5(7):1127–1135, 1988.

[18] T. Jost and H. Hugli. A multi-resolution ICP with heuristic closest point search for fast and robust 3D registration of range images. In *Proceedings of Fourth International Conference on 3-D Digital Imaging and Modeling (3DIM)*, pages 427–433, 2003.

[19] B.U. Lee, C.M. Kim, and R.H. Park. An orientation reliability matrix for the iterative closest point algorithm. *IEEE Transactions on Pattern Analysis and Machine Intelligence*, 22(10):1205–1208, 2000.

[20] A. Lorusso, D.W. Eggert, and R.B. Fisher. A comparison of four algorithms for estimating 3d rigid transformations. In *Proceedings of British Machine Vision Conference*, pages 237–246, 1995.

[21] G. McNeill and S. Vijayakumar. Part-based probabilistic point matching. In *Proceedings of 18th International. Conference on Pattern Recognition (ICPR)*, pages 382–386, 2006.

[22] A. Myronenko and X. Song. Point-set registration: Coherent point drift. Technical Report, 2009.

[23] A. Myronenko, X. Song, and M. Carreira-Perpinan. Non-rigid point set registration: Coherent point drift. In *Proceedings of Advances in Neural Information Processing Systems (NIPS)*, pages 1009–1016, 2006.

[24] A. Nuchter, K. Lingemann, and J. Hertzberg. Cached K-D tree search for ICP algorithms. In *Proceedings of Sixth International Conference on 3-D Digital Imaging and Modeling(3DIM)*, pages 419–426, 2007.

[25] N. Paragios, M. Rousson, and V. Ramesh. Non-rigid registration using distance functions. *Computer Vision and Image Understanding*, 89(2–3):142–165, 2003.

[26] B.M. Planitz, A.J. Maeder, and J.A. Williams. The correspondence framework for 3D surface matching algorithms. *Computer Vision and Image Understanding*, 97(3):347–383, 2005.

[27] S. Rusinkiewicz and M. Levoy. Multi-scale EM-ICP: A fast and robust approach for surface registration. In *Proceedings of 3rd International Conference on 3-D Digital Imaging and Modeling*, pages 145–152, 2001.

[28] G.C. Sharp, S.W. Lee, and D.K. Wehe. ICP registration using invariant features. *IEEE Transactions on Pattern Analysis and Machine Intelligence*, 24(1):90–102, 2002.

[29] L. Silva, O.R. Bellon, and K.L. Boyer. Precision range image registration using a robust surface interpenetration measure and enhanced genetic algorithms. *IEEE Transactions on Pattern Analysis and Machine Intelligence*, 27(5):762–776, 2005.

[30] C.V. Stewart, E.R. Smith, B.J. King, and R.J. Radke. Registration of combined range-intensity scans: Initialization through verification. *Computer Vision and Image Understanding*, 110(2):226–244, 2008.

[31] S. Umeyama. Least-squares estimation of transformation parameters between two point patterns. *IEEE Transactions on Pattern Analysis and Machine Intelligence*, 13(4):376–380, 1991.

[32] G. Wahba. *Spline Models for Observational Data*, SIAM, 1990.

[33] M.W. Walker, L. Shao, and R.A. Volz. Estimating 3-D location parameters using dual number quaternions. *CVGIP: Image Understanding*, 54(3):358–367, 1991.

[34] H. Zha, M. Ikuta, and T. Hasegawa. Registration of range images with different scanning resolutions. In *Proceedings of IEEE International Conference on Systems, Man, and Cybernetics*, Vol. 2, pages 1495–1500, 2000.

[35] Z. Zhang. Iterative point matching for registration of free-form curves and surfaces. *International Journal of Computer Vision*, 13(2):119–152, 1994.

[36] T. Zinßer, J. Schmidt, and H. Niemann. Point set registration with integrated scale estimation. In *Proceedings of International Conference on Pattern Recognition and Information Processing*, pages 116–119, 2005.

28

Image Semantic Feature Analysis*

Ping Guo[1,2] and Xinyu Chen[1]

[1]*Image Processing and Pattern Recognition Lab, Beijing Normal University, Beijing 100875, China*
[2]*School of Computer, Beijing Institute of Technology, 5 South Zhongguancun Street, Beijing 100081, China; E-mail: pguo@ieee.org*

Abstract

Considerable research has been devoted to the problem of image semantic representation. However, the semantic gap between low-level features and high-level human interpretation has not been filled yet. This chapter reviews currently employed approaches in image feature representation, especially the Bag of Visual-words method and probability latent models. Some learning methods to map image representation with its corresponding semantic contents are also discussed.

Keywords: semantic feature, Bag of Visual-words, probability latent model.

28.1 Introduction

The proliferation of digital images demands effective management of such tremendous images. To accomplish accurate object recognition [1, 39], automatic image annotation (or image categorization) [11], and precise image retrieval [38], we need to identify image contents (e.g., sky, sea, sand, etc.) to understand the meaning of an image. Using a set of text keywords is

*The research described in this chapter was fully supported by a grant from the National Natural Science Foundation of China (Project No. 90820010).

Patrick Shen-Pei Wang (Ed.), Pattern Recognition and Machine Vision – In Honor and Memory of Professor King-Sun Fu, 435–447.

a straightforward approach to represent image semantics [7, 24]. With text labels, users could query their interested visual objects through natural language description. Thus, an image retrieval problem becomes a text query processing procedure. However, manually annotating images is expensive, labor-intensive, time-consuming, and error-prone, especially with the large and constantly-growing images. Moreover, the text keywords themselves imply a semantic hierarchy, i.e., keywords indicate semantic granularity. For example, the keyword "scene" represents coarse-granularity, while "lake", "lawn", "mountain", etc., are finer granularity semantics. So if a user tries to retrieve scene images, which results will be satisfied by the user? This problem is still a challenge in natural language understanding because it is hard to build a universal mathematical model for semantic hierarchy. Therefore, here we concentrate on extracting image semantic from image itself, not on the basis of its annotated texts.

Extracting semantic information from images is defined as mapping low-level visual features to high-level semantic concept classes. Formally, if we have an image database $\mathbb{I} = \{I_1, I_2, \ldots, I_N\}$ and a set of image semantic concept classes $\mathbb{C} = \{C_1, C_2, \ldots, C_M\}$, each image semantic class will be defined as a function summarizing the low-level features over all or some images in the database as $C_m = \Phi_m(\mathbb{I})$. Here the challenge is to find proper Φ_m that is invariant enough to generalize across naturally-occurring intra-class variations and yet discriminative enough to distinguish between different classes [43]. However, the available low-level visual features are based on pixel data, such as color [12], shape [4], and texture [13], etc. Due to some conflicts in translating between these low-level visual features and high-level semantic interpretations, there is a mismatch which is often called semantic gap. And even a simple semantic representation of color or shape, such as red or square, requires quite different description methods. In addition, different illumination, scales, and viewpoints may produce different images for objects of the same class. Even worse, intrinsic visual difference also creates dissimilar representations for an object class. Furthermore, some objects are deformable, for example, horses may be in sitting or standing position, and partial occlusions may also introduce difficulties in recognizing objects [43]. Therefore, image semantic feature representation, extraction, and analysis are big challenges in computer vision [42].

The image semantic mapping Φ_m could be used in many image related applications. Given an image I_i, image annotation is a process to obtain a set of image-class pairs $\{(I_i, C_{l_1}), (I_i, C_{l_2}), \ldots, (I_i, C_{l_j})\}$, or we could create a binary vector with length M, in which each bit denotes whether this image

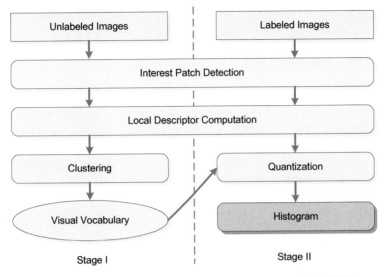

Figure 28.1 Typical procedure of representing images with the BOV

belongs to a corresponding semantic class or not when the value is 1 or 0. Thus, the task of object recognition is to find a given object in an image using its description.

The content of this chapter is organized as calculating feature vectors to represent an image and learning image semantic mappings. First we will present a counting approach by computing local features to represent an image, called Bag of Visual-words.

28.2 The Bag of Visual-Words Approach

Recently, the Bag of Visual-words (BOV)[1] approach is being used extensively to work as the feature representation of images [8, 9, 14, 16, 20, 22, 25, 26, 35, 41, 44], which is inspired by Bag of Words (BOW) used in text categorization. In the BOW representation, a text document is denoted as a histogram with the number of occurrences of each word. In a similar way, the BOV employs visual-words instead of text words. However, there is no given visual vocabulary (the set of visual-words) for the BOV as compared to the BOW, which should be learned automatically from training images. Its typical procedure is

---

[1] Many similar terms have been used in the literature, such as keypoints [9], visterms [35], keyblocks [44], codewords [26], regions [16], etc.

shown as Figure 28.1, which is composed by two stages: The first stage is to generate a visual vocabulary using unlabeled training images, and the second stage is to construct a histogram for each image or image segment.

28.2.1 Generating Visual Vocabulary

As image semantics is very abundant, it is difficult to develop semantic representation with raw vocabulary. Therefore, in the first stage, we need to build a visual vocabulary which is a finite set of visual-words learned from all the unlabeled training images. After that every image could be described with visual-words from this vocabulary.

The BOV approach extracts low-level features locally, that is, the features are extracted only from local interest patches. The underlying reasons to employ local features are: First, they are invariant to some geometric and illumination transforms, and this invariance ensures that the same image points will be extracted from a given image and its transformed pair [30], thus the same image representation will be obtained [35]; Second, objects, such as car, bike, cow, etc., usually appear at a small portion of an image [42]; In addition, the locality provides some robustness to partial visibility and occlusion [14]; and finally, the significance of such special patches lies in their compact representation of important image regions, leading to efficient and good discriminative power [10].

Several interest patch detection algorithms have been employed, such as Harris affine detector [9], difference-of-Gaussian point detector [35], saliency detector [15, 26], etc. Detailed comparison among different patch detections are discussed in [27, 31]. Instead of with interest patch, low-level visual feature vectors may also be calculated on the basis of regular grids [42], template matching [40], at random [26, 29], or even every pixel to avoid early removal of potentially useful patches [43].

Over the limited spatial support identified by local interest patches, local descriptors could then be calculated. Scale Invariant Feature Transform (SIFT) is a usually utilized descriptor [28], which has been identified as the best [30]. SIFT descriptors are Gaussian derivatives computed at 8 orientation directions over a grid of 4×4 spatial locations. Thus the dimension size of the feature vector is 128, which trade-offs between discriminative representation and descriptor size. The SIFT descriptor reduces the effects of illumination change; therefore, it is invariant to affine change in illumination. Furthermore, the distance in descriptor space should tend to coincide roughly with the distinguishability with respect to viewpoints [20]. Dense color SIFT is also used in [6]. Note that invariance is an important characteristic to be captured by local descriptors; however, increasing the degree of invariance may remove some information which is critical for discrimination [35].

Visual-words are cluster centers by clustering the invariant descriptors of image interest patches, which simulates semantic classification as perceived by human. However, it may not necessarily have visual consistency [42], i.e., each visual-word may or may not contain a high-level concept and each high-level concept may be contained in one or more visual-words. The commonly used clustering algorithm is K-means, in which Mahalanobis distance [43] or Euclidean distance may be used. The difference between these two distances is that Mahalanobis distance takes into consideration the correlations of the local descriptors and it is scale-invariant. The size of the visual vocabulary is predetermined and usually from hundreds to thousands. The computational cost will be increased and the description space will be sparse with large K, and too small K will lead to confuse different concept objects in images thus losing the description accuracy. The information bottleneck method has been utilized to reduce the number of visual-words by a pair-wise merging operation [43]. Multi-level vocabularies could also be constructed with different values of K [36]. Some authors also tried other clustering approaches to deal with unbalanced clusters, such as agglomerative clustering [1], mean-shift and on-line clustering [20], etc. Instead of using clustering to identify visual-words, Gaussian Mixture Model (GMM) has been exploited in which each mixture component is treated as a visual-word [14]. The GMM approach allows an interest patch to contribute to more than one visual-word if necessary; on the contrary, with clustering an interest patch can only belong to one cluster center, i.e., visual-word.

28.2.2 Constructing Histogram

With a visual vocabulary, the next stage is to construct a representation (feature vector) for an image. First, each image still needs to undergo detecting interest patches and computing local descriptors for those detected interest patches. Next, the local descriptors should be replaced with their nearest corresponding visual-words learned from the first stage, which we call it quantization. This step is quite similar to the stemming preprocessing step of text documents by replacing all words with their stems [35]. Therefore, it has two indications: first it provides some robustness to descriptor variations as similar patch descriptors are mapped to the nearest visual-words; and second, as all images are mapped using the same set of visual-words, it provides a fixed length description of the input set of patch descriptors [14]. Multi-level descriptor quantization also has been proposed in [36].

By calculating the count with which each visual-word occurs in an image, we obtain a histogram in which the frequency of each visual-word is presented. Thus the histogram for image I_i is a vector defined as

$$\vec{h}(I_i) = [n(I_i, w_1), n(I_i, w_2), \ldots, n(I_i, w_T)],$$

in which $n(I_i, w_t)$ is the number of occurrence of visual-word w_t and T is the size of visual vocabulary. This histogram is the BOV representation of an image. With such representation, the information of an image's geometrical configuration is lost; however, it provides some additional robustness to viewpoint changes. A binary indicator vector could also be constructed in which the presence of a visual word in the image or its count passing a threshold testing is treated as 1, otherwise 0 [20]. By partitioning an image into increasingly fine grids and calculating histograms of interest patches found in each grid cell, the spatial information would be encoded [25]. After segmenting an image into regions, the spatial dependency also are considered by grouping pair-wise regions with their vertical position relationship [16]. Instead of calculating the number of occurrence of each visual-word, [33] calculates the Euclidean distances of each local descriptor of an image to all visual-words in the visual vocabulary, and the feature vector of the image is formed by keeping the smallest distance for each visual-word.

28.3 Extensions to the BOV

There are some limitations of the BOV approach, therefore some extensions have been investigated.

28.3.1 Probabilistic Latent Models

Visual-words in the visual vocabulary are not uniquely defined, but on which image representation and then classification performance strongly depend [35]. And the histogram representation of an image contains only the frequency of each visual-word appearing in this image. It brings some synonymy and polysemy ambiguities [35]. To disambiguate this BOV representation, probabilistic latent space models have been adapted to learn semantic topics in images, such as Probability Latent Semantic Analysis (pLSA) [6, 35, 42] and Latent Dirichlet Allocation (LDA) [3, 22, 26]. These models derive a set of latent variables that encode hidden states of the semantic world, in which each state induces a joint distribution on the space of visual-words.

The pLSA is proposed to automatically learn topics from text documents [18]. While adapting to images, it similarly assumes that each visual-word is generated by a combination of hidden topics z_j:

$$P(w_t \mid I_i) = \sum_{j=1}^{J} P(w_t \mid z_j, I_i) P(z_j \mid I_i),$$

where J is the number of hidden topics, the conditional probability $P(w_t \mid z_j, I_i)$ denotes the probability of observing the visual-word w_t given topic z_j and image I_i, and $P(z_j \mid I_i)$ is the conditional probability of topic z_j given I_i. In addition, the pLSA assumes that the conditional probability of generating a visual-word by a specific topic is independent from the image, i.e., $P(w_t \mid z_j, I_i) = P(w_t \mid z_j)$ [42]. The model parameters, $P(w_t \mid z_j)$ and $P(z_j \mid I_i)$, can be learned and optimized by the Expectation-Maximization (EM) algorithm [18]. Since each topic is characterized by a multinomial distribution of visual-words, it groups them by the co-occurrence relationship. As a result, the pLSA provides more stable image representation [35].

The LDA is a generative probabilistic model for collections of discrete data, which is also proposed for inferring topics from text documents [5]. Modeling an image with the LDA assumes that it is built according to the following generative process [22]: To create an image, first we select its class label; Next, we draw a probability vector for patches in the image to determine the mixture of possible topics; After that, for each patch we decide its particular topic out of the co-occurrence of topics; Finally when a topic is selected, we construct it with visual-words [26]. Thus the LDA is a Bayesian hierarchy model, and the EM or the Gibbs sampling could be employed to estimate its parameters. A similar generative model, called Continuous-space Relevance Model (CRM), is proposed in [23], which makes no assumptions about topological structure.

28.3.2 Vocabulary Issue

Visual vocabulary is an artificial concept as it is generated by clustering of local descriptors; hence only a few of visual-words are informative and not all of them have a clear semantic interpretation [22]. The efficiency of vocabularies estimated without any consideration with the following classification task and image modeling process should be questioned. In addition, the aforementioned method to build visual vocabulary is generative, which means that it captures overall image statistics but may not be discriminative for object classification [20]. One more problem is that the visual vocabulary is build with training images, a different training set may produce different visual-words, thus leading to a different vocabulary [34].

Instead of training each class vocabulary independently, creating a universal vocabulary with all considered object classes and then adapting each class specific vocabulary from the universal vocabulary has been exploited [34]. With multiple vocabularies, an image is characterized by a set of histograms called bipartite, one for each class. One half of the bipartite is a histogram with class-specific vocabulary and the other with universal vocabulary. Therefore, it emphasizes the differences between a specific class and a general one. Such approach could process multiple semantic classes.

By viewing the visual vocabulary as a built-in component of a classification model, the authors in [22] combine the learning processes of the visual vocabulary and other parameters of the classification model, and propose a latent mixture vocabulary model. Images are considered as distributions over topics, topics as combinations of visual-words, and visual-words as Gaussian mixture densities over visual descriptors.

28.4 Other Feature Representation Approaches

Inspired by the feedforward hierarchy visual processing in primate visual cortex, a hierarchical model has been proposed to extract image semantic features [32, 37], which is composed by five layers: the first layer is a grayscale image pyramid with 10 scales; the second layer is computed by centering 2D Gabor filters with 4 orientations at each possible position and scale to increase object selectivity; the third layer is called local invariance layer by pooling nearby units with the same orientation through a maximum operation, thereby introducing gradual invariance to scale and translation; the fourth layer is the intermediate feature layer by performing template matching with d prototype patches at every position and scale; the top layer is the global invariance layer, a d-dimensional vector, in which each element is produced by selecting the maximum response to the corresponding prototype patch over anywhere in the image.

A region based approach exploited in [17] demonstrates the success of applying regions in representing image features. Regions are extracted and constructed as a hierarchical structure called region tree with the generic grouping algorithm proposed in [2]. And then a region is described with evenly subdivided cells and each cell encodes region cues, such as contour shape, edge shape, color, and texture, as histograms. This region descriptor embeds the scale invariant nature and the hierarchy structure enables image representation in different granularity levels.

28.5 Image Semantic Mapping

Once images have been represented by feature vectors (such as histograms of the BOV approach, in which each visual-word is a feature), the next step is to derive image semantic concept, such as recognizing objects in images, classifying images, etc. Although we cannot directly obtain image semantic concept by analyzing image features, we can make use of statistical features of a large amount of images. The assumption is that similar distributions of features apply to similar concept objects.

A straightforward non-parametric approach is to describe a semantic concept as a set of feature vectors which are associated with the same class label. With an input test image, the nearest-neighbor classification is applied to it and the class label of its closet neighbor is returned [43]. It can be extended to return multiple labels by selecting K nearest neighbors [6].

By splitting the image database into two parts: images with and without the concept of interest, a binary classifier could be trained to detect that concept. A random variable ω_l is introduced to indicate whether a concept class C_l presents or not in image I: $\omega_l = 1$ when presenting; otherwise $\omega_l = 0$. The interest concept is declared as be present if

$$P(I \mid \omega_l = 1)P(\omega_l = 1) \geq P(I \mid \omega_l = 0)P(\omega_l = 0).$$

Then for multi-class classification, we could adopt a one-against-all approach by training M classifiers, in which M is the number of semantic class concepts. Each classifier is trained to differentiate images of one class from images of all other classes. A test image will be attached to all the class labels of classifiers if it satisfies their corresponding decision functions. The commonly trained classifier is Support Vector Machine (SVM) [6,9,32,33,35,37]. The Naïve Bayes approach [9] to annotate image I with class C_l on the basis of the BOV representation is

$$P(C_l \mid I) \propto P(I \mid C_l)P(C_l) = P(C_l) \prod_{t=1}^{T} P(w_t \mid C_l)^{n(I, \, w_t)}.$$

It produces a ranking of semantic labels, thus it can also be used to annotate an image with multiple labels.

Furthermore, we can estimate the joint probability distribution of the visual features of image I and its semantic concepts, $P(I, C_{l_1}, C_{l_2}, \ldots, C_{l_j})$, by searching the set of concept labels that maximizes the joint probability. Having the joint probability distribution allows us to calculate a conditional likelihood $P(C_{l_1}, C_{l_2}, \ldots, C_{l_j} \mid I)$ which then can be employed to guess the most likely semantic labels [23].

A Multiple Instance Learning (MIL) approach is discussed in [8]. This approach is motivated by the fact that in the process of estimating the probability density of a specific semantic class, the representation of the feature vector of an image contains various features from other semantic classes due to without previous semantic segmentation.

An application of Self-Organizing Map (SOM) has been exploited [21], which is a type of artificial neural network by mapping image descriptor space to a 2D grid where similar images are located near each other. The distribution of feature vectors over the map forms a 2D discrete probability density. With separate features, multiple SOMs could be derived and different SOMs impose different similarity relations on the images to model various semantic concepts.

An extension to the 2D Hidden Markov Model (HMM), called Dependency-Tree HMM (DTHMM) is utilized to extract contextual information [19], in which an image is decomposed into non-overlapping blocks forming a regular grid and each block depends on its top or left neighbor with equal probability. By regarding these blocks as a sequence of observations, the DTHMM builds a dependency-tree which keeps local context information in an image as pixels with reasonable resolutions usually have spatial dependencies.

28.6 Conclusion

Two steps in obtain image semantics are described in this chapter: image feature representation and image semantic mapping. Images are represented with features extracted from low-level image content, a typical local feature representation approach called BOV has been discussed in detail, which exploits local statistics in images and introduces an intermediate representation. Image semantic concepts are learned on the basis of the combination of feature representations. This mapping procedure is achieved by various computational intelligence techniques, such as Bayesian approach, GMM, SVM,

MIL, SOM, HMM, etc. However, the semantic and conceptual gap between human and computer still exists. Knowledge-based approaches and computational intelligence techniques are required to be further investigated to bridge this gap. Therefore, a more general framework of image representation and semantic mapping is needed.

References

[1] S. Agarwal, A. Awan, and D. Roth. Learning to detect objects in images via a sparse, part-based representation. *IEEE Transactions on Pattern Analysis and Machine Intelligence*, 26(11):1474–1490, September 2004.

[2] P. Arbelaez, M. Maire, C. Fowlkes, and J. Malik. From contours to regions: An empirical evaluation. In *Proceedings of IEEE Conference on Compter Vision and Pattern Recognition (CVPR'09)*, pages 2294–2301, June 2009.

[3] K. Barnard, P. Duygulu, D. Forsyth, N. de Freitas, D.M. Blei, and M.I. Jordan. Matching words and pictures. *Journal of Machine Learning Research*, 3:1107–1135, 2003.

[4] I. Bartolini, P. Ciaccia, and M. Patella. WARP: Accurate retrieval of shapes using phase of Fourier descriptors and time warping distance. *IEEE Transactions on Pattern Analysis and Machine Intelligence*, 27(1):142–147, January 2005.

[5] D.M. Blei, A.Y. Ng, and M.I. Jordan. Latent Dirichlet allocation. *Journal of Machine Learning Research*, 3:993–1022, 2003.

[6] A. Bosch, A. Zisserman, and X. Munoz. Scene classification using a hybrid generative/discriminative approach. *IEEE Transactions on Pattern Analysis and Machine Intelligence*, 30(4):712–727, April 2008.

[7] H. Buxton and A. Mukerjee. Conceptualizing images. *Image and Vision Computing*, 12(2):79, 2000.

[8] G. Carneiro, A.B. Chan, P.J. Moreno, and N. Vasconcelos. Supervised learning of semantic classes for image annotation and retrieval. *IEEE Transactions on Pattern Analysis and Machine Intelligence*, 29(3):394–410, March 2007.

[9] G. Csurka, C.R. Dance, L. Fan, J. Willamowski, and C. Bray. Visual categorization with bags of keypoints. In *Proceedings of ECCV Workshop on Statistical Learning for Computer Vision*, pages 1–22, 2004.

[10] R. Datta, J. Li, and J.Z. Wang. Context-based image retrieval – Approaches and trends of the new age. In *Proceedings of the 7th ACM SIGMM International Workshop on Multimedia Information Retrieval*, pages 253–262, November 2005.

[11] S. Deb and Y. Zhang. An overview of content-based image retrieval techniques. In *Proceedings of the 18th International Conference on Advanced Information Networking and Applications (AINA'04)*, Vol. 1, pages 59–64, March 2004.

[12] Y. Deng, B.S. Manjunath, C. Kenney, M.S. Moore, and H. Shin. An efficient color representation for image retrieval. *IEEE Transactions on Image Processing*, 10(1):140–147, January 2001.

[13] M.N. Do and M. Vetterli. Wavelet-based texture retrieval using generalized Gaussian density and Kullback–Leibler distance. *IEEE Transactions on Image Processing*, 11(2):146–158, February 2002.

[14] J.D.R. Farquhar, S. Szedmak, H. Meng, and J. Shawe-Taylor. Improving "bag-of-keypoints" image categorisation: Generative models and PDF-kernels. Technical Report, University of Southampton, February 2005.

[15] R. Fergus, P. Perona, and A. Zisserman. Object class recognition by unsupervised scale-invariant learning. In *Proceedings of IEEE Conference on Compter Vision and Pattern Recognition (CVPR'03)*, Vol. 2, pages 264–271, June 2003.

[16] D. Gokalp and S. Aksoy. Scene classification using bag-of-regions representation. In *Proceedings of IEEE Conference on Compter Vision and Pattern Recognition (CVPR'07)*, pages 1–8, June 2007.

[17] C. Gu, J.J. Lim, P. Arbelaez, and J. Malik. Recognition using regions. In *Proceedings of IEEE Conference on Compter Vision and Pattern Recognition (CVPR'09)*, pages 1030–1037, June 2009.

[18] T. Hofmann. Unsupervised learning by probabilistic latent semantic analysis. *Machine Learning*, 41(1–2):177–196, January/February 2001.

[19] J. Jiten, B. Merialdo, and B. Huet. Semantic feature extraction with multidimensional hidden Markov model. In E.Y. Chang et al. (Eds.), *Proceedings of SPIE Conference on Multimedia Context Analysis, Management and Retrieval*, Vol. 6073, pages 211–221, January 2006.

[20] F. Jurie and B. Triggs. Creating efficient codebooks for visual recognition. In *Proceedings of the 10th IEEE International Conference on Computer Vision (ICCV'05)*, Vol. 1, pages 504–610, October 2005.

[21] M. Koskela and J. Laaksonen. Semantic annotation of image groups with self-organizing maps. In W.-K. Leow et al. (Eds.), *Image and Video Retrieval*, Lecture Notes in Computer Science, Vol. 3568, pages 518–527. Springer, Berlin, 2005.

[22] D. Larlus and F. Jurie. Latent mixture vocabularies for object categorization and segmentation. *Image and Vision Computing*, 27(5):523–534, April 2009.

[23] V. Lavrenko, R. Manmatha, and J. Jeon. A model for learning the semantics of pictures. In *Proceedings of the 17th Annual Conference on Neural Information Processing Systems (NIPS'03)*, Vol. 16, pages 553–560, December 2003.

[24] J.A. Lay and L. Guan. Semantic retrieval of multimedia by concept languages: Treating semantic concepts like words. *IEEE Signal Processing Magazine*, 23(2):115–123, March 2006.

[25] S. Lazebnik, C. Schmid, and J. Ponce. Beyond bags of features: Spatial pyramid matching for recognizing natural scene categories. In *Proceedings of IEEE Conference on Computer Vision and Pattern Recognition (CVPR'06)*, Vol. 2, pages 2169–2178, June 2006.

[26] F.-F. Li and P. Perona. A Bayesian hierarchical model for learning natural scene categories. In *Proceedings of IEEE Conference on Computer Vision and Pattern Recognition (CVPR'05)*, Vol. 1, pages 524–531, June 2005.

[27] J. Li and N.M. Allinson. A comprehensive review of current local features for computer vision. *Neurocomputing*, 71(10–12):1771–1787, June 2008.

[28] D.G. Lowe. Distinctive image features from scale-invariant keypoint. *International Journal of Computer Vision*, 60(2):91–110, November 2004.

[29] R. Marée, P. Geurts, J. Piater, and L. Wehenkel. Random subwindows for robust image classification. In *Proceedings of IEEE Conference on Computer Vision and Pattern Recognition (CVPR'05)*, Vol. 1, pages 34–40, June 2005.

[30] K. Mikolajczyk and C. Schmid. A performance evaluation of local descriptors. *IEEE Transactions on Pattern Analysis and Machine Intelligence*, 27(10):1615–1630, October 2005.

[31] K. Mikolajczyk, T. Tuytelaars, C. Schmid, A. Zisserman, J. Matas, F. Schaffalitzky, T. Kadir, and L. Van Gool. A comparison of affine region detectors. *International Journal of Computer Vision*, 65(1–2):43–72, November 2005.

[32] J. Mutch and D.G. Lowe. Multiclass object recognition with sparse, localized features. In *Proceedings of IEEE Conference on Computer Vision and Pattern Recognition (CVPR'06)*, Vol. 1, pages 11–18, June 2006.

[33] P. Mylonas, E. Spyrou, Y. Avrithis, and S. Kollias. Using visual context and region semantics for high-level concept detection. *IEEE Transactions on Multimedia*, 11(2):229–243, February 2009.

[34] F. Perronnin. Universal and adapted vocabularies for generic visual categorization. *IEEE Transactions on Pattern Analysis and Machine Intelligence*, 30(7):1243–1256, July 2008.

[35] P. Quelhas, F. Monay, J.-M. Odobez, D. Gatica-Perez, and T. Tuytelaars. A thousand words in a scene. *IEEE Transactions on Pattern Analysis and Machine Intelligence*, 29(9):1575–1589, September 2007.

[36] P. Quelhas and J.-M. Odobez. Multi-level local descriptor quantization for bag-of-visterms image representation. In *Proceedings of the 6th ACM International Conference on Image and Video Retrieval*, pages 242–249, July 2007.

[37] T. Serre, L. Wolf, and T. Poggio. Object recognition with features inspired by visual cortex. In *Proceedings of IEEE Conference on Computer Vision and Pattern Recognition (CVPR'05)*, Vol. 2, pages 994–1000, June 2005.

[38] W. Shao, G. Naghdy, and S.L. Phung. Automatic image annotation for semantic image retrieval. In G. Qiu et al. (Eds.), *Advances in Visual Information Systems*, Lecture Notes in Computer Science, Vol. 4781, pages 369–378. Springer, Berlin, 2007.

[39] Y. Tong, W. Liao, and Q. Ji. Facial action unit recognition by exploiting their dynamic and semantic relationships. *IEEE Transactions on Pattern Analysis and Machine Intelligence*, 29(10):1683–1699, October 2007.

[40] A. Torralba, K.P. Murphy, and W.T. Freeman. Sharing features: Efficient boosting procedures for multiclass object detection. In *Proceedings of IEEE Conference on Computer Vision and Pattern Recognition (CVPR'04)*, Vol. 2, pages 762–769, June 2004.

[41] M. Varma and A. Zisserman. A statistical approach to texture classification from single images. *International Journal of Computer Vision*, 62(1–2):61–81, April/May 2005.

[42] Y. Wang, T. Mei, S. Gong, and X.-S. Hua. Combining global, regional and contextual features for automatic image annotation. *Pattern Recognition*, 42(2):259–266, February 2009.

[43] J. Winn, A. Criminisi, and T. Minka. Object categorization by learned universal visual dictionary. In *Proceedings of the 10th IEEE International Conference on Compter Vision (ICCV'05)*, Vol. 2, pages 1800–1807, October 2005.

[44] L. Zhu, A. Rao, and A. Zhang. Theory of keyblock-based image retrieval. *ACM Transactions on Information Retrieval*, 20(2):224–257, April 2002.

Subject Index

Author Index

451

RIVER PUBLISHERS SERIES IN INFORMATION SCIENCE AND TECHNOLOGY